HAZARDOUS MATERIALS AND HAZARDOUS WASTE MANAGEMENT

HAZARDOUS MATERIALS AND HAZARDOUS WASTE MANAGEMENT

Second Edition

GAYLE WOODSIDE

JOHN WILEY & SONS, INC.
New York • Chichester • Weinheim • Brisbane • Singapore • Toronto

This book is printed on acid-free paper. ♾

Library of Congress Cataloging-in-Publication Data:

Woodside, Gayle
Hazardous materials and hazardous waste management / Gayle Woodside. — 2nd ed.
p. cm.
Includes bibliographical references and index.
ISBN 0-471-17449-1 (alk. paper)
1. Hazardous wastes—Management. 2. Hazardous substances—Management. I. Title.
TD1030.W66 1999
604.7—dc21 98-33647

Printed in the United States of America.

10 9 8 7 6 5 4 3 2 1

CONTENTS

PREFACE

This book covers aspects of managing hazardous materials and hazardous waste that an engineer, scientist, or other professional who works in the field will encounter when working in industry or government. Included are regulatory references and applications for the various topics associated with the hazardous materials and hazardous waste management field. Tables have been included whenever possible to give the reader information in a summary form, and figures have been added for clarity. The book is structured by discrete topic, with inclusive information in each chapter, for ease of use.

This second edition includes updated information on all the topics in the first edition, and enhanced information about waste treatment/disposal technologies, air emissions assessment and control, and pollution prevention. In addition, several new topics have been added, including voluntary standards and initiatives, hazardous materials transportation, workplace management of radiation exposure, water quality assessment and management, and process safety management.

The intent of writing on the subjects of hazardous materials and hazardous waste (and not just on one topic or the other) is to give those who are learning about or working in the field a sense of the interconnection between hazardous materials and hazardous waste regulations and management practices. Traditionally, reference and text books have addressed only one topic. In today's world of industry downsizing, professionals need to be knowledgeable in more than one area. In the hazarous materials and hazardous waste fields, the professional needs to be conversant not only in the regulations, but also in other areas, such as exposure assessment, emergency planning, transportation of hazardous materials and waste, process safety, administrative requirements, and many other topics. I have addressed those topics that will be beneficial to both the professional just entering the field and professionals involved in developing and evolving hazardous materials and hazardous waste management regulations and management practices.

There are many people over the years who have contributed technical information and management knowledge to me during the writing of this second edition. In particular I would like to thank Dianna Kocurek, John Woodside, John Prusak, David Dalke, Mike Tomme, and Lial Tischler. A special thanks goes to Millie Torres, Associate Managing Editor, who has worked with me untiringly on three books. Finally, I would like to thank my husband and friend, Bruce Almy, for his unwavering support.

HAZARDOUS MATERIALS AND HAZARDOUS WASTE MANAGEMENT

PART I

HAZARDOUS MATERIALS AND HAZARDOUS WASTE: AN OVERVIEW

1

A REGULATORY OVERVIEW

Regulations pertaining to environmental and workplace hazards have been in place for over three decades. The National Environmental Policy Act of 1969 became the basis for environmental protection in the United States, and all environmental laws enacted thereafter—including those pertaining to air quality, water quality, and hazardous waste—were an outgrowth of the policies and goals set forth in this act. The Occupational Safety and Health Act of 1970 and subsequent amendments have been the basis for regulation of hazardous materials in the workplace as well as other aspects of industrial safety and health. Similarly, the Atomic Energy Act of 1954 and subsequent acts set forth the basis for regulation of radioactive source materials and waste. This chapter provides an overview of the regulatory process and a summary of key regulations that govern hazardous materials and hazardous waste.

THE PROCESS OF REGULATING

The law provides protection for the environment and human health through several mechanisms. These include congressional, state, or local acts; federal, state, and local regulations; court decisions interpreting laws and regulations; and constitutional law. In the following paragraphs we discuss in detail the more important points associated with the process of regulating hazardous materials and hazardous waste.

Rulemaking

The process of regulating starts with acts of Congress in the form of bills, which must be passed by the House of Representatives and Senate and signed into law by the President (or at least not vetoed by the President for 10 days). This process is lengthy and complicated and typically requires hearings, followed by amendments or modifications to the original piece of legislation. To become law, a bill must receive majority approval by both chambers of the legislature. Many bills get tabled or otherwise stalled during the process and are never enacted into law. An act can take several years to pass, with modifications and amendments introduced throughout the process.

Acts typically embody a concept or goal of the legislature and set forth a time frame in which these goals are to be achieved. Acts also empower the executive branch of the government, through an agency such as the Environmental Protection Agency (EPA), Occupational Safety and Health Administration (OSHA), Nuclear Regulatory Commission (NRC), or Department of Transportation (DOT), to create regulations that take the concept and refine it into specific rules that are meant to accomplish the goals set forth in the act.[1]

In some instances the executive agency lacks scientific or other information required to craft regulations properly, and will commission studies before proposing specific standards or requirements. These studies then become the technical basis of the regulation or future agency policies. In the area of hazardous materials and hazardous waste, the studies may be used to set requirements for permissible exposure limits for a material, treatment standards for a waste, or allowable emissions for a particular pollutant.

When the executive agency begins drafting a regulation, the rule-making procedure must follow the terms defined in the Administrative Procedure Act. These include providing adequate notice, allowing public comment, publishing substantive rules at specified times, and giving interested parties the right to petition for issuance, amendment, or repeal.

The vehicle used for providing notice of proposed rule making is the *Federal Register*.[2] Once the notice is published, the agency is required to give interested persons an opportunity to participate through submission of written data, views, or arguments about the proposed rule. The opportunity for oral presentation is normally part of the process but is not required under the statute. This public comment period on proposed legislation is usually 60 to 90 days. The basic steps generally followed in the rule-making process for hazardous materials and hazardous waste are:

- *Advanced notice of proposed rulemaking* (*ANPR*). With this published notice, the regulating agency signals its intent to begin studying key aspects of a regulatory objective. The agency solicits from the public input on how best to achieve its regulatory goal.

[1]The President may also empower an executive agency to promulgate regulations through an executive order.

[2]The *Federal Register* system of publication was originally established by the Federal Register Act. Its function was amended and expanded by the Administrative Procedure Act. The *Federal Register* is published daily and contains federal agency regulations, proposed rules, meetings and proceedings, and other documents of the executive branch.

- *Proposed rule.* A draft regulation, or proposed rule, is published for public comment. Any significant input from the public received during the ANPR comment period, along with comments of agency rationale for including or excluding specific aspects from the rule, is published in the preamble. Once again, written comments are solicited from the public, and public hearings about the rule may be held at strategic locations in regions throughout the nation.

  The overall intent of the rule normally is not changed through public comments, although definition of terms, applicability, and administrative aspects of the rule, such as record-keeping requirements and reporting deadlines, may get modified. In some cases, however, the development and publication of the final rule may be postponed for a significant period of time because of adverse public opinion about the rule, a change in rule-making priorities at the agency, or for other reasons.
- *Final rule.* This rule can become legally enforceable 30 days after publication in the *Federal Register,* with some exceptions allowing for earlier implementation. It is not uncommon, however, for the agency to set the effective date at six months to one year after publication of the rule. This is meant to give the regulated community time to understand the rule and come into compliance. Final rules are published in the *Code of Federal Regulations.*[3]

State Statutes and Regulations Implementing the Federal Statutes

Many federal regulations establish federal–state regulatory programs—programs that give states the opportunity to enact and enforce laws. These laws must meet federal minimum criteria and must achieve the regulatory objectives established by Congress. Examples of these federal state programs include regulations promulgated under the Clean Air Act, the Clean Water Act, and the Resource Conservation and Recovery Act. Essentially, under these programs, states are delegated primary permitting and enforcement functions, subject to federal intervention only if they do not meet federal enforcement and some other guidelines.

Permitting

Once regulations are finalized, the executive agency may establish a permits program to ensure compliance with the new regulations by the regulated com-

[3]Rules and regulations are codified in the *Code of Federal Regulations* (CFR), which is updated annually. The CFR is divided into 50 titles, encompassing broad subject categories such as labor (29 CFR), which includes occupational safety and health; energy (10 CFR), which includes radioactive materials and waste; protection of environment (40 CFR); and transportation (49 CFR). The *Federal Register* serves as a daily update to the CFR.

munity. An industrial facility might need a permit if the facility uses, produces or generates, treats, stores, emits or discharges, or disposes of a hazardous material or hazardous waste.

Most of the permits necessary for hazardous materials and hazardous waste management are required by EPA. If EPA issues a permit to a facility, it is issued by a regional office. EPA regions and states and U.S. territories that are served by each are presented in Table 1.1.

As mentioned earlier, the EPA can delegate its permitting authority to a state if that state has an acceptable implementation plan for the permitting process and regulates with at least as much restriction as EPA. If the state has not received permitting authority from the EPA, the state may require a separate permit in addition to the EPA permit for specific industrial activities, such as wastewater discharge or hazardous waste treatment, storage, and disposal. In some instances, local governments may also require a separate permit.

OSHA has outlined record-keeping and other responsibilities in its regulations for facilities that use chemicals, but individual facility permits are not presently required. The agency maintains the right to inspect a facility's operations and records at any time for compliance with OSHA regulations. Like EPA, OSHA can delegate its inspection authority and other regulatory aspects to a state. States and U.S. territories that have OSHA-approved plans are presented in Table 1.2.

TABLE 1.1 EPA Regions and States and U.S. Territories Served

Region 1: Boston, MA
Connecticut, Maine, Massachusetts, New Hampshire, Rhode Island, Vermont
Region 2: New York, NY
New Jersey, New York, Puerto Rico, Virgin Islands
Region 3: Philadelphia, PA
Delaware, District of Columbia, Maryland, Pennsylvania, Virginia, West Virginia
Region 4: Atlanta, GA
Alabama, Florida, Georgia, Kentucky, Mississippi, North Carolina, South Carolina, Tennessee
Region 5: Chicago, IL
Illinois, Indiana, Michigan, Minnesota, Ohio, Wisconsin
Region 6: Dallas, TX
Arkansas, Louisiana, New Mexico, Oklahoma, Texas
Region 7: Kansas City, KS
Iowa, Kansas, Missouri, Nebraska
Region 8: Denver, CO
Colorado, Montana, North Dakota, South Dakota, Utah, Wyoming
Region 9: San Francisco, CA
Arizona, California, Hawaii, Nevada, American Samoa, Guam
Region 10: Seattle, WA
Alaska, Idaho, Oregon, Washington

TABLE 1.2 States and U.S. Territories That Have OSHA-Approved Plans

Alaska	Michigan	Tennessee
Arizona	Minnesota	Utah
California	Nevada	Vermont
Connecticut	New Mexico	Virgin Islands
Hawaii	New York	Virginia
Indiana	North Carolina	Washington
Iowa	Oregon	Wyoming
Kentucky	Puerto Rico	
Maryland	South Carolina	

The permitting process begins when a facility files an application for a specific permit. Examples of permit applications include Resource Conservation and Recovery Act (RCRA) waste management permit, Toxic Substance Control Act (TSCA) premanufacture notice, National Pollution Discharge Elimination System (NPDES) wastewater discharge permit, EPA Title V air emissions permit, and NRC permit (license) to dispose of radioactive waste. Examples of facilities that typically are required to apply for and receive permits include chemical-producing facilities; manufacturing facilities; hazardous waste treatment, storage, and disposal facilities; industrial wastewater treatment facilities; publicly owned treatment works (POTWs); and others.

The application usually requires general information pertaining to the facility, such as name, location, standard industrial classification, and type of manufacturing or operational activities engaged in at the facility. Information specific to the type of permit is also required. The completed application can range from a short application of several pages to a complex application with volumes of information included. Table 1.3 gives examples of activities that typically require permitting.

After the permit application is submitted, the agency reviews it for administrative completeness and technical soundness. Where applicable, the agency relies on health studies, risk assessments, demonstrated technology, established standards (regulatory and nonregulatory), and other engineering data, such as modeling, to determine the acceptability of the facility's application.

If the specifics of the application are approved by the agency, a draft permit will be issued and public notice will be made to solicit comments on the permit, as required. At this point, the public has an opportunity to ask questions about the specifics of the permit or request a public hearing to oppose granting the permit. If a citizen or other group does not request a public hearing within a specified time frame—usually 30 to 60 days—the permit is granted by the agency.

If a public hearing is requested, opponents of the permit application are given a chance to explain their reasons for opposition. The agency defends the technical basis of the draft permit and the applicant provides input on its own behalf. An unbiased hearing examiner from the agency listens to all

TABLE 1.3 Selected Examples of Activities That Typically Require Permitting

Permitted Activity	Requirements	Comments
Treatment, storage, or disposal of hazardous waste	Notify EPA or state of activity; obtain an EPA identification number.	Facility must meet interim status or permitted standards.
	Apply for and obtain a RCRA Part B permit.	Hazardous waste can be stored for up to 90 days without a permit; waste can be treated during this period.
Off-site discharge of wastewater or stormwater to land or stream from municipal or industrial facility.	Apply for an obtain NPDES and/or state permit.	Stormwater discharges can be permitted under the general permit system.
Wastewater discharge to publicly owned treatment works (POTWs)	If required, apply for a permit or authorization from the POTW.	Discharge must meet applicable EPA effluent guidelines and pretreatment standards.
Emissions of air pollutants with published air quality standards or that are hazardous air pollutants	Rederal permit required for major sources. State permit required in states with approved state implementation plans.	Title V permitting and state implementation plan evaluation are in process.
Manufacture or importation of chemicals	Premanufacture notice must be submitted to EPA under TSCA.	Health/environmental effects must be submitted for each new chemical.
Disposal of radioactive waste	Apply for an obtain NRC license.	Requirements for disposal vary depending on class of radioactive waste.
Formulation/manufacture of a new pesticide	Apply for an receive registration and classification from EPA for the pesticide.	Test data must be submitted to EPA before registration can be obtained.
Transport of hazardous waste	Apply for and obtain EPA transporter number.	Transporters must meet EPA manifest and DOT shipping requirements.

arguments and makes the final determination as to whether the permit will be granted or whether specific permit provisions require modification.

Once a permit is issued to regulate certain facility activities, the permit provisions become the facility's enforceable standards. Generally, standards set in the permit remain the facility's enforceable standards until the permit expires. Any changes to the standards developed by the agency typically are incorporated into the permit during the permit renewal process, although some new or revised standards may become enforceable earlier.

Monitoring and Record Keeping

Almost all hazardous materials and hazardous waste regulations require some type of monitoring or record keeping. Additionally, a permit can specify a monitoring or inspection schedule to verify that permit provisions are being met. Depending on permit requirements, the data or records are submitted to the agency on a periodic basis such as monthly, quarterly, annually, biennially, or when an exception occurs.

Once submitted to an agency, monitoring records and other data become public record. Data that are maintained at the facility are subject to audit by the agency but generally are not considered public information. Under Title III of the Superfund Amendments and Reauthorization Act (SARA), the citizens of a community have a right to know what chemicals and wastes are stored at a facility and in what quantities. Additionally, citizens have access to reports detailing releases of certain chemicals to the environment.

Enforcement

Like permitting, enforcement of the regulations is the responsibility of the executive agency. As part of its enforcement program, the agency has the right to audit records and inspect the regulated facility for compliance. An audit can include items such as records, data, processes, operating parameters, material use and handling methods, storage practices, and other facility operations.

When noncompliance issues arise, the agency has the power to begin enforcement proceedings. Generally, enforcement tools include civil penalties; administrative orders to respond to or abate the noncompliance; civil action for relief, such as prohibition or injunction; and criminal sanctions against corporations or individuals, usually for known violations or negligence. Most of the enforcement suits (except administrative orders) are processed through the judicial branch. In many cases, the agency will levy fines for administrative noncompliance, which vary depending on the nature of the violation or discrepancy and which can range from a few hundred dollars to $25,000 per day per violation. Criminal penalties may include personal fines and (for the most grievous violations) jail sentences.

Judicial Branch Involvement

In addition to responsibility for enforcement suits, the judicial branch of the government must handle law suits that are filed by the public or the community regulated. Citizens or public organizations can bring a lawsuit against the executive agency if they can substantiate that the regulations do not meet the legislative intent required by the act. This can include missed deadlines, lack of standards, and the exemption of the regulations for certain practices or industries.

Similarly, the regulated community can bring a lawsuit against the executive agency if it has a strong case indicating that the agency has overstepped its regulatory bounds. This can include promulgating rules not specified by the act or basing rules on flawed studies. For these cases, the role of the judicial body is to determine the intent of the act of Congress and the facts regarding the actions and regulations of the executive agency. Based on this review, the agency may have to recraft its regulations or the regulations may be allowed to stand.

The Rights of the Regulated Community

Search Warrants. Collection of evidence is necessary for any civil or criminal enforcement program. The issue of whether or not an agency needs a search warrant to search corporate premises arises continuously. In numerous cases[4] the courts have upheld the requirement (under the Fourth Amendment of the Constitution) for agencies to obtain a search warrant for inspections, even those of routine nature. Despite this fact, it is usually considered imprudent for a business to demand a warrant for most inspections, especially for those that are considered routine, since it gives the appearance that the business has something to hide.

Due Process. The government cannot enforce any law that abridges the privileges or immunities of its citizens; not can it deprive any person of life, liberty, or property without due process of law. These provisions are part of the Fifth and Fourteenth Amendments. The government does have the inherent right, however, to pass laws for the protection of the health, welfare, morals, and property of people within its jurisdiction, provided that these laws are reasonable.

Acquisition of Information

Under the Freedom of Information Act, an agency is required to make available to the public substantive rules along with statements of general policy or interpretations of general applicability formulated and adopted by the agency. This is accomplished primarily through the *Federal Register.* An agency is also required to make available, for a reasonable fee, any guidance documents that the agency staff may develop or contract. These documents are generally available through the National Technical Information Service

[4]*Camara v. Municipal Court of San Franscisco*, 387 U.S. 523 (1967) and *Marshall v. Barlow's Inc.*, 436 U.S. 307 (1978).

(NTIS), which is part of the Department of Commerce,[5] or can be ordered through the EPA Web page on the Internet. Numerous private firms also sell regulations and updates to the regulated community.

In addition, the public has a right to view permits, enforcement actions, and other agency documents. These documents can be viewed at the agency. Copies can be requested and will be provided for a reasonable fee. To make information easily accessible, database systems containing regulatory compliance information have been developed by EPA and other agencies. These systems include information on releases of oil and hazardous substances, toxic release inventory data, enforcement actions, and TSCA test data. Like the agency manuals, these databases are generally available through NTIS.

KEY HAZARDOUS MATERIALS AND HAZARDOUS WASTE REGULATIONS

National Environmental Policy Act (NEPA) of 1969

The NEPA Act established the Council on Environmental Quality and set lofty policies and goals for environmental protection. Included in the act, as set forth in 42 U.S.C. §4331, is a charge to the federal government:

1. To fulfill the responsibilities of each generation as trustee of the environment for succeeding generations
2. To assure for all Americans safe, healthful, productive, and aesthetically and culturally pleasing surroundings
3. To attain the widest range of beneficial uses of the environment without degradation, risk to health or safety, or other undesirable and unintended consequences

The policies set forth in this act gave Congress the initiative for subsequent legislation and research and development in the environmental area.

Occupational Safety and Health (OSH) Act of 1970

The OSH Act and its amendments pertain mainly to workplace safety. The Occupational Safety and Health Administration, part of the Department of Labor, has promulgated a comprehensive set of regulations that set standards for hazardous materials management, electrical safety, fire and life safety, and other areas of workplace safety.

In terms of hazardous materials management, OSHA regulates numerous safety and health aspects, including flammable and compressed gas storage,

[5]The mailing address for NTIS is 5285 Port Royal Road, Springfield, VA 22161; the telephone number is 703-487-4650.

material labeling and information communication, personal protective equipment, workplace monitoring, medical surveillance, management of ionizing and nonionizing radiation sources, and training requirements. Additionally, numerous chemicals are regulated individually. Most regulations pertaining to hazardous materials used in the industrial sector are defined in 29 CFR Parts 1900–1910; hazardous materials used in the construction industry are delineated in 29 CFR §1926. All industrial facilities that use chemicals are subject to all or part of these regulations.

***Hazard Communication Standard* (*Effective 1986*).** A key OSHA rule that has affected facilities that manage hazardous materials and hazardous waste is the Hazard Communication Standard. Regulations driven by the law are defined in 29 CFR §1910.1200. Included are requirements for training employees about the hazards associated with chemicals used in the workplace. The training program must be established in writing and kept at the facility. Additionally, material safety data sheets must be available at the workplace and accessible to employees at all times, and chemical containers, including stationary containers, must be labeled appropriately.

***Hazardous Waste Operations and Emergency Response* (*Effective 1990*).** Another key rule promulgated and administered by OSHA (as issued under the authority of the Superfund Amendments and Reauthorization Act of 1986) sets forth the requirements for hazardous waste operations at Superfund and other cleanup sites. Hazardous materials emergency response training and other requirements are also outlined in this rule, which is documented in 29 CFR §1910.120.

***Occupational Exposure to Hazardous Chemicals in Laboratories* (*Effective 1992*).** This standard sets forth requirements to pertaining to hazardous chemical exposure in laboratories, as defined in 29 CFR §1910.1450. The rule requires the employer to develop and carry out the provisions of a written chemical hygiene plan, including designating a chemical hygiene officer and, if appropriate, a chemical hygiene committee. The chemical hygiene plan must be capable of protecting employees from health hazards associated with hazardous chemicals in the laboratory, including keeping exposures below the limits specified under 29 CFR §1910, Subpart Z, as well as specifying standard operating procedures relevant to safety and health considerations. Other requirements include monitoring, employee information and training, medical consultation and examinations, appropriate labeling, use of respirators (if appropriate), and record-keeping requirements.

***Standard for Process Safety Management of Highly Hazardous Chemicals* (*Effective 1992*).** This rule, published in 29 CFR §1910.119, lists over 100 highly hazardous chemicals and threshold quantities ofthese chemicals. If the chemical is manufactured, stored, or used in an amount greater than the threshold quantity, the law requires the facility to conduct a process hazard analysis for all associated process equipment. In addition, employees are required to receive training on process hazards associated with the equipment.

Personal Protective Equipment Enhanced Standard (Effective 1994). This rule, defined under 29 CFR §1910, Subpart I, specifies requirements pertaining to personal protective equipment (PPE). Included are requirements to assess hazards and select appropriate PPE based on those hazards, to communicate the selection decisions to affected employees, and to provide training to employees required to use PPE. Specific requirements are detailed for eye and face protection; respiratory protection; head protection; foot protection; electrical protective equipment; and hand protection.

Clean Air Act (CAA) Amendments of 1970, 1977, and 1990

The CAA Act originated in 1963, with most of the key aspects of the law as we know it today defined by the 1970, 1977, and 1990 CAA Amendments. Regulations based on the CAA Amendments govern emissions of pollutants into the atmosphere from industrial and commercial activity. The regulations, delineated in 40 CFR Parts 0–99, focus on categories of industrial processes as well as specific hazardous and toxic pollutants. Definitions of what constitutes a major source, definitions of air quality standards, and control requirements based on these definitions are detailed in the rules. Methods for sampling various constituents coming from stacks and vents are also set forth in the regulations.

Among other things, the 1970 CAA Amendments required EPA to set National Ambient Air Quality Standards (NAAQS) for six air pollutants that posed a significant threat to human health. Congress set 1975 as the deadline for achieving these goals, but this deadline was not met. In an effort to achieve NAAQS and other national air quality goals, the CAA Amendments of 1977 required that new processes or sources be installed with best available control technology (BACT). Areas that exceeded (or did not attain) NAAQS were considered nonattainment areas. As such, these areas required additional controls on sources in order to achieve attainment of the standards, and proportional decreases (or offsets) for emissions were required for each emission increase.

National standards were also set for specific pollutant sources, such as dry cleaners and plastics manufacturers, and were termed new source performance standards (NSPS). New sources constructed or modified after NSPS are proposed are required to meet these standards. Additionally, numerous hazardous air pollutants were listed and regulated via individual standards.

The CAA Amendments of 1990 altered significantly the national air pollution control strategy. While all new sources in attainment areas are still required to meet BACT, new sources in nonattainment areas are required to meet the lowest achievable emission rate (LAER) for the pollutants that exceed NAAQS. Existing sources in nonattainment areas are required to retrofit emission sources with reasonably available control technology (RACT). Additionally, decreases in emissions from sources at ratios greater than 1:1 will have to be demonstrated before a new source will be permitted in a nonattainment area.

The new CAA Amendments of 1990 also affected significantly the regulation of hazardous air pollutants. Rather than regulating these pollutants on a specific basis, EPA established source categories, such as refineries and ethylene manufacturers, and regulates emissions of hazardous air pollutants by these source categories. Maximum achievable control technology (MACT) is required for these sources. Sources are expected to retrofit abatement equipment to meet MACT within three years of promulgation of a standard. With MACT, it is expected that hazardous air pollutants can be reduced significantly.

Clean Water Act (CWA), Formerly the Federal Water Pollution Control Act of 1972 Amended 1977 and 1987

The basic statutory structure for water quality management was set in place with the Federal Water Pollution Control Act in 1972. This act set up the framework for the establishment of minimum acceptable requirements for water quality and wastewater management. In 1977 the Act was renamed the Clean Water Act and was enhanced to include increased controls on toxic pollutants. In 1987, the Water Quality Act provided further amendments and enhancements to the control of toxic pollutants.

Point source discharges affected by this act include wastewater discharges and stormwater discharges from industrial and commercial facilities, municipalities and POTWs, private treatment plants, and other sources. Regulations governing these discharges are found in several places in the Code of Federal Regulations, depending on the particular focus of regulation. Effluent guidelines are defined in 40 CFR Parts 401–471, general pretreatment standards are provided in 40 CFR Part 403, and regulations pertaining to permitting are located in 40 CFR Parts 122–125. Water quality standards are given in 40 CFR Part 131.

The effluent guidelines for wastewater discharges cover a limited, but diverse number of specific industrial and commercial categories, such as hospitals, food processors, metal finishers, semiconductor manufacturers, and chemical manufacturers. Both facilities that discharge directly to water bodies and those that discharge to POTWs are covered by effluent guidelines.

The general pretreatment standards regulate nondomestic discharges to POTWs. These pretreatment standards detail responsibilities for POTWs with flows of greater than 5 million gallons per day that receive nondomestic discharges. The standards also set out requirements for those categorical discharges covered by effluent guidelines, discharges that are considered to have significant impacts on the POTW, and all other nondomestic discharges to POTWs.

Permitting of wastewater and stormwater discharges that go directly to water bodies is administered under the NPDES process. Discharges that require permitting include wastewater discharges covered by effluent guidelines, wastewater discharges from POTWs and from privately owned treatment facilities, and stormwater discharges from urban municipalities, construction sites, and certain industries.

Water quality standards define the quality goals of a water body. They designate the use or uses to be made of the water (i.e., for the protection and propagation of fish or for recreational purposes), and they set criteria necessary to protect these uses. Once established, the standards become the basis for the establishment of water quality–based treatment controls and other strategies for water quality protection.

Federal Insecticide, Fungicide, and Rodenticide Act (FIFRA) of 1972

Although actually created in 1947, the "modern" FIFRA was enacted through the Federal Environmental Pesticide Control Act (FEPCA) of 1972, which amended (and virtually rewrote) the FIFRA law. Provisions within this act included strengthened enforcement and a focus on effects on health and environment. Subsequent amendments were enacted in 1975, 1978, 1980, and 1988.

A key aspect of FIFRA is the pesticide registration program. In essence, the law requires that before a pesticide can be manufactured, distributed, or imported, it must be approved by EPA. Data gathering and material testing can take years (and cost millions of dollars), and subsequent evaluation by EPA can take equally long. Thus the process is laborious and not a simple notification process. In addition to registration, FIFRA also regulates pesticide use, in that some categories of pesticides can be applied only by certified applicators. The numerous requirements of FIFRA are presented in 40 CFR Parts 150–189.

Safe Drinking Water Act (SDWA) of 1974 and Amendments of 1986

The SDWA authorized EPA to regulate contaminants in public drinking water systems by establishing national standards to be implemented and enforced by EPA or authorized states. Large and small municipal systems alike are regulated under SDWA, which is delineated in 40 CFR Parts 140–149. The 1986 amendments strengthened the act in several key ways. It directed EPA to set maximum concentration levels for 83 priority contaminants in three years, and 25 contaminants every three years thereafter; it established requirements for certain types of treatment, such as disinfection and filtration; it strengthened the statute's enforcement authorities; it expanded groundwater protection provisions; and it required public water systems to monitor unregulated contaminants.

In addition to regulating public water systems, SDWA also has authority regulating underground injection. Pursuant to SDWA, EPA established the underground injection control (UIC) program, which regulates underground injection by class of well. The five well classes are: Class I, industrial and municipal waste disposal wells injecting fluids below the lowermost drinking water formations; Class II, wells associated with the oil and gas industry, including saltwater injection wells, enhanced recovery wells, wells injecting liquid hydrocarbons for storage, and certain wells at natural gas plants; Class III, wells for injected fluids for the extraction of minerals; Class IV, wells that inject hazardous waste or radioactive waste into a formation $\frac{1}{4}$ mile from an underground source of drinking water; and Class V, all other injection wells.

Toxic Substance Control Act (TSCA) of 1976

The intent of TSCA is to assure that chemicals manufactured or imported into the United States are registered and listed in the TSCA registry. Rules for this act are defined in 40 CFR Parts 700–799. To receive authorization to manufacture or import for commercial purposes a new chemical, a premanufacture notice (PMN) must be filed with EPA (some exceptions apply). Through this registration process, environmental fate and health effects studies must be developed and submitted. Other aspects of the regulation require periodic reporting of production and importation quantities and record keeping and reporting of allegations of health and environmental effects and associated investigations.

Chemical manufacturers are most affected by this law and its regulations. They are required to file the PMN and to study and report health and environmental data about each new chemical. Research and development facilities are exempt from filing a PMN if certain requirements are met. Facilities that use chemicals in their processes must ensure that appropriate chemical information accompanies the chemical and that the chemical is registered under TSCA before it is received. These facilities are also subject to record-keeping and reporting requirements related to health and environmental allegations.

In addition to requirements pertaining to new chemicals, TSCA regulates polychlorinated biphenyls (PCBs) and asbestos. Included in these regulations are labeling requirements, record-keeping requirements, and other requirements. The requirements for PCBs are published under 40 CFR Part 761. Asbestos requirements are given in 40 CFR Part 763.

Resource Conservation and Recovery Act (RCRA) of 1976

EPA promulgated rules under RCRA to regulate the management of hazardous solid waste from generation to final disposal, also termed "cradle to grave." These rules are defined in 40 CFR Parts 260–270. Under these regulations the term *solid waste* includes waste solids, sludges, liquids, and containerized gases.

The regulations include criteria for defining a hazardous waste. The definitions for hazardous waste include hazardous characteristics such as ignitability or toxicity, wastes from nonspecific sources such as certain spent halogenated and nonhalogenated solvents, and wastes from specific sources such as wastewater treatment sludges generated from a specific process. In addition, the rules list several hundred toxic and acutely hazardous chemicals that if discarded or spilled, become a hazardous waste.

The regulations also set forth standards for hazardous waste generators, transporters, and owners and operators of treatment, storage, and disposal facilities. Included in the standards are facility-permitting requirements, transporter identification and tracking requirements, record-keeping and inspection requirements, requirements for financial bonding, and other requirements.

Hazardous and Solid Waste Amendments (HSWA) of 1984. These amendments were enacted and subsequent regulations were developed to strengthen the RCRA regulations. These regulations provide technical standards for landfill disposal, leak detection systems, and underground storage of petroleum products and CERCLA hazardous substances. Additionally, the regulations prohibit specified hazardous wastes from land disposal unless certain treatment standards are met. HSWA also regulates continuing releases from solid waste management units.

Medical Waste Tracking Act of 1988. This act, aimed specifically at medical waste management, required EPA to set up a demonstration program for characterizing and tracking of medical waste, and evaluating treatment techniques for this waste. The specifics of the demonstration program and other aspects of medical management regulation are published in 40 CFR Part 259. Information from the demonstration program may be used to draft additional regulations for this type of solid waste.

Pollution Prevention Act of 1990. This act has changed the focus of cradle-to-grave management of hazardous waste to "cradle-to-cradle" management. The new cradle-to-cradle concept emphasizes prevention of waste through recycling, source reduction, elimination of toxic materials, and other methods to attain environmentally conscious manufacturing. In essence, the act required EPA to develop programs to minimize pollution and to develop a list of priority chemicals that will be the target of minimization (and in some cases elimination) programs.

Comprehensive Environmental Response, Compensation, and Liability Act of 1980 (CERCLA or Superfund Act) and Amendments

This act—termed CERCLA or Superfund interchangeably—and several other acts passed under the Superfund umbrella provided for liability, compensation, cleanup, and emergency response for hazardous substances released into the environment. Regulations pertaining to the naming and cleanup of Superfund sites, which are designated in the National Priority List, are found in 40 CFR Part 300. There are presently 1000 sites, with over 100 federal facilities on the list. The 1986 amendments to the act provided for $8.5 billion to be allocated for cleanup of these sites over a five-year period, as well as the requirement for potentially responsible parties to share in cleanup and associated costs.

Another aspect of this act is spill reporting. Spill reporting requirements are published in 40 CFR Part 302 and include a list of hundreds of hazardous materials and reporting quantities. The National Response Center must be notified immediately if the material is released to the environment without a permit in amounts greater than the reporting quantity. This requirement for reporting allows EPA to track hazardous materials and hazardous waste incidents. Once reported, these incidents become part of the public record.

Emergency Planning and Community Right-to-Know Act (EPCRA) of 1986. EPCRA is often referred to as SARA Title III since the provisions of the act are incorporated in Title III of the Superfund Amendments and Reauthorization Act (SARA) of 1986. Regulations resulting from this act are delineated in 40 CFR Parts 355, 370, and 372.

The act required the establishment of local emergency planning and release notification, which were accomplished under 40 CFR Part 355. The rule requires each state to set up local emergency planning committees (LEPCs) to collect local or regional data for hazard assessment. The LEPCs were required to develop an emergency response plan by October 1988, which incorporated the hazards found in the local area. This plan was to be provided to the state. Also, release notification requirements are outlined in this part of the regulations. Any off-site release of a listed chemical above a specified quantity is required to be reported to the LEPC and the State Emergency Response Commission (SERC).

Another key aspect of the act was to provide knowledge to the community about chemicals stored and released from facilities in the area. This is provided by two reports. The first report, known as the tier I/tier II report and defined in 40 CFR §§370.40 and 370.41, requires aggregate information pertaining to hazard type (tier I) or a list of chemicals stored in threshold amounts at the facility, including locations, amounts, and hazards (tier II). The second report, the toxic release inventory (TRI) report, is defined in 40 CFR §372.85. This report is required by facilities that manufacture or process certain chemicals in quantities over 25,000 pounds or that use these chemicals in quantities over 10,000 pounds. In this report, a facility specifies information pertaining to the chemical. Before 1991, required information included estimates or actual data pertaining to on-site waste treatment and treatment efficiencies, releases to wastewater and air, and on- and off-site disposal of certain chemicals. As of 1991, the report must include data pertaining to off- and on-site waste recycling, energy recovery from disposal of materials, and source reduction activities used at the facility. EPA can use these data to perform technology and other assessments and to recommend best management practices for specific chemicals.

Radon Gas and Indoor Air Quality Research Act of 1986. Incorporated into the SARA amendments was the provision under Title IV for setting up a research program for radon gas and indoor air quality. The program is intended to identify, characterize, and monitor the sources and levels of indoor air pollution, particularly radon. Control technologies and other mitigation measures to prevent or abate radon gas and other indoor air pollution are to be researched and developed. It is likely that regulations pertaining to indoor air quality will be promulgated after the research is completed.

Oil Pollution Act (OPA) of 1990

The OPA addresses oil pollution liability and prevention comprehensively. The act establishes the federal liability scheme for vessels and facilities that spill oil on waters subject to U.S. jurisdiction, including enumerating compensable damages and financial responsibility requirements. Also addressed are U.S. requirements for prevention and removal, including licensing requirements for drug and alcohol testing, tank vessel manning, double-hull requirements for tank vessels, and marine casualty reporting. Requirements for international oil pollution prevention and removal include participation in an international prevention and compensation regime that is at least as effective as domestic law.

Hazardous Materials Transportation Act (HMTA) of 1975

The HMTA made the Department of Transportation responsible for regulating the transportation of hazardous materials. Regulations pertaining to packaging, container handling, labeling, vehicle placarding, and other safety aspects are detailed in 49 CFR Parts 171–180. Also included are requirements for reporting accidents involving hazardous materials and hazardous waste. Some DOT regulations overlap with regulations from other agencies, such as NRC packaging requirements and EPA hazardous waste manifest requirements.

In late 1990, the act was amended by the Hazardous Materials Transportation Uniform Safety Act of 1990, commonly termed HM 181, so-called for its assigned docket number. The new regulations aligned the categories of hazardous materials to international regulations based on UN recommendations, and addressed outdated nonbulk packaging terminology and specifications and other topics.

Atomic Energy Act of 1954

The Atomic Energy Act was the original act governing the production and use of source materials, special nuclear materials, and by-product materials for defense and peaceful purposes. The Atomic Energy Commission (AEC) was set up to license the processing and use of these nuclear materials. Additionally, the act granted to the AEC the responsibility for regulating health and safety (and the environment) at nuclear facilities.

Energy Reorganization Act of 1974. This act divided the AEC into two agencies, the agency that is now known as the Department of Energy (DOE), and the Nuclear Regulatory Commission (NRC). The act made DOE responsible for energy development and defense production activities, while the NRC was given licensing authority for civilian nuclear energy activities and certain defense activities.

Uranium Mill Tailings Radiation Control Act of 1978. This act established the basis for regulation of uranium and thorium mill tailings separately from other radioactive materials and wastes. In general, the act deals with the control and stabilization of these wastes for protection of public health and the environment. Regulations were promulgated with respect to this act by EPA and NRC under 40 CFR Part 192 and 10 CFR Part 40, respectively. EPA standards govern control and cleanup of residual radioactive materials at inactive uranium-processing sites. NRC regulations establish technical criteria for siting and design of disposal facilities for protection of the groundwater.

Low-Level Radioactive Waste Policy Act of 1980 and 1985 Amendments. This act defined low-level radioactive wastes and made each state responsible for providing disposal capacity for these wastes generated within its border. It encouraged regional compacts and allowed the compacts ratified by Congress to exclude waste generated outside their borders beginning January 1, 1986.

When it became evident that regions without waste sites would be unable to have facilities operating by the deadline of 1986, the Low-Level Radioactive Waste Policy Act Amendments of 1985 were enacted. These amendments extended the deadline for facility sitings to 1993 while providing a series of specific dates for progress toward new facility construction. The amendments also specified which categories of low-level radioactive waste are a state responsibility, established volume ceilings for individual nuclear reactors and for operating disposal sites, and set forth other requirements.

Standards that regulate the disposal of low-level radioactive waste are found in 10 CFR Part 61. Included are performance objectives, technical requirements for land disposal facilities and financial assurances. Licensing requirements for disposal sites and administrative requirements such as record keeping, reporting, and inspections are addressed.

Packaging and transport of radioactive materials, including low-level radioactive waste, is regulated in 10 CFR Part 71. These packaging and transport standards address NRC and DOT requirements.

Nuclear Waste Policy Act of 1982. This act established a program for the disposal of civilian spent fuel and high-level waste in geologic repositories.

The facilities are operated by DOE and licensed by NRC. High-level waste is defined by NRC in 10 CFR Part 60. Included in these regulations are licensing criteria for geologic repositories. Management and disposal of this type of waste are also regulated by EPA under 40 CFR Part 191. Included are standards for public protection, containment requirements, qualitative assurance requirements, and groundwater protection requirements.

REFERENCES

Code of Federal Regulations, 10 CFR Parts 40, 60, 61, and 71, Nuclear Regulatory Commission, Washington, DC.

———, 29 CFR Parts 1900–1910, Occupational Safety and Health Administration, U.S. Department of Labor, Washington, DC.

———, 40 CFR Parts 0–99, U.S. Environmental Protection Agency, Washington, DC.

———, 40 CFR Parts 122–125, 131, and 401–471, U.S. Environmental Protection Agency, Washington, DC.

———, 40 CFR Parts 140–149, U.S. Environmental Protection Agency, Washington, DC.

———, 40 CFR Parts 150–189, U.S. Environmental Protection Agency, Washington, DC.

———, 40 CFR Parts 191–193, U.S. Environmental Protection Agency, Washington, DC.

———, 40 CFR Parts 259–270, U.S. Environmental Protection Agency, Washington, DC.

———, 40 CFR Parts 355, 370, and 372, U.S. Environmental Protection Agency, Washington, DC.

——— 40 CFR Parts 700–799, U.S. Environmental Protection Agency, Washington, DC.

———, 40 CFR Parts 171–180, Department of Transportation, Washington, DC.

Environmental Law Deskbook, Environmental Law Institute, Washington, DC, 1989.

Environmental Law Handbook, 13th ed., Government Institutes, Rockville, MD, 1995.

Environmental Statutes, 1996 ed., Government Institutes, Rockville, MD, 1996.

5 U.S.C. §§550–559, Administrative Procedure Act.

——— §552, Freedom of Information Act.

7 U.S.C. §136, Federal Insecticide, Fungicide, and Rodenticide Act.

15 U.S.C. §§2601–2671, Toxic Substances Control Act (TSCA).

——— §§2601–2629, TSCA Control of Toxic Substances.

——— §2605e, TSCA Regulation of Polychlorinated Biphenyls (PCBs).

——— §§2641–2655c, TSCA Asbestos Hazard Emergency Response.

——— §§2661–2671, TSCA Indoor Radon Abatement.

29 U.S.C. §§651–678, Occupational Safety and Health Act.

——— §§653, 655, 657, Process Safety Management of Highly Hazardous Chemicals.

——— §655g, Hazard Communication Standard.

——— §§655 and 657, Hazardous Waste Operations and Emergency Response Standard.

33 U.S.C. §§1251–1387, Federal Water Pollution Control Act (Clean Water Act).

42 U.S.C. §§300f–300j-26 Safe Drinking Water Act.

——— §§2011–2021, 2022–2286i, Atomic Energy Act.

——— §§2021b–2021j, Low-Level Radioactive Waste Policy Act.

——— §§4321–4370a, National Environmental Policy Act.

——— §§6901–6992k, Resource Conservation and Recovery Act (RCRA).

——— §§6921–6939b, RCRA Hazardous Waste Management.

——— §§6991–6991i, RCRA Regulation of Underground Storage Tanks.

——— §§6992–6992k, RCRA Demonstration Medical Waste Tracking Program.

——— §§7401–7626, Clean Air Act, as amended.

——— §7401, Sec. 401–405, Radon Gas and Indoor Air Quality Research Act (Title IV of the Superfund Amendments and Reauthorization Act).

——— §7918, Uranium Mill Tailings Radiation Control Act.

——— §§9601–9675, Comprehensive Environmental Response, Compensation, and Liability Act, as amended.

——— §§10101–10270, Nuclear Waste Policy Act of 1982.

——— §§ 11001–11050, Emergency Planning and Community Right-to-Know Act (Title III of the Superfund Amendments and Reauthorization Act).

49 U.S.C. §§1801–1813, Hazardous Materials Transportation Act.

BIBLIOGRAPHY

Bierlein, L. (1987). *Red Book on Transportation of Hazardous Materials,* 2nd ed., Van Nostrand Reinhold, New York.

Burns, M. E. (1987). *Low-Level Radioactive Waste Regulation,* Lewis Publishers, Boca Raton, FL.

Carpenter, D. A., R. F. Cushman, and B. W. Roznowski (1991). *Environmental Dispute Handbook: Liability and Claims,* Wiley, New York.

Chemical Manufacturers Association (1987). *A Manager's Guide to Title III,* CMA, Washington, DC.

——— (1991). *NPDES Discharge Permitting and Compliance Issues Manual,* CMA, Washington, DC.

——— (1989). *Overview of the Resource Conservation and Recovery Act,* CMA, Washington, DC, videotape.

Cooper Musselman, V. (1989). *Emergency Planning and Community Right-to-Know,* Van Nostrand Reinhold, New York.

Department of Energy (1991). *OSHA Training Requirements for Hazardous Waste Operations,* DOE/EH-0227P, DE92 004780, Office of Environment, Safety, and Health, U.S. DOE, Washington, DC.

Environmental Protection Agency, (1989). *CERCLA Compliance with Other Laws Manual: Parts I and II,* OSWER Directives 9234.1-01 and 9234.1-02, Office of Solid Waste and Emergency Response, U.S. EPA, Washington, DC.

——— (1991). *Land Disposal Restriction: Summary of Requirements,* OSWER 9934.0-1A, Office of Waste Programs Enforcement, Office of Solid Waste and Emergency Response, U.S. EPA, Washington, DC.

Environmental Resource Center (1989). *How to Comply with the OSHA Hazard Communication Standard,* Van Nostrand Reinhold, New York.

Fire, F. L., N. K. Grant, and D. H. Hoover (1989). *SARA Title III,* Van Nostrand Reinhold, New York.

Gershey, E. L., R. C. Klein, and A. Wilkerson (1990). *Low-Level Radioactive Waste: Cradle to Grave,* Van Nostrand Reinhold, New York.

Gibson, M. M. (1993). *Environmental Regulation of Petroleum Spills and Wastes,* Wiley, New York.

Government Institutes (1991). *Environmental Law Handbook,* 11th ed., Government Institutes, Inc., Rockville, MD.

Hartl, A. V. (1994). *Transporting Hazardous Materials: Law and Compliance,* Wiley, New York.

Keith, L. H., and D. B. Walters (1992). *The National Toxicology Program's Chemical Data Compendium,* Vol. 3, "Standards and Regulations," Lewis Publishers, Boca Raton, FL.

National Governors' Association (1986). *The Low-Level Radioactive Waste Handbook: A User's Guide to the Low-Level Radioactive Waste Policy Amendments Act of 1985,* NGA 444, Center for Policy Research, NGA, Washington, DC.

Stensvaag, J.-M. (1989). *Hazardous Waste Law and Practice,* Vol. 2, Wiley, New York.

Stensvaag, J.-M. (1991). *Clean Air Act 1990 Amendments: Law and Practice,* Wiley, New York.

Wagner, T. P. (1989). *The Hazardous Waste Q & A,* Van Nostrand Reinhold, New York.

2

VOLUNTARY MANAGEMENT STANDARDS AND INITIATIVES

Traditionally, the United States has regulated industry through command-and-control regulations such as those detailed in Chapter 1. Recently, voluntary management standards and initiatives have gained credibility with OSHA, EPA, industry, and the general public. In this chapter we focus on two of the most important of the voluntary programs: ISO 14001, the international environmental management system standard, and OSHA's voluntary protection programs (VPPs). Benefits of subscribing to these voluntary standards and initiatives are presented in Table 2.1.

ISO 14001: ENVIRONMENTAL MANAGEMENT SYSTEM SPECIFICATION

The International Organization for Standardization (ISO) has been a leader in the international technical arena for a half century. ISO is commonly used when referring to the organization and its standards, but the term *iso* is actually a Greek word meaning "equal." Technically speaking, ISO is primarily a body for developing products and safety standards; however, in the late 1980s, it journeyed off its traditional path by developing a series of quality management standards, the ISO 9000 series. These quality management standards are process standards (i.e., they specify a process and not an end goal), the first of their kind for ISO. The ISO 9000 series brought world recognition to the organization as a leader in developing these types of standards. Further, although all ISO standards are termed voluntary, in some countries registration to the appropriate ISO 9000 standard(s) has become a requirement for trade.

With the resounding success of the ISO 9000 series, in mid-1991 ISO once again embarked on a new journey—to environmental management standards. Like the ISO 9000 series, these standards are process standards. They are not intended to set environmental performance goals but rather, to specify the

TABLE 2.1 Benefits of Subscribing to ISO 14001 and Voluntary Standards and Initiatives

- Positions facility as a leader in hazardous materials and hazardous waste management
- Provides a framework for establishing an integrated approach to hazardous materials and hazardous waste management
- Promotes a positive image in the community and with regulating agencies
- Involves commitment and focus from top management
- Integrates the concepts of hazardous materials and hazardous waste management into the fabric of the organization
- Provides an impetus for continual improvement of the hazardous materials and hazardous waste management system

Source: Adapted from Woodside et al. (1998).

elements of a management system that provides a framework for an organization to develop and maintain a reliable process that consistently meets environmental obligations and commitments.

ISO 14001—the environmental management system (EMS) specification—which was finalized and issued as a first edition on September 1, 1996, is the most widely recognized environmental management standard. It is a specification standard, which means that organizations which conform to its requirements can become registered to the standard. ISO 14001 was written as a consensus standard with nearly 50 countries participating. The standard is applicable to all types and sizes of organizations, and it accommodates diverse geographical, cultural, and social conditions. It can be applied to all parts or any single part of an organization and/or its activities, products, and services. A basic model of the standard is depicted in Figure 2.1. A discussion of each of the elements of ISO 14001 follows.

Figure 2.1 Environmental management system model. (Adapted from Woodside et al., 1998.)

Environmental Policy (Section 4.2)

It is top management's responsibility to set the environmental policy. This policy forms the foundation of the entire EMS. There are certain requirements that the environmental policy must meet, including the following:

- It must be appropriate to the nature, scale, and environmental impacts of the organization's activities, products, and services.
- It must include a commitment to continual improvement of prevention of pollution.
- It must include a commitment to comply with relevant environmental legislation and regulations and with other (voluntary) requirements to which the organization subscribes.
- It must provide a framework for setting and reviewing environmental objectives and targets.
- It must be documented, implemented, and maintained.
- It must be communicated to all employees.
- It must be available to the public.

The organization should consider it own situation and what it can strive for with this environmental policy. In addition to minimizing use of raw materials and minimizing waste generation, the organization might consider goals such as being a responsible neighbor, use of recycled materials and products, as appropriate, and a commitment to sustainable development.

Examples of how to communicate the policy to employees, contractors, and the general public are presented in Table 2.2.

Planning (Section 4.3)

The planning section of the ISO 14001 includes four elements: environmental aspects, legal and other requirements, objectives and targets, and the environ-

TABLE 2.2 Environmental Policy Communication Methods

• New employee training
• Contractor training
• Inclusion in contracts
• Open house for external parties
• Employee newsletters
• Earth Day activities
• Tent cards in the cafeteria
• Posting at copy and fax machines
• Posting near recycling bins

mental management program. The following paragraphs describe each of these elements.

Environmental Aspects (Section 4.3.1). Perhaps one of the critical elements of the ISO 14001 standard, this section requires the organization to consider the inputs and outputs of its activities and to evaluate which of its environmental aspects are significant. Environmental aspects are "those elements of an organization's activities, products, and services that can impact the environment" (ISO 14001, Section 3.3). Examples of aspects and significant aspects could include air emissions, chemical use, noise, unplanned releases, wastewater discharges, and other such elements. A sample schematic of inputs and outputs and the environmental aspects of a typical manufacturing company is presented in Figure 2.2.

Legal and Other Requirements (Section 4.3.2). This element of ISO 14001 is succinct and to the point. It requires an organization to establish and maintain a procedure for identifying and providing access to legal requirements applicable to its activities, products, and services. Further, if the organization subscribes to other requirements—such as voluntary agency or industry requirement or internal company mandate—these requirements must also be part of the procedure.

Objectives and Targets (Section 4.3.3). The ISO 14001 standard requires an organization to establish and maintain documented objectives and targets. Elements to consider when setting these objectives and targets include:

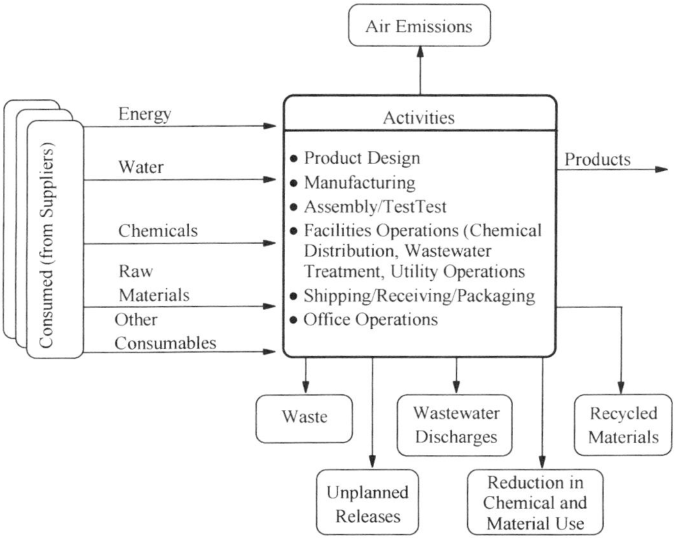

Figure 2.2 Environmental aspects process. (Adapted from Woodside et al., 1998.)

- Relevant legal requirements and other requirements to which the organization subscribes
- Significant environmental aspects of the organization's activities, products, and services
- Technological options available to the organization
- The organization's financial, operational, and business requirements
- The views of interested parties

This list—deceivingly short—covers a broad range of topics and takes into account legal and other requirements, the organization's specific business situation, and the views of interested parties, the latter being "an individual or group concerned with or affected by the environmental performance of the organization" (ISO 14001, Section 3.11). Examples of such could include a variety of groups—neighborhood, environmental, citizen, politcal—as well as employees, stockholders, or customers. Examples of objectives and targets that might be set by a manufacturing facility, based on identified significant aspects, are presented in Table 2.3.

Environmental Management Program (Section 4.3.4). Defining the environmental management program is the next step after setting objectives and targets. Essentially, this program describes the implementation process with respect to the objectives and targets in that it describes the "who," the "when," and the "how." The standard specifically requires the organization to define:

TABLE 2.3 Examples of Objectives and Targets Based on Significant Aspects Identified

Significant Aspect	Objective	Target
Chemical use	Reduce chemical use over time.	Evaluate chemical use in major manufacturing processes. Set reduction targets based on results of evaluation.
Energy use	Reduce energy consumption.	Reduce consumption 5% over the preceding year.
Wastewater discharge	Reduce the amount of copper in effluent.	Reduce copper in effluent by 20%.
Waste	Reduce the amount of waste requiring disposal.	Reduce the amount of hazardous waste generated by 4%. Reduce the amount of nonhazardous waste sent to the landfill by 10%.

- The designation of responsibility for achieving objectives and targets at each relevant function and level of the organization
- The time frame by which they are to be achieved
- The means by which they are to be achieved

The environmental management program must be updated as new developments or modifications to activities, products, and services warrant.

Implementation and Operation (*Section 4.4*)

Once the planning elements of the standard have been accomplished, it is time to proceed with implementation and operation of the EMS. Elements that are part of this section of the standard include structure and responsibility; training, awareness, and competence; communication; EMS documentation; document control; operational control; and emergency preparedness and response.

***Structure and Responsibility* (*Section 4.4.1*).** Requirements pertaining to this element are as follows:

- Roles, responsibilities, and authorities must be defined, documented, and communicated.
- Management must provide resources essential to the implementation and control of the EMS.
- Resources should include human resources and specialized skills, technology, and financial resources.
- Top management must appoint representative(s) to establish, implement, and maintain the EMS.
- Top management must appoint representative(s) to report on the performance of the EMS to them for review of the EMS's suitability, adequacy, and effectiveness and as a basis for improvement of the EMS.

Essentially, this section of the standard ensures that those persons who are responsible for the EMS are formally defined and that all roles and responsibilities are communicated at all levels of the organization.

***Training, Awareness, and Competence* (*Section 4.4.2*).** Training of relevant personnel is integral to proper functioning of the EMS. Since training is dependent on the organization's activities, products, and services, the organization must identify training needs. All personnel whose work may affect the environment must receive appropriate training. In addition, the organization must establish and maintain procedures in order to make employees or members at each relevant level aware of:

- The importance of conformance with the environmental policy and procedures and with the requirements of the EMS.
- The significant environmental impacts, actual or potential, or their work activities and the environmental benefits of improved personal performance.
- Their roles and responsibilities in achieving conformance with the environmental policy and procedures and with the requirements of the EMS, including emergency preparedness and response requirements.
- The potential consequences of departure from specified operating procedures.

In addition, the standard requires that personnel performing tasks that could cause a significant impact must be competent to perform that task, based on education, training, or experience.

Communication (Section 4.4.3). This is perhaps one of the most important elements of the standard. This element requires the organization to establish and maintain procedures for both internal and external communications about significant aspects and the EMS. Internal communication is expected to be multidirectional: that is, not just from the top down, but also from the bottom up and throughout relevant levels and functions of the organization. Examples of internal communications include the following:

- Department or functional meetings that review significant environment aspects and progress toward achieving objectives and targets
- Employee newsletters or brochures that communicate element of the EMS
- Internal Web page
- Periodic internal reports about the status of the EMS
- A publicized internal phone number to provide information about the EMS and/or to allow feedback or recommendations for improvements
- Area wall charts that depict environmental measurements
- Ongoing communication between the environmental staff and relevant functions/levels with respect to environmental management program and efforts toward prevention of pollution and continual improvement

Examples of external communication methods include the following:

- External environmental performance reports
- Communication through stockholder reports
- An external communication hotline
- External Web page

- Presentations at industry and/or governmental meetings about the organization's EMS and environmental performance

***Environmental Management System Documentation* (*Section 4.4.4*).** This section of the standard requires the organization to establish and maintain information that describes the core elements of the EMS and their interaction, and to provide direction to related documentation. The information can be in paper or electronic form. Examples of core elements could include the environmental policy, the EMS manual, and facility-wide operating procedures. Supporting (related) documentation might be department procedures or desk procedures.

***Document Control* (*Section 4.4.5*).** This section of the standard should not be confused with the extensive and rigid requirements of ISO 9000 (the quality management system standards). Although the name of the section implies compatibility with ISO 9000, in fact ISO 14001 is somewhat vague about what is required with respect to this element. Essentially, the organization must establish procedures to ensure the following:

- Documents can be located.
- Documents are reviewed periodically, revised as necessary, and approved for adequacy by authorized personnel.
- Current versions of the documents are available where needed.
- Obsolete documents are removed promptly or otherwise assured against unintended use.
- Obsolete documents retained for legal and/or knowledge-preservation purposes are suitably identified.
- Responsibilities are established concerning the creation and modification of the various types of documents.

An environmental documentation pyramid (which includes records) is presented in Figure 2.3. As shown in the figure, there can be numerous levels of documents. The organization is free to use varied levels of document control for each level of documents as long as they are specified in the document control procedure.

***Operational Control* (*Section 4.4.6*).** This section requires the organization to do the following:

- Establish and maintain documented procedures to cover situations where their absence could lead to deviations from the environmental policy and objectives and targets.
- Stipulate operating criteria in the procedures.

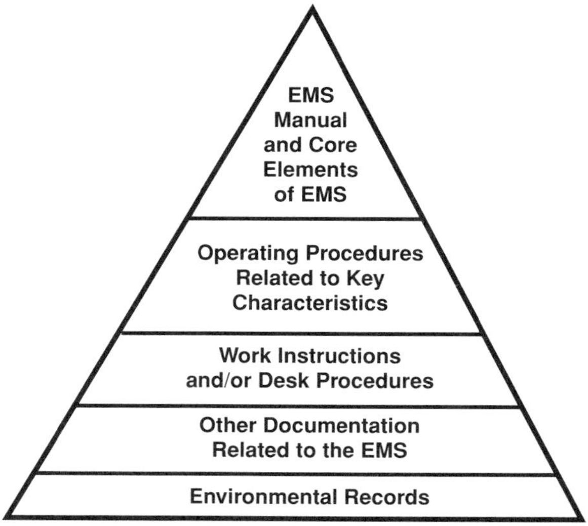

Figure 2.3 Environmental management system documentation pyramid.

- Establish and maintain procedures related to identifiable significant aspects of goods and services used by the organization.
- Communicate relevant procedures and requirements to suppliers and contractors.

Procedures that control operations have long been part of most EMSs. Thus organizations can take advantage of what is in place that the facility can use to meet the first part of this requirement.

In addition, organizations must ensure that significant aspects of good and services used by the organization are covered by procedures, and these procedures must be communicated to applicable suppliers and contractors. These procedures can be communicated in the contract, through contractor training, or by other means.

Emergency Preparedness and Response (Section 4.4.7). ISO 14001 requires an organization to establish and maintain procedures to identify and respond to emergency situations and to prevent and mitigate environmental impacts associated with them. In the United States, emergency preparedness and response programs and plans are typically mature. Types of information covered in these plans include:

- General information, such as an overview of operations and potential emergency scenarios and designated personnel who are to respond during emergencies.

- Emergency notification information, such as facility coordinators, community response personnel, emergency hotline numbers, and names and numbers of regulatory agency contacts.
- Reportable information, such as information about the facility, time and place of incident, name and quantity of material involves, possible hazards to human health and/or the environment, and potential impact outside the facility.
- Available resources listings, such as spill equipment list and list of nearby medical facilities and fire stations.
- Emergency-specific information, such as weather emergencies, fire and explosion emergencies, chemical-release emergencies, and evacuation procedures.
- Site information, such as evacuation routes, location of safety equipment, and location of chemical-spill supply carts.

Checking and Corrective Action (Section 4.5)

***Monitoring and Measurement* (*Section 4.5.1*).** This element of the ISO 14001 standard requires that the organization establish and implement procedures to monitor and measure, on a regular basis, key characteristics of its operations and activities that could have a significant impact on the environment. When conforming to this requirement, the organization must record information to track performance, relevant operational control, and conformance with the organization's environmental objectives and targets. In addition to monitoring and measuring environmental objectives and targets, the organization must also monitor and measure other activities related to significant aspects. Thus the total monitoring and measurement program could include tracking of chemical use, water use, wastewater quality, energy use, waste generation, recycling activities, and other key characteristics.

A second part of this element of the standard requires the organization to calibrate monitoring equipment which is part of the EMS, and to record calibrations. When determining which pieces of equipment are part of the EMS, the organization should consider those pieces of equipment used for reporting compliance, as well as those used to measure objectives and targets and to control key operations.

The final part of this section of the standard specifies that the organization must put in place a documented procedure for periodically evaluating compliance with relevant environmental legislation and regulations. This could include self-assessments of compliance, peer reviews, or external evaluations of compliance.

***Nonconformance and Corrective and Preventive Action* (*Section 4.5.2*).** ISO 14001 requires that the organization establish and maintain procedures for defining responsibility and authority for handling and investigating nonconformances, for taking action to mitigate any environmental impacts, and for initiating and completing corrective and preventive action. Following are some nonconforming activities that might be identified within the EMS:

- Activities or operations do not support the environmental policy.
- Employees whose job function could affect legal requirements do not have access to the requirements.
- Significant aspects have not been identified.
- Views of interested parties were not considered when setting objectives and targets.
- Environmental management program does not specify responsible person, time frame, and/or means.
- Defined roles, responsibilities, and authorities have not been defined and/or communicated to relevant employees.
- Training needs have not been identified and/or the training plan is not being followed.
- Relevant external communications are not documented.
- Directions to those documents that interact with the core elements of the EMS are not provided.
- Obsolete documents were not removed promptly from points of use.
- Contractors were not aware of the organization's procedures related to the EMS.
- An equipment calibration schedule was not followed.
- Responsibility for handling nonconformances to the EMS was not defined.
- Records were not easily retrievable.
- The EMS audit schedule was not defined.
- Management review was not documented.

Records (Section 4.5.3). Records are not the same as documents. Documents include procedures, work tasks, instructions, manuals, and other forms of documentation that are used to manage the EMS. In contrast, records are evidence that something has been accomplished, such as inspection, equipment calibration, and training. In this section of the standard, the organization is required to establish procedures for the maintenance and disposition of records. In addition, records should be:

- Legible
- Identifiable and traceable to the activity, product, or service involved
- Easily retrievable
- Protected against damage, deterioration, or loss
- Retained according to established and recorded retention times

EMS Audit (Section 4.5.4). The EMS audit is an audit performed by an unbiased and objective persons or persons to assess whether or not the EMS conforms to planned arrangements and has been properly implemented and maintained. The standard requires that the organization establish and maintain procedures and a program(s) for EMS auditing. The procedures must cover audit scope, audit frequency, audit methodologies, and responsibilities and requirements for conducting audits and reporting results. The frequency of the audits must be based on the environmental importance of the activity being audited and the results of previous audits.

Management Review (Section 4.6)

The final element of the ISO 14001 standard is that of management review. This section requires top management of the organization to review the EMS at specified intervals to ensure its continuing suitability, adequacy, and effectiveness. (*Note:* This determination is not made during the EMS audit; only top management can make this determination based on the audit results and other information presented.) Each organization can determine its own meaning for this collection of terms, but it should consider this meaning early in the process so that the management review can align with the ISO 14001 standard. Examples of information that might be included during management review include environmental progress toward achieving objectives and targets, data about key characteristics, plans for technological upgrades, permit compliance records, results of EMS audits, and changes in activities, products, or services that might require changes to the EMS.

OSHA'S VOLUNTARY PROTECTION PROGRAMS

Voluntary protection programs (VPPs) are authorized under the OSH Act and are intended to encourage employers and employees to reduce the number of occupational safety and health hazards at their places of employment and to stimulate the improvement of programs for providing safe and healthful working conditions. These programs recognize and encourage work sites that have lower than average workplace injury rates and that have adopted comprehensive safety and health program management approaches that are considered the best in the industry. VPPs recognize that good workplace safety and health programs go well beyond OSHA standards.

In addition to recognizing excellent safety and health programs, VPPs supplement OSHA's enforcement effort in that VPP work sites are not subject to the same inspection requirements to which other sites are subject. Once certified as a VPP site, participants are exempted from OSHA general inspections (from one to three years, depending on the level of certification), although employees' rights to contact OSHA and to file complaints remain the same.

TABLE 2.4 Elements Common to Programs in Approved VPP Work Sites

Safety and Health Program

- Documented required safety programs, such as hazard communication program, medical surveillance program, process safety management program, laser safety program, personal protective equipment program, lockout/tag-out program, others
- Specific safety and health training, including process equipment safety, electrical safety, personal protective equipment training, confined space training, training on processes containing highly hazardous chemicals, hazard communication training, and other task-related training, as required
- Safety manual accessible to managers and employees at work site
- Job- or department-specific documented safety plans kept in the work area
- Total safety and health program evaluation system that provides for annual reports and includes evaluation of safety staffing, training, hazard assessment, safety rules and procedures, employee participation, and special emphasis programs
- Lower than national average for total injury rates and lost-workday case rates for all employees averaged over three years

Systematic Assessment and Control of Hazards

- Identification of system for identifying and controlling workplace hazards
- Program for conducting monthly self-assessments
- Documentation of findings during self-assessment and correction of hazards
- Documentation of environmental monitoring programs, including those for hazardous substances, noise, radiation, and others
- Training of personnel who conduct environmental sampling; verification that testing and analysis are performed in accordance with nationally recognized procedures
- System of communication to management of employee observations of unsafe or unhealthful conditions
- Accident investigation program, with documented results
- Preventive maintenance program
- Program for correcting identified hazards in timely fashion
- Medical program, including access to physician services and adequate first-aid provisions

Training

- General periodic safety and health training, such as safe lifting, good housekeeping practices, general chemical safety, hazard identification, and other general safety training
- Training program for new hires
- Management training on safety and health issues, policies, and procedures
- Training for emergency response and building or site evacuation

TABLE 2.4 (*Continued*)

Safety Planning, Work Procedures, and Rules

- Safety and health planning as integrated in the overall management system
- Documented safety rules and enforcement of safety rules and procedures
- Routine analysis of jobs, processes, or construction crafts for hazards
- Written personal protective equipment procedures
- Emergency planning documentation
- Identification of supervisor's role in safety and health program

Management Commitment

- Written management commitment to worker safety and health protection
- Top management involvement in safety and health program
- Clear system of accountability for safety at the line level and clear assignment of safety responsibility
- Clearly established safety and health policies and results-oriented objectives
- Resources authorized for safety and health and certified safety professionals, industrial hygiene professionals, and medical personnel available on-site or through corporate or other means
- Safety and health protection for contract workers equivalent to that of regular employees

Employee Participation

- Employee pride in the safety and health programs
- Active employee involvement with safety and health programs

Source: Woodside and Kocurek (1997).

There is no one "best" design of a safety management system, since operations and workplace hazards of individual work sites are varied. Elements common to programs in approved VPP work sites include a comprehensive safety and health program, systematic hazard assessment and control, training programs, clear safety planning, work procedures and rules, management commitment, and employee participation. Additional detail for each of these elements is presented in Table 2.4.

REFERENCES

International Organization for Standardization (1996). *Environmental Management Systems Specification,* ISO 14001, ISO, Geneva, Switzerland.

Woodside, G., P. Aurrichio, and J. Yturri (1998). *ISO 14001 Implementation Guide,* McGraw-Hill, New York.

Woodside, G., and D. Kocurek (1997). *Environmental, Safety, and Health Engineering,* Wiley, New York.

3

DEFINING A HAZARDOUS MATERIAL OR WASTE

Regulators have spent nearly three decades defining what makes a material or waste hazardous. Generally, a *hazardous chemical* or *waste* is a material that is potentially dangerous to human health or the environment. In this chapter we review some of the more important definitions that are currently being used for regulating hazardous materials and hazardous waste.

OSHA HAZARDOUS CHEMICAL DEFINITIONS

OSHA defines hazardous chemicals in terms of health hazards and physical hazards. Presented in Table 3.1 are characteristics related to health and physical hazards that can make a chemical hazardous under 29 CFR §1910.1200.

Generally, a health hazard is assessed as either chronic or acute. A *chronic health hazard* occurs as a result of long-term exposure. An *acute health hazard* occurs rapidly as a result of a short-term exposure. In addition to chronic or acute hazard classifications, there are specific health hazards classified in the regulations. These include carcinogens, toxics, corrosives, irritants, sensitizers, and health hazards affecting human organs.

A chemical is considered to be a *carcinogen,* or cancer-causing agent, under the OSHA regulations if it has been evaluated by the International Agency for Research on Cancer (IARC)[1] and is listed in IARC's latest edition of monographs as a carcinogen or potential carcinogen. Additionally, if it is listed as a carcinogen or potential carcinogen in the latest annual report on carcinogens published by the National Toxicology Program,[2] OSHA regulates

[1]The International Agency for Research on Cancer is part of the World Health Organization.
[2]The National Toxicology Program is part of the National Center for Toxicological Research, Department of Human Health and Human Services.

TABLE 3.1 Characteristics That Can Make a Chemical Hazardous Under OSHA Regulations[a]

	Health Hazards
Carcinogen	Toxic or highly toxic chemical
Reproductive toxin	Irritant
Hepatoxin (liver)	Corrosive
Neurotoxin (nervous system)	Sensitizer
Nephrotoxin (kidney)	Agent that damages the blood, lungs, eyes, or skin
	Physical Hazards
Combustible liquid	Water reactive
Flammable	Organic peroxide
Explosive	Oxidizer
Pyrophoric	Compressed gas

[a]OSHA has found that health hazards are difficult to define; thus OSHA has issued guidelines for health hazard definitions under 29 CFR §1910.1200, Appendixes A and B.

the chemical as a carcinogen. OSHA also has the right to list and regulate other chemicals as carcinogens as the agency deems appropriate.

Toxic chemicals are defined as either highly toxic or toxic. *Highly toxic chemicals* are defined as follows:

- The chemical has a *median lethal dose* (LD_{50}) of 50 milligrams or less per kilogram of body weight when administered orally to albino rats weighing between 200 and 300 grams each.
- The chemical has a median lethal dose (LD_{50}) of 200 milligrams or less per kilogram of body weight when administered by continuous contact for 24 hours (or less if death occurs before 24 hours) with the bare skin of albino rabbits weighing between 2 and 3 kilograms each.
- The chemical has a *median lethal concentration* (LC_{50}) in air of 200 parts per million by volume or less of gas or vapor, or 2 milligrams per liter or less of mist, fume, or dust, when administered by continuous inhalation for 1 hour (or less if death occurs before 1 hour) to albino rats weighing between 200 and 300 grams each.

A chemical is *toxic* if it meets any of the following criteria:

- The chemical has a median lethal dose (LD_{50}) of more than 50 milligrams per kilogram but not more than 500 milligrams per kilogram of body weight when administered orally to albino rats weighing between 200 and 300 grams each.
- The chemical has a median lethal dose (LD_{50}) of more than 200 milligrams per kilogram but not more than 1000 milligrams per kilogram of body weight when administered by continuous contact for 24 hours (or

less if death occurs within 24 hours) with the bare skin of albino rabbits weighing between 2 and 3 kilograms each.

- The chemical has a median lethal concentration (LC_{50}) in air of more than 200 parts per million but not more than 2000 parts per million by volume of gas or vapor, or more than 2 milligrams per liter but not more than 20 milligrams per liter of mist, fume, or dust when administered by continuous inhalation for 1 hour (or less if death occurs within 1 hour) to albino rats weighing between 200 and 300 grams each.

Corrosive chemicals cause visible destruction of living tissue. The regulations under DOT in 49 CFR Part 173, Appendix A, delineate a test method acceptable to OSHA for defining the term *corrosive*. The test exposes albino rats to a chemical for 4 hours. If, after the exposure period, the chemical has destroyed or changed irreversibly the structure of the tissue at the site of contact, the chemical is considered corrosive.

Chemicals that are *irritants* cause a reversible inflammatory effect on skin or eyes. Tests for determining whether the chemical is an irritant are defined in 16 CFR §1500.41. Tests for eye irritants are defined in 16 CFR §1500.42.

A chemical is defined as a *sensitizer* if a large number of exposed people or animals develop an allergic reaction in normal tissue after repeated exposure to the chemical. The effects typically are reversible once the exposure ceases.

Chemicals are also considered a health hazard if exposure causes damage to any of the human organs. The damage can take many forms. Examples include liver enlargement, kidney disease or malfunction, excessive nervousness or decrease in motor functions, decrease in lung capacity, damage to cornea, damage to the reproductive organs, and other organ damage.

Selected examples of chemicals that create the health hazards discussed in this section are presented in Table 3.2. Signs and symptoms from overexposure to these chemicals are presented in Chapter 4. Further, since many chemicals are associated with more than one health hazard, information pertaining to each chemical should be reviewed thoroughly before the chemical is used in the workplace.

Physical Hazards

A chemical is defined as *physical hazard* if there is scientific evidence that it is a combustible liquid, flammable, explosive, pyrophoric, or unstable (reactive). Additionally, a chemical is deemed *hazardous* if it is a compressed gas, an organic peroxide, or an oxidizer.

A *combustible liquid* is any liquid having a flash point at or above 100°F but below 200°F. A *flammable material* can be any of the following:

- An aerosol that when tested by the method described in 16 CFR §1500.45 yields a flame projection exceeding 18 inches at full valve opening, or a flashback at any degree of valve opening.

TABLE 3.2 Selected Examples of Chemicals that Create Health Hazards

Health Hazard	Chemicals That Create the Hazard
Carcinogen	Aldrin, formaldehyde, ethylene dichloride, methylene chloride, dioxin
Toxic	Xylene, phenol, propylene oxide
Highly toxic	Hydrogen cyanide, methyl parathion, acetonitrile, allyl alcohol, sulfur dioxide, pentachlorophenol
Reproductive toxin	Methyl cellosolve, lead
Corrosive	Sulfuric acid, sodium hydroxide, hydrofluoric acid
Irritant	Ammonium solutions, stannic chloride, calcium hypochlorite, magnesium dust
Sensitizer	Epichlorohydrin, fiberglass dust
Hepatotoxin	Vinyl chloride, malathion, dioxane, acetonitrile, carbon tetrachloride, phenol, ethylenediamine
Neurotoxin	Hydrogen cyanide, endrin, mercury, cresol, methylene chloride, carbon disulfide, xylene
Nephrotoxin	Ethylenediamine, chlorobenzene, dioxane, hexachloronaphthalene, acetonitrile, allyl alcohol, phenol, uranium
Agent that damages:	
Blood	Nitrotoluene, benzene, cyanide, carbon monoxide
Lungs	Asbestos, silica, tars, dusts
Eyes or skin	Sodium hydroxide, ethylbenzene, perchloroethane, allyl alcohol, nitroethane, ethanolamine, sulfuric acid, liquid oxygen, phenol, propylene oxide, ethyl butyl ketone

Source: Information from NIOSH (1987), NFPA (1991), and 29 CFR §1910.1200, Appendix A.

- A gas that at ambient temperature and pressure forms a flammable mixture with air at a concentration of 13% by volume or less.
- A gas that at ambient temperature and pressure forms a range of flammable mixtures with air wider than 12% by volume, regardless of the lower limit.
- A liquid having a flash point below 100°F.
- A solid, other than a blasting agent or explosive, that is likely to cause fire through friction, absorption of moisture, spontaneous chemical change, or retained heat from manufacturing or processing, or that can be ignited readily and when ignited burns so vigorously and persistently as to create a serious hazard.
- A solid that when tested by the method described in 16 CFR §1500.44 ignites and burns with a self-sustained flame at a rate greater than one-tenth of an inch per second along its major axis.

Other physical hazard definitions include:

- *Explosive.* An explosive is defined as a chemical that causes a sudden, almost instantaneous release of pressure, gas, and heat when subjected to sudden shock, pressure, or high temperature.
- *Pyrophoric.* A pyrophoric is a material that will ignite spontaneously in air at a temperature of 130°F or below.
- *Water reactive.* A chemical is water reactive if it reacts with water to release a gas that is either flammable or presents a health hazard.
- *Oxidizer.* An oxidizer is a chemical that initiates or promotes combustion in other materials, thereby causing fire either of itself or through the release of oxygen or other gases.
- *Organic peroxide.* In addition to being an oxidizer, an organic peroxide is a material that is extremely unstable. In its pure state, or when produced or transported, it will vigorously polymerize, decompose, condense, or become self-reactive when exposed to shock, heat, or friction.

Selected examples of chemicals that create physical hazards are presented in Table 3.3.

Compressed gases are considered hazardous materials under OSHA. A *compressed gas* is defined as:

- A gas or mixture of gases having, in a container, an absolute pressure exceeding 40 pounds per square inch (psi) at 70°F.
- A gas or mixture of gases having, in a container, an absolute pressure exceeding 104 psi at 130°F regardless of the pressure at 70°F.
- A liquid having a vapor pressure exceeding 40 psi at 100°F as determined by ASTM D323-72.

TABLE 3.3 Examples of Chemicals That Create Physical Hazards

Physical Hazard	Chemicals That Create the Hazard
Combustible liquid	Fuel oil, crude oil, other heavy oils
Flammable	Gasoline, isopropyl alcohol, acetone, spray cans that use butane propellants
Explosive	Dynamite, nitroglycerine, ammunition
Pyrophoric	Yellow phosphorus, white phosphorus, superheated toluene, silane gas, lithium hydride
Water reactive	Potassium, phosphorus pentasulfide, sodium hydride
Oxidizer	Sodium nitrate, magnesium nitrate, bromine, sodium permanganate, calcium hypochlorite, hydrogen peroxide
Organic peroxide	Methyl ethyl ketone peroxide, dibenzoyl peroxide, dibutyl peroxide

Source: Information from NFPA (1991) and 29 CFR §1910.1200.

Examples of compressed gases include oxygen, helium, and acetylene. Hazards associated with these materials include abnormal pressure release and possible rupture if the container is punctured. *Liquefied gases,* a subset of compressed gases, include butane, propane, and vinyl chloride. In addition to the potential rupture hazard, these three materials are also flammable. Some liquefied gases are also classified as cryogenic. Cryogenics are substances that change from a gas to a liquid at or below −200°C. Examples include liquid nitrogen and liquid oxygen. These materials can cause frostbite or burns to the skin if released.

Highly Hazardous Chemicals

OSHA defines a highly hazardous chemical as one that possesses toxic, reactive flammable, or explosive properties. These chemicals present a potential for a catastrophic event at or above certain threshold quantities. OSHA has listed over 100 highly hazardous chemicals that are regulated if maunufactured, stored, or used in quantities at or greater than threshold amounts. These chemicals are listed in 29 CFR §1910.119, Appendix A. Examples of these chemicals and respective threshold quantities are presented in Table 3.4.

OSHA-Regulated Chemicals Under 29 CFR 1910, Subpart Z

In addition to regulating chemicals that fall into the categories reviewed previously, OSHA has set regulatory standards for permissible exposure limits to workers for a number of chemicals, defined in 29 CFR 1910, Subpart Z. Under §1910.1000, approximately 500 air contaminants are regulated in terms of exposure limits. Examples of these chemicals and allowable exposures are presented in Chapter 4. In addition, approximately 30 chemicals are regulated in more detail, including (as appropriate) the delineation of respiratory protection standards, medical surveillance, periodic monitoring, and other requirements. These requirements for these chemicals are also delineated in 29 CFR 1910, Subpart Z and in 29 CFR 1926, Subpart Z, and the list and regulatory citations are presented in Table 3.5. In establishing limits for allowable worker exposure, OSHA uses data gathered by the National Institute of Occupational Safety and Health (NIOSH).

HAZARDOUS WASTE DEFINITIONS UNDER RCRA

Under RCRA regulations promulgated by EPA, a waste is hazardous if it is a solid waste and meets one of several definitions. The term *solid waste* might be slightly misleading since it might be construed as wastes that are in the solid phase. In fact, solid wastes are liquids, solids, and containerized gases. Definitions of RCRA hazardous wastes are delineated in 40 CFR Part 261, Subpart C.

TABLE 3.4 Examples of OSHA-Defined Highly Hazardous Chemicals and Respective Threshold Quantities

Chemical Name	Threshold Quantity (pounds)
Acetaldehyde	2,500
Acrolein (2-propenal)	150
Allyl chloride	1,000
Arsine	100
Bromine trifluoride	15,000
Chloropicrin	500
Dimethylamine, anhydrous	2,500
Ethyl methyl ketone peroxide	5,000
Ethylene fluorohydrin	100
Furan	500
Hydrogen chloride	5,000
Methyl chloride	15,000
Methyl chloroformate	500
Methyl isocyanate	250
Methyl vinyl ketone	100
Nitrogen dioxide	250
Ozone	100
Perchloric acid	5,000
Phosgene	100
Propyl nitrate	2,500
Sulfur dioxide (liquid)	1,000
Tetramethyl lead	1,000

A waste is a *characteristic hazardous waste,* as defined in 40 CFR §§261.21–261.24, if it has any of the following characteristics:

- *Ignitability.* A waste is hazardous if it is a liquid, other than an aqueous solution containing less than 24% alcohol by volume, and has a flash point of less than 140°F. A waste is also hazardous if it is not a liquid and is capable, under standard temperature and pressure, of causing fire through friction, absorption of moisture, or spontaneous chemical changes. Further, it is a hazardous waste if it is an ignitable compressed gas or an oxidizer.
- *Corrosivity.* A waste is hazardous if it is aqueous and its pH is less than or equal to 2 or greater than or equal to 12.5. Additionally, a waste is considered corrosive if it is a liquid and corrodes steel at a rate greater than 0.250 inch per year at 130°F.
- *Reactivity.* A waste is hazardous if it is normally unstable and readily undergoes violent change, or if it reacts violently or creates toxic fumes when mixed with water. Additionally, a waste is hazardous if it is a cyanide- or sulfide-bearing waste that can generate toxic gases or fumes when exposed to pH conditions between 2 and 12.5.

TABLE 3.5 Chemicals Regulated in Detail by OSHA Under 29 CFR Subpart Z

Chemical Name	Regulatory Citation
Asbestos	§§1910.1001 and 1926.58
Coal tar pitch volatiles	§§1910.1002 and 1926.1102
4-Nitrobiphenyl	§§1910.1003 and 1926.1103
α-Naphthylamine	§§1910.1004 and 1926.1104
Methyl chloromethyl ether	§§1910.1106 and 1926.1106
3,3′-Dichlorobenzidine (and its salts)	§§1910.1007 and 1926.1107
Bis(chloromethyl) ether	§§1910.1008 and 1926.1108
Benzidine	§§1910.1010 and 1926.1110
4-Aminodiphenyl	§§1910.1011 and 1926.1111
Ethyleneimine	§§1910.1012 and 1926.1112
β-Propiolactone	§§1910.1013 and 1926.1113
2-Acetylaminofluorene	§§1910.1014 and 1926.1114
4-Dimethylaminoazobenzene	§§1910.1015 and 1926.1115
N-Niitrosodimethylamine	§§1910.1016 and 1926.1116
Vinyl chloride	§§1910.1017 and 1926.1117
Inorganic arsenic	§§1910.1018 and 1926.1118
Lead	§§1910.1025 and 1926.62
Cadmium	§§1910.1027 and 1926.1127
Benzene	§§1910.1028 and 1926.1128
Coke oven emissions	§§1910.1029 and 1926.1129
Bloodborne pathogens	§1910.1030
Cotton dust	§1910.1043
1,2-Dibromo-3-chloropropane	§§1910.1044 and 1926.1144
Acrylonitrile	§§1910.1045 and 1926.1145
Ethylene oxide	§§1910.1047 and 1926.1147
Formaldehyde	§§1910.1048 and 1926.1148
Methylenedianline	§§1910.1050 and 1926.60

- *Toxicity.* A waste is hazardous if toxic concentrations of compounds enumerated in 40 CFR §261.24 can be leached into water, using EPA's toxicity characteristic leaching procedure (TCLP). If a waste is less than 0.5% filterable solids, it is considered hazardous if the liquid concentration exceeds the TCLP standard. These standards are set for numerous compounds, including metals, pesticides, and organics.

A waste can be deemed hazardous if it is a waste that is specifically listed as being hazardous in the RCRA regulations under 40 CFR Part 261, Subpart D. Included are wastes from specific sources and nonspecific sources. Examples of listed hazardous wastes from specific sources include specific waste from a process such as distillation bottoms from aniline production, untreated process wastewater from the production of toxaphene, and others. Wastes from nonspecific sources include spent solvents and general process wastes such as residues, heavy ends, sludges, and other wastes from processes such as electroplating operations, metal heat treating operations, and other manufacturing operations. An additional listing gives the names of several hundred specific chemicals that, when discarded, become hazardous waste. Examples include acrylonitrile, benzene, and toluene.

Any mixture of a listed hazardous waste and a nonhazardous waste renders the entire mixture hazardous. Similarly, a waste derived from a hazardous waste is considered hazardous. These rules are being reviewed for change, and another method for determining whether or not these types of wastes are hazardous may be used in the future.

DEFINITIONS FOR RADIOACTIVE WASTES

Major Categories of Radioactive Wastes

Definitions of radioactive wastes are published by the NRC. There are basically three major types of radioactive wastes: high-level radioactive wastes, transuranic radioactive wastes, and low-level radioactive wastes.

High-level radioactive wastes (HLWs) are those reprocessing wastes derived from nuclear reactors, including irradiated reactor fuel, liquid wastes resulting from the operation of the first-cycle solvent extraction system or equivalent, concentrated wastes resulting from subsequent extraction cycles or equivalent, and solids derived from conversion of high-level radioactive liquids. Weapons by-products also can contain high-level radioactivity. The wastes require permanent isolation in a geologic repository or equivalent at the time of disposal, as defined under 10 CFR Part 60.

Transuranic (TRU) *wastes* include wastes containing elements with atomic numbers greater than uranium (92), such as plutonium and curium, and that contain more than 100 nanocuries per gram of alpha-emitting transuranic isotopes with half-lives greater than 20 years. Some TRU materials are exceptionally long-lived, such as plutonium-239, which has a half-life of 24,400 years. TRU waste is generated from the reprocessing of plutonium-bearing fuel and irradiated targets and from operations required to prepare the recovered plutonium for weapons use. The waste includes TRU metal scrap, glassware, process equipment, soil, laboratory wastes, filters, and wastes contaminated with TRU materials (U.S. Congress 1991). NRC approves the disposal of these wastes, which includes disposal in a geologic repository or equivalent for most cases.

Low-level radioactive wastes (LLWs) are defined as wastes containing radioactivity that is neither high-level radioactive nor transuranic. In general, low-level radioactive waste, regulated under 10 CFR Part 61, is divided into three classes: A, B, and C, with Class C containing the highest concentration of radionuclides. Class C requires stabilization and barriers to protect intruders. Class B requires stabilization, but not barriers, to protect intruders. Class A, with the lowest concentration of radionuclides of all the classes, does not require stabilization or barriers. Examples of low-level radioactive wastes include solidified liquids, filters, and resins, and lab trash from nuclear reactors, hospitals, research institutions, and industry (NRC 1989).

Other Radioactive Wastes

In addition to the major categories of radioactive wastes, there are definitions and regulation for other types of radioactive waste. These include uranium and thorium mill tailings, waste derived from naturally occurring and accelerator-produced radioactive materials, and radioactive materials or waste mixed with hazardous waste (DOE 1989).

Uranium and thorium mill tailings are regulated separately from low-level radioactive wastes. Technical criteria are set by NRC for siting and design of disposal facilities for these wastes under 10 CFR Part 40. EPA governs the control and cleanup of residual radioactive materials from inactive uranium processing sites, as well as the management of uranium and thorium by-product materials at active sites. These regulations are found under 40 CFR Part 192.

Naturally occurring and accelerator-produced radioactive materials (NARM) are defined as any radioactive material not classified as source, special nuclear, or by-product material (DOE 1989). A waste from this material is regulated as if it were low-level radioactive wastes.

A *mixed waste* is defined as a radioactive waste that is mixed with hazardous waste. Radioactive wastes can be classified as mixed by using engineering knowledge of the hazardous characteristics of the waste. If the determination of hazardous or nonhazardous must be made through testing, a surrogate waste stream that is devoid of radioactive material, but which is equivalent otherwise, can be used.

Once a determination is made that the waste stream is a mixed waste, the waste is regulated under both EPA and NRC rules. EPA has set forth treatment standards for several types of mixed waste. Incineration is recommended for organic and ignitable low-level mixed waste. For transuranic and high-level mixed wastes, EPA has established vitrification as an acceptable treatment technology prior to long-term storage.

OTHER REGULATED CHEMICALS

Materials Regulated Under the Department of Transportation

The U.S. Department of Transportation (DOT) regulates the transportation of hazardous materials and has a structured method for categorizing chemicals by hazard class. In all, there are nine classes of hazardous materials, including explosive materials, gases, flammable liquids, flammable solids, oxidizers, poisonous materials, radioactive materials, corrosive materials, and miscellaneous hazardous materials. These classes are defined further in Chapter 10.

Chemicals Regulated Under the Clean Water Act (CWA)

Numerous chemicals are regulated by the EPA under the Clean Water Act. These include hazardous substances, toxic pollutants, conventional pollutants, priority pollutants, and water priority chemicals. Examples of these chemicals are presented in Table 3.6. Effluent guidelines and other wastewater discharge standards are set for these chemicals to ensure that all waters of the United States remain fishable and swimmable. Further, regulations promulgated under CWA outline threshold quantities of chemicals that must be reported to EPA if spilled into the waterways of the nation. This list comprises over 300 chemicals.

Additionally, the aggregate concentration of toxic chemicals in effluent can make it unacceptable for discharge even if the individual constituents are within standards or guidelines. An EPA-approved bioassay test is used for determining total effluent toxicity. The basis of the toxicity determination is the survival rate of certain aquatic species in whole or diluted effluent.

Comprehensive Environmental Response, Compensation, and Liability Act (CERCLA) Reportable Quantities

Under CERCLA, EPA regulates release reporting of chemicals in amounts that it has determined to be hazardous to human health or the environment. Chemicals regulated in this fashion include metals, solvents, pesticides, toxics, and radionuclides. The list contains approximately 2000 chemicals, and the reporting quantities range from 1 pound for very hazardous chemicals such as arsenic acid to up to 5000 pounds for less hazardous chemicals such as phosphoric acid. If these chemicals are released without a permit in amounts greater than the reportable quantity, the National Response Center must be notified. Examples of chemicals with very low reporting thresholds are presented in Table 3.7.

Pollutants Regulated Under the Clean Air Act Amendments

There are numerous chemicals regulated under the Clean Air Act. Chemicals deemed pivotal to achieving national ambient air quality standards (NAAQS) are defined and include nitrogen oxide, sulfur dioxide, carbon monoxide, ozone, particulate matter, and lead. Additionally, the regulations define separate lists of pollutants that are regulated under the CAA Amendments as hazardous air pollutants (HAPS), hazardous organic air pollutants (HONS), and synthetic organic chemical manufacturing industry (SOCMI) chemicals affected by CAAA regulations (40 CFR §60.707, Subpart RRR and §60.489, Subpart VV). Examples of these chemicals are presented in Table 3.8.

Superfund Amendments and Reauthorization Act (SARA) Title III Chemicals

Under SARA Title III, EPA has listed over 300 chemicals that are considered extremely hazardous substances. These include many of the same chemicals

TABLE 3.6 Examples of Chemicals Defined as Hazardous or Considered a Priority Under the Clean Water Act

Conventional Pollutants (40 CFR §401.16)	
Biochemical oxygen demand (BOD)	pH
Fecal coliform	Total suspended solids (nonfilterable)
Oil and grease	
Priority Pollutants (40 CFR §423, Appendix A)	
Acrolein	Ethylbenezene
Acrylonitrile	Fluorene
Antimony	Heptachlor
Arsenic	Methylene chloride
Benzidine	2-Nitrophenol
Cadmium	PCB-1242
Carbon tetrachloride	Phenol
Chloroform	Bis(2-ethylhexyl) phthalate
4,4-DDT	Pyrene
1,2-Dichlorobenzene	Trichloroethylene
Endrin	Zinc
Hazardous Chemicals (40 CFR §116.4)	
Acetaldehyde	Methoxychlor
Acetic acid	Methyl mercaptan
Acrylonitrile	Naphthalene
Aldrin	Nitrotoluene
Ammonium chromate	Phosphoric acid
Aniline	Potassium cyanide
Benzene	Silver nitrate
Butyl acetate	Sodium chromate
Cadmium chloride	Sodium selenite
Carbon tetrachloride	2,4,5-T esters
Chlorine	Toluene
Chromic acetate	Vinyl acetate
DDT	Xylene (mixed)
Diquat	Zinc borate
Ferric sulfate	Zirconium nitrate
Lead fluoride	

(*continued*)

TABLE 3.6 (*Continued*)

Water Priority Chemicals (Section 313, 57 FR 41331, September 9, 1992)	
Acrylonitrile	Lead
Allyl chloride	Mercuric nitrate
Antimony trioxide	Naphthalene
Arsenic	Nickel
Benzidine	Parathion
Beryllium fluoride	Phenol
Cadmium	Phosgene
Calcium cyanide	Selenium
Chlorine	Silver
Chromic acid	Styrene
1,2-Dichlorobenzene	Sulfuric acid
Ethylbenzene	Thallium
Heptachlor	Vinyl chloride
Hydrogen cyanide	Zinc formate

Source: Government Institutes (1994).

regulated under RCRA and CERCLA. Like CERCLA, SARA also has release reporting requirements. The regulations require notification to state and local authorities if a release leaves the facility in an amount greater than the reportable quantity.

Pesticides

EPA maintains a list of over 350 pesticides that are governed under the Federal Insecticide, Fungicide, and Rodenticide Act (FIFRA), as detailed in 40 CFR Part 180. Examples include boron, carboxin, endosulfan, ethylene glycol, lindane, and paraquat. The hazard of the pesticide is controlled by the pesticide tolerance commodity/chemical index, which specifies the allowable application for type of crop.

Safe Drinking Water Act (SWDA) Chemicals

Chemicals regulated under SWDA include organic and inorganic contaminants, metals, pesticides, radionuclides, and synthetic organic chemicals. Certain contaminants have published maximum concentration levels and maximum concentration level goals in the drinking water. Other chemicals have best technology or treatment technique identified in the standard. Examples of SWDA chemicals are presented in Table 3.9.

TABLE 3.7 Examples of CERCLA Chemicals with Low Unauthorized Release Reporting Thresholds

Chemical Name	Reporting Threshold (pounds)
Acetic acid	10
Acetone	10
Aldrin	1
Barium cyanide	1
Benzene	1
Berillium fluoride	1
Bis(2-chloroethyl)	10
Calcium cyanide	10
Coke oven emissions	1
Creosote	1
DDE	1
Diazinon	1
2,4-Dinitrophenol	10
Endrin	1
Ethylene oxide	10
Fluorine	10
Lead chloride	10
Lindane	1
Methyl isocyanate	10
Methoxychlor	1
Methylene chloride	10
Methyl isocyanate	10
Nitrogen oxide	10
N-Nitrosodiethylamine	1
Parathion	10
PCBs	1
Phosgene	10
Phosphorus	1
Potassium silver cyanide	1
Quinone	10
Selenium oxide	10
Silver cyanide	1
TDE	1
Toluenediamine	10
Trichloromethane	10
Vinyl chloride	1

Toxic Substance Control Act (TSCA) Chemicals

TSCA-regulated chemicals, once approved by EPA, receive a chemical abstract service (CAS) number, which identifies the chemical regardless of the common name used by the manufacturer. For example, isopropyl alcohol and its synonym isoproanol will both have the same CAS number: 67-63-0.

TABLE 3.8 Examples of Chemicals Determined to Be Hazardous Under the Clean Air Act Amendments

Hazardous Air Pollutants (40 CFR §63.56, Subpart C)	
Acetaldehyde	Heptachlor
Acrolein	Hexane
Cumene	Isophorone
Ethylbenzene	Lindane
Ethylene glycol	Methyl ethyl ketone
Hazardous Organic Air Pollutants (40 CFR §63, Subpart F, Table 2)	
Acrylonitrile	Methy ethyl ketone
Carbon tetrachloride	Naphthalene
Chloroform	Nitrobenzene
Chrysene	Pyrene
Dichloroethane	Quinone
Dimethyl phthalate	Stryrene
Ethylbenzene	Toluene
Glycol ethers	Vinyl chloride
Hexane	Xylenes
Methanol	
SOCMI Chemicals Affected by CAAA Regulations (40 CFR §60.707, Subpart RRR)	
Acetone	Isobutanol
Benzene	Isopropanol
1-Butene	Methyl isobutyl ketone
Cyclohexane	Octene
Diethylene glycol	Phenol
Epichlorohydrin	Styrene
Ethyl acetate	Toluene
Hexane	Xylenes
SOCMI Chemicals Affected by CAAA Regulations (40 CFR §60.489, Subpart VV)	
Acetic acid	Ethylene glycol
Allyl alcohol	Ethylene glycol monomethyl ether
Amyl alcohols	Methyl formate
Carbon disulfide	Methylene chloride
Chlorobenzoic acid	Paraldehyde
Dioxane	Phenol
Ethyl chloride	Phosgene
Ethylene	Trichlorobenzenes
	Triethylene glycol

TABLE 3.9 Examples of Chemicals Deemed Hazardous in Drinking Water Above Published Concentration Levels or Concentration Level Goals

Contaminant	Level (mg/L)
Maximum Contaminant Levels for Inorganic Chemicals (40 CFR §141.11)	
Arsenic	0.05
Barium	1
Cadmium	0.01
Lead	0.05
Mercury	0.002
Nitrate (N)	10
Selenium	0.01
Maximum Contaminant Level Goals for Organic Chemicals (40 CFR §141.50)	
Benzene	0
Chlordane	0
Diquat	0.02
Endrin	0.002
Ethylbenzene	0.7
Methoxychlor	0.04
Styrene	0.1
Toluene	1
Toxaphene	0

OVERLAPPING REGULATIONS FOR INDIVIDUAL CHEMICALS AND WASTES

There are numerous chemicals may be governed by more than one regulation. Examples of hazardous materials that fall under more than one statute, depending on the amount that is used, stored, spilled, or other factors is presented in Table 3.10. Note that the list of statutes is not all inclusive, but gives a visual indication of the complexity of regulation for hazardous materials, including chemicals and wastes.

LIST OF LISTS

Definitions of what constitutes a hazardous material or hazardous waste are numerous and overlap in many statutes. Literally, thousands of chemicals and wastes are regulated in some manner. Presented in Table 3.11 is an overview of lists of chemicals and wastes cited in the regulations under OSHA, NRC, the Department of Transportation (DOT), and EPA. EPA regulations and no-

TABLE 3.10 Examples of Hazardous Materials That May Be Regulated by More Than One Statute

	Statute						
Hazardous Material	OSHA	RCRA	CAA	CWA	FIFRA	CERCLA	SWDA[a]
Acrylonitrile	×	×	×	×		×	×
Allyl alcohol	×	×	×	×		×	
Benzene	×	×	×	×		×	×
Coke oven emissions	×		×				
Formaldehyde	×	×	×	×		×	
Isopropyl alcohol	×	[b]	[c]				
Lindane	×	×		×	×	×	×
Parathion	×	×		×	×	×	
Sulfuric acid	×	[b]		×		×	
Vinyl chloride	×	×	×	×		×	

[a]Excludes list of prohibition of underground injection of RCRA-defined wastes.
[b]Potentially, a characteristic RCRA waste.
[c]May be regulated under total volatile organics.

tations include CWA, CAA, FIFRA, SDWA, SARA, CERCLA, and TSCA. The materials in the lists presented in the table are either defined as hazardous or have been given a standard or other regulatory discharge or reporting limit.

HAZARDOUS MATERIALS USE AND WASTE GENERATION RELATED TO MANUFACTURING OPERATIONS

Hazardous materials are used in many manufacturing and nonmanufacturing operations, and hazardous waste is typically generated from these operations. In the following paragraphs we discuss major types of hazardous materials and hazardous waste associated with processes and operations found across the United States.

Chemical-Use Processes and Waste Types

Almost all manufacturing facilities use chemicals in some form in the production of marketable goods. From the manufacture of pure chemicals to the reprocessing of chemicals to produce derivative products to the use of chemicals as a cleaning agent, chemicals are a common part of the industrial sector.

There are too many chemical-use applications to present an all-inclusive list of industrial processes that use hazardous materials; thus only selected categories of regulated industrial processes and associated hazardous materials

TABLE 3.11 Summary of Lists of Regulated Materials

Regulation	Regulated Material
10 CFR Part 20, Appendix B to §§20.1001–20.2401	Annual limits of intake and derived air concentrations of radionuclides for occupational exposure; effluent concentraions; concentrations for release to sewerage (NRC)
10 CFR §61.55, Table 1	List of long-lived radionuclides used in low-level waste classification (NRC)
10 CFR §61.55, Table 2	List of short-lived radionuclides used in low-level waste classification (NRC)
29 CFR §1910.119, Appendix A	List of highly hazardous chemicals, toxics, and reactives (OSHA)
29 CFR Part 1910, Subpart Z, Tables Z-1-A, Z-2, Z-3	Limits for air contaminants (OSHA)
29 CFR §1910.1001–1048	List of OSHA specifically regulated substances (OSHA)
40 CFR §§50.4–50.12	Pollutants with national primary and secondary ambient air quality standards (CAA)
40 CFR §61.01	List of hazardous air pollutants (CAA)
40 CFR §§ 60.617, 60.667, and 60.707	Regulated emissions from the synthetic organic chemical manufacturing industry (SOCMI) (CAA)
40 CFR §61, Subparts B–FF	List of NESHAP standards (CAA)
40 CFR §63.74	List of high-risk pollutants for early reduction program (CAA)
40 CFR §68.130, Table 1	Regulated toxic substances and threshold quantities for prevention of accidental releases (CAA)
40 CFR §68.130, Table 3	Regulated flammable substances and threshold quantities for prevention of accidental releases (CAA)
40 CFR 82, Subpart A, Appendices A and B	Class I- and class II-controlled substances (CAA)
42 USC 7412; Clean Air Act, Title I, Part A, §112 (as amended 1990)	List of hazardous air pollutants (CAA)
40 CFR §116.4, Table 116.4A	List of hazardous substances by common name (CWA)
40 CFR §116.4, Table 116.4B	List of hazardous substances by CAS number (CWA)
40 CFR §117.3	Reportable quantities for hazardous substances (CWA)
40 CFR §122, Appendix D, Table II	Organic pollutants of concern in NPDES discharges (CWA)
40 CFR §122, Appendix D, Table III	Listed metals, cyanides, and phenols of concern in NPDES discharges (CWA)

TABLE 3.11 ***(Continued)***

Regulation	Regulated Material
40 CFR §122, Appendix D, Table V	Toxic pollutants and hazardous substances of con cern in NPDES discharges (CWA)
40 CFR Part 129	Toxic pollutant effluent standards (CWA)
40 CFR §141.11–141.12	Primary maximum contaminant levels for drinking water (SDWA)
40 CFR §141.15–141.16	Maximum contaminant levels for radioactivity (SWDA)
40 CFR §141.50–141.51	Maximum contaminant goals for organic and inorganic contaminants (SDWA)
40 CFR §141.61–141.62	Revised (and expanded) maximum primary contaminant levels for organic and inorganic contaminants (SDWA)
40 CFR §143.3	Secondary maximum contaminant levels for drinking water (SDWA)
40 CFR 148.10–148.17	Lists of waste prohibited from underground injection (SDWA)
40 CFR Part 180	Alphabetical listing of pesticide chemicals (FIFRA)
40 CFR §261.24	List of maximum concentration of contaminants for the toxicity characteristics—toxicity characteristic leaching procedure (TCLP) limits (RCRA)
40 CFR §261.31	List of hazardous wastes from nonspecific sources (RCRA)
40 CFR §261.32	List of hazardous wastes from specific sources (RCRA)
40 CFR §261.33 (e)	List of acutely hazardous wastes (RCRA)
40 CFR §261.33 (f)	List of toxic wastes (RCRA)
40 CFR Part 261, Appendix VII	Basis for listing wastes from specific and nonspecific sources (RCRA)
40 CFR Part 261, Appendix VIII	List of hazardous constituents (RCRA)
40 CFR Part 264, Appendix IX	Practical quantitiation limits for groundwater constituents (RCRA)
40 CFR §268.30–268.31	Prohibitions for solvent wastes and dioxin-containing wastes (RCRA)
40 CFR §268.32	California list wastes with specific prohibitions (RCRA)
40 CFR §§268.33–268.35	Lists of land restricted wastes (RCRA)
40 CFR §268.40	Treatment standards for land restricted wastes (RCRA)

(continued)

TABLE 3.11 ***(Continued)***

Regulation	Regulated Material
40 CFR §268.45	Treatment standards for hazardous debris (RCRA)
40 CFR §268.48	Universal treatment standards (RCRA)
40 CFR §302.4, Table 302.4	List of hazardous substances and reportable quantities CERCLA)
40 CFR §302.4, Appendix A	Sequential CAS registry number list of CERCLA hazardous substances (CERCLA)
40 CFR §355, Appendix A	List of extremely hazardous substances for emergency planning and notification (SARA)
40 CFR §372.65	List of toxic chemicals for toxic release inventory reporting (SARA)
40 CFR §401.15	List of toxic pollutants (CWA)
40 CFR §401.16	List of conventional pollutants (CWA)
40 CFR Part 423, Appendix A	Priority pollutants (CWA)
40 CFR §700, Index (Finders Aid)	Toxic substances CAS number/chemical index (TSCA)
49 CFR §172.101	Hazardous materials table (DOT)
49 CFR §172.101, Appendix	List of hazardous substances and reportable quantities per CERCLA (DOT)
47 FR 9352	83 contaminants required to be regulated under SDWA Amendments of 1986 (SDWA)
55 FR 1470, January 14, 1991	Priority list of drinking water contaminants
57 FR 41331, September 9, 1992	Section 313 water priority chemicals
57 FR 42102, September 14, 1992	OSHA specifically regulated substances
59 FR 4495, January 31, 1994	List of regulated toxic substances and threshold quantities for accidental release prevention

are presented in Table 3.12. Some of these processes and associated materials are currently the focus of EPA as candidates for chemical source reduction and waste minimization.

Waste types generated by processes within industry include bottoms or residues, dusts, discarded or off-spec chemicals or by-products, lab packs, slags, sludges or slurries, spent liquors, waste packages, and wastewaters. Selected examples of manufacturing operations that generate these types of waste are detailed in Table 3.13.

TABLE 3.12 Examples of Manufacturing Operations or Processes That Use Hazardous Materials

Manufacturing Category	Examples of Materials Potentially Used
Chemical reprocessing operations	Nonhalogenated solvents, cupric chloride pyrophosphate, acids, caustics, others
Coking operations	Ammonia, benzene, phenols, cyanide
Degreasing operations	Perchloroethylene, trichloroethylene, methylene chloride, 1,1,1-trichloroethane, carbon tetrachloride, chlorinated fluorocarbons
Distillation operations	Chlorobenzene, trichloroethylene, perchloroethylene, aniline, cumene, *ortho*-xylene, naphthalene, others
Electroplating processes	Cyanides, nickel, copper, acids, chrome, cadmium, gold
Ink formulation	Solvents, caustics, chromium- or lead-containing pigments and stabilizers
Leather tanning	Tannic acid, chromium
Painting operations	Methylene chloride, trichloroethylene, toluene, methanol, turpentine
Petroleum processes	Arsenic, cadmium, chromium, lead, halogenated solvents, flammable oils, distillate products
Primary metal processes	Cyanides, salt baths, heavy metals such as chromium and lead
Pulp and paper operations	Chlorine, sodium sulfite, sodium hydroxide, dioxins, furans, phenols
Textile finishing	Solvents, solutions of dyes
Weapons manufacture	Trinitrotoluene (TNT), nitroglycerin, uranium alloys, plutonium
Wood-preserving processes	Creosote, pentachlorophenol, other creosote and chlorophenolic formulations, copper, arsenic, chromium

Some of these wastes are listed as hazardous in the RCRA regulations. Others are hazardous as a result of having hazardous characteristics, such as ignitability, corrosivity, reactivity, or toxicity as defined in EPA's toxicity characteristic leaching procedure. Others may be regulated as hazardous by another agency or by the state.

TABLE 3.13 Waste Types and Examples of Operations That Generate the Waste Type

Waste Type	Examples of Operations That Generate the Waste Type
Bottoms or residues	Solvent and petroleum distillation operations, chemical purification operations, decant and separation techniques, cleaning operations
Discarded or off-spec chemicals or by-products	Chemical manufacturing operations, bench- or pilot-scale testing, operations, lab analysis activities
Dusts	Grinding operations, brushing operations, packaging operations, asbestos removal and other demolition activities, mining activities
Lab packs	Lab experiments, bench-scale studies, hazardous waste analysis, process analysis
Slags	Metal heat-treating operations, metal-processing operations
Sludges or slurries	Wastewater treatment operations, decantation operations; other separation techniques, such as filtration or gravity settling
Spent liquors	Metal-finishing operations, wood-preserving operations, pulping operations
Waste packages	Nuclear material processing, weapons manufacturing
Wastewaters, inorganic	Chemical products manufacturing, electroplating processes, petroleum processes, primary metal processing, coking processes
Organic	Chemical products manufacturing, cleaning and rinsing operations, other products manufacturing operations

REFERENCES

Department of Energy (1989). *Waste Classification: History, Standards, and Requirements for Disposal,* DE89 013705, prepared by D. C. Kocher, Oak Ridge National Laboratory, Oak Ridge, TN.

Government Institutes (1994). *Book of Lists for Regulated Hazardous Substances,* GI, Rockville, MD.

National Fire Protection Association (1991). *Fire Protection Guide for Hazardous Materials,* 10th ed., NFPA, Quincy, MA.

National Institute for Occupational Safety and Health (1987). *Pocket Guide to Chemical Hazards,* DHHS (NIOSH) Publication 85-114, U.S. Department of Health and Human Services, Public Health Service, Centers for Disease Control, Washington, DC.

Nuclear Regulatory Commission, *Regulating the Disposal of Low-Level Radioactive Waste: A Guide to the Nuclear Regulatory Commission's 10 CFR Part 61,*

NUREG/BR-0121, Office of Nuclear Material Safety and Safeguards, NRC, Washington, DC.

U.S. Congress (1991). *Long-Lived Legacy: Managing High Level and Transuranic Waste at the DOE Nuclear Weapons Complex,* OTA-BP-O-83, Office of Technology Assessment, Congress of the United States, U.S. Government Printing Office, Washington, DC.

BIBLIOGRAPHY

Alliance of American Insurers (1983). *Handbook of Hazardous Materials, Fire, Safety, Health,* 2nd ed., AAI, Schaumberg, IL.

American Conference of Governmental Industrial Hygienists (1998). *Threshold Limit Values for Chemical Substances and Physical Agents and Biological Exposure Indices,* Cincinnati, OH.

American Petroleum Institute (1988). *Naturally Occurring Radioactive Material (NORM) in Oil and Gas Production Operations,* API, Washington, DC, videotape.

Bretherick, L. (1990). *Handbook of Reactive Chemical Hazards,* 4th ed., Butterworth-Heinemann, Newton, MA.

Clayton, G. D., and F. E. Clayton, eds. (1993). *Patty's Industrial Hygiene and Toxicology,* Vol. 2, "Toxicology," 3rd ed., Wiley-Interscience, New York.

Department of Energy (1988). *Review of EPA, DOE, and NRC Regulations on Establishing Solid Waste Performance Criteria,* DE88 015331, ORNL/TM-9322, prepared by A. J. Mattus, T. M. Gilliam, and L. R. Dole, Oak Ridge National Laboratory, Oak Ridge, TN.

Department of Labor (1990). *OSHA Regulated Hazardous Substances: Health, Toxicity, Economic and Technological Data,* Vols. 1 and 2, Noyes Data Corporation, Park Ridge, NJ.

Environment Protection Agency (1988). *Extremely Hazardous Substances: Superfund Chemical Profiles,* Vols. 1 and 2, Noyes Data Corporation, Park Ridge, NJ.

Friedman, D., ed. (1988). *Waste Testing and Quality Assurance,* Special Technical Publication 999, American Society for Testing and Materials, Philadelphia.

International Labour Office (1991). *Occupational Exposure Limits for Airborne Toxic Substances,* 3rd ed., ILO, New York.

Keith, L. H., and D. B. Walters (1992). *The National Toxicology Program's Chemical Data Compendium,* Vol. 2, "Chemical and Physical Properties," Lewis Publishers, Boca Raton, FL.

Lewis, R. J., Sr. (1991). *Hazardous Chemicals Desk Reference,* 2nd ed., Van Nostrand Reinhold, New York.

National Fire Protection Association (1990). *Identification of the Fire Hazards of Materials,* NFPA 740, Quincy, MA.

Perry, R. (1997). *Perry's Chemical Engineers' Handbook,* 7th ed., McGraw-Hill, New York.

Sax, N. I., and R. J. Lewis, Sr. (1987). *Hawley's Condensed Chemical Dictionary,* 11th ed., Van Nostrand Reinhold, New York.

——— (1992). *Dangerous Properties of Industrial Materials,* 8th ed., Van Nostrand Reinhold, New York.

Sittig, M. (1991). *Handbook of Toxic and Hazardous Chemicals and Carcinogens,* Noyes Data Corporation, Park Ridge, NJ.

Weiss, G., ed. (1986). *Hazardous Chemicals Data Book,* 2nd ed., Noyes Data Corporation, Park Ridge, NJ.

Windholz, M., ed. (1989). *The Merck Index,* 11th ed., Merck & Co., Rahway, NJ.

Wolman, Y. (1988). *Chemical Information: A Practical Guide to Utilization,* 2nd ed., Wiley, New York.

PART II

WORKPLACE MANAGEMENT OF HAZARDOUS MATERIALS AND HAZARDOUS WASTE

4

UNDERSTANDING EXPOSURES FROM HAZARDOUS MATERIALS AND HAZARDOUS WASTES

Exposures from chemicals and waste in industrial applications can be evaluated in several ways. In this chapter we discuss methods typically used by industry to provide workers with information or real-time data about exposures from workplace hazardous materials and hazardous waste. These methods include the use of material safety data sheets, documented exposure limits, testing for toxicity characteristics, and medical surveillance programs. The use of chemical and waste labels, which also provides workers with information about materials in the workplace, is discussed in Chapter 9. Workplace and personal monitoring, which is integral to evaluating exposure, is addressed in Chapter 5.

MATERIAL SAFETY DATA SHEETS

OSHA Requirements

All chemicals that are manufactured, imported, sold, or used in a manufacturing process must be accompanied by a material safety data sheet (MSDS), as defined in 29 CFR §1910.1200. Chemical manufacturers or importers must provide the MSDS to distributors or employers with their initial shipment of chemicals and with the first shipment after data on the MSDS have been updated. Distributors who sell to other distributors or employers must also provide the MSDS and updates to their customers. An MSDS or equivalent must be available for review by employees in the workplace.

ANSI has provided recommendations for formatting information found in an MSDS in its standard Z400.1. The standard specifies a 16-part document that is now widely used in industry; mandatory and optional information is specified. An example of this 16-part MSDS provided in the ANSI standard is shown in Figure 4.1. An MSDS checklist based on the ANSI standard and 29 CFR §1910.1200 is presented in Table 4.1.

EXPOSURE LIMITS

Exposure Limits Defined by OSHA and ACGIH

OSHA has published exposure limits for hundreds of hazardous airborne contaminants in terms of permissible exposure limits (PELs) as time-weighted averages (TWAs). These limits, which are enforceable standards and based on information provided by NIOSH, define the maximum time weighted exposure over an 8-hour work shift of a 40-hour work week that should not be exceeded. These limits are expressed in parts per million (ppm) and/or milligrams per cubic meter (mg/m^3) and are published in 29 CFR §1910.1000, Table Z-1.

The American Conference of Governmental Industrial Hygienists (ACGIH) defines airborne contaminant exposure similarly, and terms this exposure concentration the threshold limit value–time weighted average (TLV-TWA). TLV-TWAs are defined as a concentration to which nearly all workers can be repeatedly exposed over a normal 8-hour workday and a 40-hour workweek, without adverse affects. Like OSHA, ACGIH has documented TLV-TWAs for hundreds of chemicals in the annual publication *Threshold Limit Values for Chemical Substances and Physical Agents and Biological Exposure Indices.* Although not legally enforceable, ACGIH standards are considered industry standards.

In addition, NIOSH has recommended exposure limits (RELs) that are based on the TWAs. Although NIOSH provides information to OSHA, OSHA standards are not identical to NIOSH's. Also, in some instances, industry standards established by ACGIH may differ from OSHA and/or NIOSH standards. These differences are typically because OSHA standards must be adjusted through regulatory amendments, which is a lengthy and time-consuming process. ACGIH and NIOSH standards are updated annually based on the latest scientific data, although interpretation of these data by the two organizations may differ. Thus, to ensure adequate protection of workers, it is considered a good management practice for employers to abide by the lowest standard practicable, even if it is lower than the OSHA standard. If the OSHA standard is lower, the OSHA standard must be met.

For some chemicals, ACGIH, NIOSH, and OSHA have documented additional exposure concentrations. These include a short-term exposure limit (STEL) and a ceiling limit (C). The STEL is typically a 15-minute time-weighted average exposure limit that should not be exceeded, although for some substances such as asbestos, the STEL is defined as a 30-mintue exposure limit. This limit is not a stand-alone limit and must be included in the 8-hour time weighted average. The STEL generally applies to a substance whose toxic effects are considered to be chronic but for which there may be

RESIN MATERIAL SAFETY DATA SHEET

1. CHEMICAL PRODUCT AND COMPANY IDENTIFICATION

Plastics Company	1000 Main Street	Uptown, NJ 87654
Emergency Phone	123-456-7890 24-hours	3/7/97
Effective Date:	2/23/97 Print Date:	3/7/97 MSD #01234

PRODUCT NAME: ABC Resin
PRODUCT CODE: 1234

2. COMPOSITION/INFORMATION ON INGREDIENTS

Chemical Ingredients (% by wt.)

Component A/B/C Resin	CAS# XXXXXX-XX-X	90–99%
Component D	CAS# XXXXXX-XX-X	0–2%
Component E	CAS# XXXXXX-XX-X	0–2%
Component F	CAS# XXXXXX-XX-X	0–2%
Impurity C	CAS# XXXXXX-XX-X	2000 ppm max.

(See Section 8 for exposure guidelines)

3. HAZARDS IDENTIFICATION

EMERGENCY OVERVIEW

Odorless white or colored powder. Can burn in a fire. Slippery, can cause falls if walked on.

POTENTIAL HEALTH EFFECTS

EYE: Solid or dusts may cause irritation or scratch the surface of the eye.

SKIN CONTACT: Not irritating. Rubbing may cause irritation similar to sand or dust.

SKIN ABSORPTION: Unlikely to occur. Material is a dry solid.

INGESTION: The material is believed to present very little hazard if swallowed.

INHALATION: Exposure to dust is not expected to present a hazard.

CHRONIC EFFECTS/CARCINOGENICITY:

Impurity C is listed as a potential carcinogen by IARC. Release of ingredient C from the product may occur in small quantities during processing of the product, but is not expected to present a significant hazard.

4. FIRST-AID MEASURES

EYES: Immediately flush with water for at least 5 minutes.

SKIN: Wash off in flowing water or shower.

INGESTION: No adverse effects anticipated by swallowing.

INHALATION: No adverse effects anticipated by breathing small amounts during proper industrial handling.

5. FIREFIGHTING MEASURES

FLAMMABLE PROPERTIES

Figure 4.1 Example MSDS Using 16-Part ANSI Format.

FLASH POINT: None

METHOD USED: Not applicable

FLAMMABLE LIMITS

LFL: Not applicable

UFL: Not applicable

EXTINGUISHING MEDIA: Water fog, foam, CO_2, dry chemicals.

FIRE & EXPLOSION HAZARDS: Dense smoke emitted when burned without sufficient oxygen. Possible dust explosion.

FIRE-FIGHTING EQUIPMENT: Wear full bunker gear, including a positive pressure self-contained breathing apparatus in any closed space.

6. ACCIDENTAL RELEASE MEASURES

Vacuum or sweep material and place in a disposal container.

7. HANDLING AND STORAGE

Avoid contact with eyes. Avoid breathing dust. Ground and bond containers when transferring material to prevent dust explosion. Minimize dust generation and accumulation.

8. EXPOSURE CONTROLS/PERSONAL PROTECTION

RESPIRATORY PROTECTION: For most conditions, no respiratory protection should be needed; however, in dusty atmospheres, use an approved dust respirator.

SKIN PROTECTION: No precautions other than clean body-covering clothing should be needed.

EYE PROTECTION: Use safety glasses. If there is a potential for exposure to particles that could cause mechanical injury to the eye, wear chemical goggles.

EXPOSURE GUIDELINES(s): None established for Resin ABC. Impurity C: ACGIH TLV and OSHA PEL are 50 ppm TWA, 100 ppm STEL.

ENGINEERING CONTROLS: Provide general and/or local exhaust ventilation to control airborne levels below the exposure guidelines. Good general ventilation should be sufficient for most conditions.

9. PHYSICAL AND CHEMICAL PROPERTIES

APPEARANCE: Milky white or colored powder
ODOR: None
BOILING POINT: Not applicable
VAPOR PRESSURE: Not applicable
VAPOR DENSITY: Not applicable
SOLUBILITY IN WATER: Not applicable
SPECIFIC GRAVITY: 1.05 @ 25/25°C (°F)
FREEZING POINT: Not applicable
pH: Not relevant
VOLATILE: Not applicable

Figure 4.1 *(Continued)*

10. **STABILITY AND REACTIVITY**

STABILITY: (CONDITIONS TO AVOID) At temperatures over 572°F (300°C), highly toxic cyanide fumes are released.
INCOMPATIBILITY: (SPECIFIC MATERIALS TO AVOID) Oxidizing materials.
HAZARDOUS DECOMPOSITION PRODUCTS: Compound G and hydrogen cyanide.
HAZARDOUS POLYMERIZATION: Will not occur.

11. **TOXICOLOGICAL INFORMATION**

For detailed toxicological information, write to the address listed in Section 1 of this MSDS.

12. **ECOLOGICAL INFORMATION**

See Technical Bulletin 25 on ecological toxicity of ABC Resin. Copies are available from the address in Section 1 of this MSDS.

13. **DISPOSAL CONSIDERATIONS**

Waste Management Information (Disposal): Burn in an adequate incinerator or bury in landfill in accordance with all applicable regulations. Any disposal practice must be in compliance with local, state, and federal laws and regulations (contact local or state environmental agency for specific rules).

14. **TRANSPORT INFORMATION**

TRANSPORTATION AND HAZARDOUS MATERIALS DESCRIPTION:
Not a hazardous material for DOT shipping.

15. **REGULATORY INFORMATION**

OSHA HAZARD COMMUNICATION RULE, 29 CFR 1910.1200: Impurity C is considered hazardous if present in more than trace amounts.

CERCLA/SUPERFUND, 40 CFR §117, 302

This product contains no reportable quantity (RQ) substances.

SARA HAZARD CATEGORY: This product has been reviewed according to the EPA Hazard Categories promulgated under Sections 311 and 312 of the Superfund Amendment and Reauthorization Act of 1986 (SARA Title III) and is considered, under applicable definitions, to meet the following categories:

Not to have met any hazard category.

SARA 313 INFORMATION:
This product contains the following substances subject to the reporting requirements of section 313 of Title III of the Superfund Amendments and Reauthorization Act of 1986 and 40 CFR Parts 372:

CHEMICAL NAME	ASNUMBER	CONCENTRATION
Impurity C	XXXXXX-XX-X	2000 ppm max. 0.2%

TOXIC SUBSTANCES CONTROL ACT (TSCA): The ingredients of this product are all on the TSCA inventory list.

CALIFORNIA PROPOSITION 65: The following statement is made to comply with the California Safe Drinking Water and Toxic Enforcement Act of 1986. This product contains Impurity C, a chemical known to the state of California to cause cancer.

16. **OTHER INFORMATION**

MSDS STATUS: Revised Regulatory Information Section.

Figure 4.1 (*Continued*)

TABLE 4.1 MSDS Checklist Based on ANSI Standard Z400.1-1993 and 29 CFR §1910.1200

Section 1: Chemical Product and Company Identification

_____ Identifies product name found on label
_____ Includes both generic chemical name and specific name
_____ Gives name, address, and telephone number of the manufacturer, distributor, employer, or other responsible party
_____ Provides date of preparation of the MSDS or date of last change

Section 2: Composition, Information on Ingredients

_____ Provides chemical and common name(s) of the ingredients contributing to known hazards (trade secrets can be indicated as such, in lieu of identifying ingredient)
_____ For untested mixtures, provides the chemical and common name of ingredients at 1% or more that present a health hazard and those that present a physical hazard in the mixture
_____ Provides ingredient at 0.1% or greater, if ingredient is a listed carcinogen
_____ *Optional:* Provides additional information such as percentages or percentage range of each component; CAS registry numbers; exposure limits; components regulated under federal, state, or local regulations

Section 3: Hazard Identification

_____ Provides health hazards, including acute and chronic effects and target organs or systems affected
_____ Gives signs and symptoms of exposure
_____ Gives medical conditions aggravated by exposure
_____ Gives primary routes of exposure
_____ If listed as a carcinogen by OSHA, IARC, or NTP, provides this information
_____ *Optional:* Provides emergency overview information such as physical color and form; odor; flammable, combustible, and explosive properties; reactive properties; other information that would aid in an emergency situation

Section 4: First-Aid Measures

_____ Provides emergency and first-aid procedures
_____ *Optional:* Provides a note to physicians, including additional information on potential health effects and recommended specific procedures for treatment

Section 5: Firefighting Measures

_____ Gives pertinent physical information such as flash point; upper and lower flammable (explosive) limits; autoignition temperature; hazardous combustion products; conditions of flammability; explosion data; fire extinguishing methods
_____ *Optional:* Provides additional information such as potential for dust explosion; reactions that release flammable gases or vapors; release of invisible flammable vapors; nonflammable that could contribute to unusual hazard to a fire, such as strong oxidizing and reducing agents

TABLE 4.1 (*Continued*)

Section 6: Accidental Release Measures

_____ Provides procedures for cleanup of leaks and spills
_____ *Optional:* Provides specific federal, reporting requirements

Section 7: Handling and Storage

_____ Gives safe handling and storage procedures

Section 8: Exposure Control, Personal Protection

_____ Provides specific engineering controls
_____ Specifies PPE
_____ Provides exposure limits

Section 9: Physical and Chemical Properties

_____ Provides physical and chemical properties such as appearance, odor threshold, physical state, pH, vapor pressure, boiling point, freezing point, specific gravity, evaporation rate, other

Section 10: Stability and Reactivity

_____ Provides information on reactivity, including unstable properties and water or air reactive properties

Section 11: Toxicological Information

_____ *Optional:* Provides information on toxicity testing of the material, its components, or both

Section 12: Ecological Information

_____ *Optional:* Provides information on the environmental impact of the material if released to the environment

Section 13: Disposal Considerations

_____ *Optional:* Provides information that may be useful in proper disposal of the material

Section 14: Transport Information

_____ *Optional:* Provides basic shipping classification information

Section 15: Regulatory Information

_____ *Optional:* Provides information on regulatory status of materials, including the components; addresses regulations such as OSHA, TSCA, FIFRA, CERCLA, EPCRA (SARA Title III), CAA, CWA, SDWA, and others

Section 16: Other Information

_____ *Optional:* Provides any additional pertinent information about the material or its components recognized acute effects at a particular limit. ACGIH recommends that exposure at this limit not be repeated more than four times daily and that at least a 60-minute rest period between exposures should be allowed.

A ceiling limit is a maximum concentration that should not be exceeded for any period of time. Ceiling limits are established for chemicals which, at certain concentrations, could produce acute poisoning during very short exposures.

Since OSHA, ACGIH, and NIOSH exposure limits are based on an 8-hour workday, 40-hour workweek time-weighted average, worker exposure may safely exceed published exposure limits for periods of time during the workday as long as they are compensated with periods of time when exposures are below the limit, so that the 8-hour time-weighted average is not exceeded. However, all factors related to a chemical exposure—including cumulative effects, frequency and duration of excursions, nature of contaminant, and others—should be considered before this type of exposure is allowed. In some cases, excursions above the permissible exposure limit or threshold limit value may not be acceptable.

The STEL and ceiling limits should not be exceeded at any time during the workday. For many chemicals, there is not enough toxicological data available for ACGIH, NIOSH, or OSHA to publish an STEL or ceiling limit. When an STEL has not been published, ACGIH recommends that worker exposure at levels of three times the TLV-TWA should last for no more than 30 minutes during a workday. Additionally, if there is no ceiling limit, a concentration level of five times the TLV-TWA should never be exceeded. Consideration should also be given to areas that contain chemical mixtures. When calculating the exposure limit of chemical mixtures in the air, additive effects for individual components should be used when the components have similar toxicological effects.

Selected examples of chemicals that have published OSHA, ACGIH, and NIOSH exposure limits are presented in Table 4.2. Examples of allowable exposure patterns based on the time weighted average are shown in Figure 4.2.

Odor threshold is a physical property of a chemical and has no relation to acceptable exposure concentrations. There are numerous chemicals that are toxic at levels below the odor threshold. Odor has been detected at varying levels during tests conducted over the years, so there are no absolute numeric criteria for odor threshold. In addition, human differences in ability to smell play a factor in the recognition of a chemical through the olfactory sense. Examples of chemical odor thresholds (in ranges) in comparison to exposure concentrations are presented in Table 4.3.

Odor—or the absence of odor—should never be used in determining acceptable workplace concentrations. In cases where workers are using air-purifying (cartridge) respirators, odor (and taste) is considered a warning

TABLE 4.2 Examples of Documented OSHA, ACGIH, and NIOSH Exposure Limits for Selected Chemicals[a]

Chemical Compound	PEL/TLV/REL (ppm)	STEL (ppm)	Ceiling (ppm)
Acetic acid	10/10/10	–/15/15	–/–/–
Benzyl chloride	1/1/–	–/–/–	–/–/1
Fluorine	0.1/1/0.1	–/2/–	–/–/–
Hexachloroethane	1/1/1	–/–/–	–/–/–
Iodine	–/–/–	–/–/–	0.1/0.1/0.1
Isopropyl alcohol	400/400/400	–/500/500	–/–/–
Methylamine	10/5/10	–/15/–	–/–/–
Naphthalene	10/10/10	–/15/15	–/–/–
Phenol	5/5/5	–/–/–	–/–/15.6[b]
Sulfur dioxide	2/2/2	–/5/5	–/–/–
Vinyl chloride	1/5/1	5/–/5	–/–/–
Xylenes	100/100/100	–/150/150	–/–/–

[a]Values in terms of milligrams per cubic meter (mg/m^3) may exist but are not included.
[b]15-minute ceiling
Source: Information from ACGIH (1996a,b), and 29 CFR §1910.1000, Table Z-1.

property indicating breakthrough of the filter material.[1] When noticed, the work area should be evacuated immediately.

Exposure Limits for Asbestos

Asbestos is regulated under 29 CFR §1910.1001 (general industry) and 29 CFR §1926.58 (construction industry). Asbestos is the name of a class of magnesium-silicate minerals that occur in fibrous form. Minerals that are included in this group are chrysotile, crocidolite, amosite, anthophylite asbestos, tremolite asbestos, and actinolite asbestos. Revised in 1994, asbestos exposure limits include an 8-hour time-weighted average of 0.1 fiber per cubic centimeter (f/cm^3) and a 30-minute short-term exposure limit (excursion limit) of 1 f/cm^3.

Exposure Limits for Radioactive Materials

NRC has codified the practice of maintaining all radiation exposures to workers and the general public as low as reasonably achievable (known as the *ALARA concept*). Public and worker protection standards for specific radio-

[1]In general, cartridge respirators are not allowed for chemicals that have permissible exposure limits below the odor threshold.

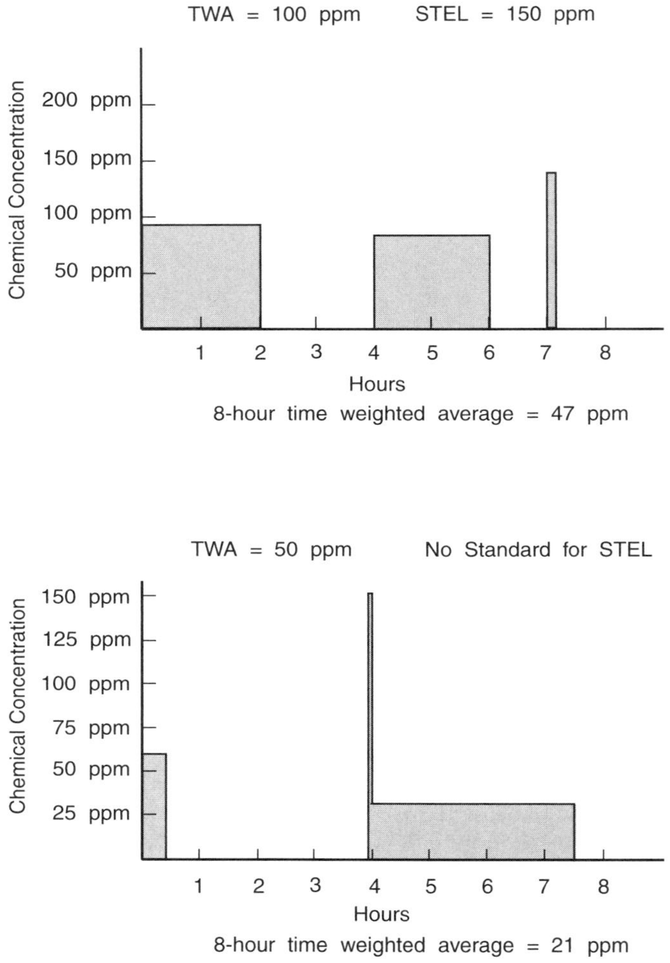

Note: A trained specialist should evaluate thoroughly factors such as the TWA and chemical additive effects to determine the need for respiratory protection.

Figure 4.2 Allowable Chemical Exposure Patterns Based on Time-Weighted Average.

active materials and wastes have been established by OSHA, EPA, NRC, Mine Safety and Health Administration (MSHA), and other agencies (Oak Ridge Associated Universities 1988). Worker limiting requirements are categorized by body part and organ and are defined in numerous regulations, including 10 CFR Part 20, Subpart C, 29 CFR §1910.96, 40 CFR Part 191, 30 CFR Part 57, and other citations. The dose requirements of OSHA and NRC are summarized in Table 4.4.

TABLE 4.3 Examples of Chemical Odor Thresholds Compared to Exposure Limits

Chemical	ACGIH TLV-TWA (ppm)	OSHA PEL-TWA (ppm)	Odor Threshold[b] (range in ppm)
Acetone	750	1000	3.6–653
Allyl alcohol	2	2	1.4–2.1
Carbon tetrachloride	5	10	140–584
Chloroform	10	50 (C)[a]	133–276
Cyclohexanone	25	50	0.052–219
Ethyl acetate	400	400	6.4–50
Isopropyl alcohol	400	400	37–610
Methyl Cellosolve	5	25	0.096–61
Methyl ethyl ketone	200	200	2–85
Propylene oxide	20	100	10–199

[a]This value is the ceiling value.
[b]Odor threshold ranges are detection levels from critiqued sources deemed acceptable by AIHA.
Sources: AIHA (1989), ACGIH (1996a,b), and 29 CFR 1910, Subpart Z.

TABLE 4.4 Worker-Limiting Dose Requirements of OSHA and NRC

OSHA Requirements

- Whole body (head and trunk; active blood-forming organs; lens of eyes; or gonads)—$1\frac{1}{4}$ rems per calendar quarter
- Hand and forearms; feet and ankles—$18\frac{3}{4}$ rems per calendar quarter
- Skin of entire body—$7\frac{1}{4}$ rems per calendar quarter

NRC Requirements

- An annual limit, which is the more limiting of:
  —The total effective dose equivalent being equal to 5 rems; or
  —The sum of the deep-dose equivalent and the committed dose equivalent to any individual organ or tissue other than the lens of the eye being equal to 50 rems
- The annual limits to the lens of the eye, to the skin, and to the extremities, which are as follows:
  —An eye dose equivalent of 15 rems; and
  —A shallow-dose equivalent of 50 rems to the skin
  —A shallow-dose equivalent of 50 rems to any extremity

Source: 29 CFR §1910.96 and 10 CFR §20.1202.

TOXICITY CHARACTERISTIC LEACHING PROCEDURE

EPA's toxicity characteristic leaching procedure (TCLP), designated as Method 1311 and presented in 40 CFR §261, Appendix II of the RCRA regulations, is a solid waste extraction procedure used to identify hazardous characteristics of solid waste contaminated with listed toxic chemicals, metals, and pesticides. Chemicals regulated as toxic by TCLP testing are listed in 40 CFR §261.24 and are presented in Table 4.5. Regulatory limits in milligrams per liter (mg/L) are set for each chemical. If the TCLP analysis shows concentrations above the regulatory limits, the waste must be managed as hazardous.

MEDICAL SURVEILLANCE

Medical surveillance programs offer information about the biological effects that chemicals may be having on an individual worker. Medical surveillance is required by OSHA for workers who are exposed routinely to certain concentrations of regulated chemicals or dusts, such as lead, benzene, formaldehyde, acrylonitrile, asbestos, and coke oven emissions. Additionally,

TABLE 4.5 Chemicals Regulated as Toxic by TCLP Testing

Chemical Category	Chemicals Regulated
Metals	Arsenic, barium, cadmium, chromium, lead, mercury, selenium, silver
Pesticides	Chlordane, 2,4-D, endrin, heptachlor (and its epoxide), lindane, methoxychlor, toxaphene, 2,4,5-TP (Silvex)
Cresols	Cresol, *m*-cresol, *o*-cresol, *p*-cresol
Phenols	Pentachlorophenol, 2,4,5-trichlorophenol, 2,4,6-trichlorophenol
Benzenes	Benzene, chlorobenzene, 1,4-dichlorobenzene, 2,4-dinitrotoluene, hexachlorobenzene, nitrobenzene
Chlorinated	Carbon tetrachloride, chloroform, 1,2-dichloroethane, compounds
	1,1-dichloroethylene, hexachlorobutadiene, hexachloroethane, tetrachloroethylene (perchloroethylene), trichloroethylene, vinyl chloride
Other toxic compounds	methyl ethyl ketone, pyridine

medical surveillance is required for workers who routinely wear a respirator or who take part in hazardous waste operations and emergency response activities. The employer is responsible for administering and paying for the medical examinations and other medical surveillance activities. Tests included in a typical medical surveillance program can vary depending on specific chemical exposures.

A medical questionnaire is mandatory for workers who are exposed to asbestos and is generally included in some (less exhaustive) form in most medical surveillance programs. The information required in the asbestos questionnaire (parts 1 and 2) is presented in Figure 4.3.

Other aspects of the medical surveillance program could include a physical exam, general health survey, blood count, chemistry screen, urinalysis, spirometry test, chest x-ray, audiogram, and electrocardiogram. Breath analysis may be included for workers exposed to chemicals that can be detected through this type of screening.

All exams should be performed on a regularly scheduled basis, with the time between exams depending on types of exposures. The time of testing, such as at the end of the shift or the end of the week, often can be critical for observing accurate effects on blood or urine, so this should be taken into account when establishing the examination schedule.

Biological effects of chemical overexposure are documented by ACGIH and can be seen in blood, urine, and respiratory functions. Effects on blood can be seen directly or indirectly through blood counts and blood screening. Indirect indicators are red and white blood cells, plasma, and other blood components. Examples of chemicals that affect the blood composition include parathion, carbon monoxide, aniline, and nitrobenzene.

Overexposure can be ascertained directly by the presence of a specific chemical in the blood stream at elevated levels. Chemicals that can be tested in this manner include cadmium, lead, toluene, and perchlorethylene.

In addition to blood screening, chemistry screening can provide information on cholesterol, triglycerides, total protein, blood urea nitrogen, calcium, phosphorus, and other chemical components. Urinalysis is also a very useful screening test. This test can provide direct detection of numerous chemicals in the urine. Examples include mercury, cadmium, lead, methyl ethyl ketone, and phenol.

Breath analysis can be used to detect the presence of some chemicals. Standards for acceptable concentrations of chemicals in the breath have been developed by ACGIH for chemicals such as benzene, carbon monoxide, toluene, and trichloroethylene.

Decreases in lung capacity can be determined with the spirometry test. When necessary, this test can be coupled with chest x-rays. In some cases, respiratory capacity tests may be performed more often than other parts of the medical surveillance tests, if exposure warrants.

Part 1
INITIAL MEDICAL QUESTIONNAIRE

1. Name ____________________

2. Social Security No. ___ ___ ___ ___ ___ ___ ___ ___ ___
1 2 3 4 5 6 7 8 9

3. Clock No. ___ ___ ___ ___ ___ ___
10 11 12 13 14 15

4. Present occupation ____________________

5. Plant ____________________

6. Address ____________________

7. ____________________
(zip code)

8. Telephone No. ____________________

9. Interviewer ____________________

10. Date ____________________ ___ ___ ___ ___ ___ ___
16 17 18 19 20 21

11. Date of birth ____________________ ___ ___ ___ ___ ___ ___
Month Day Year 22 23 24 25 26 27

12. Place of birth ____________________

13. Sex
1. Male ___
2. Female ___

14. What is your marital status?
1. Single ___
2. Married ___
3. Widowed ___
4. Separated/ Divorced ___

15. Race
1. White ___
2. Black ___
3. Asian ___
4. Hispanic ___
5. Indian ___
6. Other ___

16. What is the highest grade completed in school? ____________________
(For example, 12 years is completion of high school)

OCCUPATIONAL HISTORY

17A. Have you ever worked full time (30 hours per week or more) for 6 months or more? 1. Yes ___ 2. No ___
If YES to 17A:

17B. Have you ever worked for a year or more in any dusty job? 1. Yes ___ 2. No ___
3. Does not apply___
Specify job/industry ____________ Total years worked ________
Was dust exposure: 1. Mild ___ 2. Moderate ___ 3. Severe ___

17C. Have you ever been exposed to gas or chemical fumes in your work? 1. Yes ___ 2. No. ___
Specify job/industry ____________ Total years worked ________
Was exposure: 1. Mild ___ 2. Moderate ___ 3. Severe ___

FIGURE 4.3 Example Health Questionnaire.

17D. What has been your usual occupation or job—the one you have worked at the longest?

1. Job occupation __________
2. Number of years employed in this occupation __________
3. Position/job title __________
4. Business, field, or industry __________

(Record on lines the years in which you have worked in any of these industries, e.g., 1960–1969)

Have you ever worked:

	YES	NO
E. In a mine?	☐	☐
F. In a quarry?	☐	☐
G. In a foundry?	☐	☐
H. In a pottery?	☐	☐
I. In a cotton, flax, or hemp mill?	☐	☐
J. With asbestos?	☐	☐

18. **PAST MEDICAL HISTORY**

	YES	NO
A. Do you consider yourself to be in good health? If NO, state reason__________	☐	☐
B. Have you any defect of vision?__________ If YES, state nature of defect__________	☐	☐
C. Have you any hearing defect?__________ If YES, state nature of defect__________	☐	☐
D. Are you suffering from or have you ever suffered from:		
a. Epilepsy (or fits, seizures, convulsions)?	☐	☐
b. Rheumatic fever?	☐	☐
c. Kidney disease?	☐	☐
d. Bladder disease?	☐	☐
e. Diabetes?	☐	☐
f. Jaundice?	☐	☐

19. **CHEST COLDS AND CHEST ILLNESSES**

19A. If you get a cold, does it usually go to your chest? (Usually means more than $\frac{1}{2}$ the time.) 1. Yes _____ 2. No_____ 3. Don't get colds_____

Figure 4.3 (*Continued*)

20A. During the past 3 years, have you had any chest illnesses that have kept you off work, indoors at home, or in bed?	1. Yes_____ 2. No_____
If YES to 20A:	
20B. Did you produce phlegm with any of these chest illnesses?	1. Yes_____ 2. No_____ 3. Does not apply_____
20C. In the last 3 years, how many such illnesses with (increased) phlegm did you have which lasted a week or more?	Number of illnesses _____ No such illnesses _____
21. Did you have any lung trouble before the age of 16?	1. Yes_____ 2. No_____
22. Have you ever had any of the following?	
1A. Attacks of bronchitis?	1. Yes_____ 2. No_____
If YES to 1A:	
1B. Was it confirmed by a doctor?	1. Yes_____ 2. No_____ 3. Does not apply_____
1C. At what age was your first attack?	Age in years _____ Does not apply _____
2A. Pneumonia (include bronchopneumonia)?	1. Yes_____ 2. No_____
If Yes to 2A:	
2B. Was it confirmed by a doctor?	1. Yes_____ 2. No_____ 3. Does not apply _____
2C. At what age did you first have it?	Age in years __________ Does not apply __________
3A. Hay fever?	1. Yes_____2. No_____
If YES to 3A:	
3B. Was it confirmed by a doctor?	1. Yes_____ 2. No_____ 3. Does not apply_____
3C. At what age did it start?	Age in years _____ Does not apply_____
23A. Have you ever had chronic bronchitis?	1. Yes_____ 2. No_____
If YES to 23A:	
23B. Do you still have it?	1. Yes_____ 2. No_____ 3. Does not apply_____
23C. Was it confirmed by a doctor?	1. Yes_____ 2. No_____ 3. Does not apply_____
23D. At what age did it start?	Age in years_____ Does not apply_____

Figure 4.3 (*Continued*)

24A. Have you ever had emphysema? 1. Yes_____ 2. No_____
If YES to 24A:

24B. Do you still have it? 1. Yes_____ 2. No_____
3. Does not apply_____

24C. Was it confirmed by a doctor? 1. Yes_____ 2. No_____
3. Does not apply_____

24D. At what age did it start? Age in years _____
Does not apply_____

25A. Have you ever had asthma? 1. Yes_____ 2. No_____
If YES to 25A:

25B. Do you still have it? 1. Yes_____ 2. No_____
3. Does not apply_____

25C. Was it confirmed by a doctor? 1. Yes_____ 2. No_____
3. Does not apply_____

25D. At what age did it start? Age in years _____
Does not apply_____

25E. If you no longer have it, at what age did it stop? Age in years _____
Does not apply _____

26. Have you ever had:

26A. Any other chest illness? 1. Yes_____ 2. No_____
If yes, please specify____

26B. Any chest operations? 1. Yes_____ 2. No_____
If yes, please specify____

26C. Any chest injuries? 1. Yes_____ 2. No_____
If yes, please specify____

27A. Has a doctor ever told you that you had heart trouble? 1. Yes_____ 2. No_____
If YES to 27A:

27B. Have you ever had treatment for heart trouble in the past 10 years? 1. Yes_____ 2. No_____
3. Does not apply _____

28A. Has a doctor ever told you that you had high blood pressure? 1. Yes_____ 2. No_____
If YES to 28A:

28B. Have you had any treatment for high blood pressure (hypertension) in the past 10 years? 1. Yes_____ 2. No_____
3. Does not apply_____

29. When did you last have your chest X-rayed? (Year) _____ _____ _____ _____
25 26 27 28

30. Where did you last have your chest X-rayed? _____________________________
What was the outcome? ____

Figure 4.3 (*Continued*)

FAMILY HISTORY

31. Were either of your natural parents ever told by a doctor that they had a chronic lung condition such as:

		FATHER			MOTHER		
		1. Yes	2. No	3. Don't know	1. Yes	2. No	3. Don't know
A.	Chronic bronchitis?	_____	_____	_____	_____	_____	_____
B.	Emphysema?	_____	_____	_____	_____	_____	_____
C.	Asthma?	_____	_____	_____	_____	_____	_____
D.	Lung cancer?	_____	_____	_____	_____	_____	_____
E.	Other chest conditions?	_____	_____	_____	_____	_____	_____
F.	Is parent currently alive?	_____	_____	_____	_____	_____	_____

G. Please specify

Father	Mother
_____ Age if living	_____ Age if living
_____ Age at death	_____ Age at death
_____ Don't know	_____ Don't know

H. Please specify cause of death

_______________________________ _______________________________

COUGH

32A. Do you usually have a cough? (Count a cough with first smoke or on first going out of doors. Exclude clearing of throat.) (If no, skip to 32C.) 1. Yes_____ 2. No_____

32B. Do you usually cough as much as 4 to 6 times a day 4 or more days out of the week? 1. Yes_____ 2. No_____

32C. Do you usually cough at all on getting up or first thing in the morning? 1. Yes_____ 2. No_____

32D. Do you usually cough at all during the rest of the day or at night? 1. Yes_____ 2. No_____

If YES to any of above (32A, B, C, or D), answer the FOLLOWING questions. If NO to all, check Does not apply and skip to 33A.

32E. Do you usually cough like this on most days for 3 consecutive months or more during the year? 1. Yes_____ 2. No_____

3. Does not apply _____

Figure 4.3 (*Continued*)

32F.	For how many years have you had the cough?	Number of years_____ Does not apply _____
33A.	Do you usually bring up phlegm from your chest? (Count phlegm with the first smoke or on first going out of doors. Exclude phlegm from the nose. Count swallowed phlegm.) (If no, skip to 33C.)	1. Yes_____ 2. No_____
33B.	Do you usually bring up phlegm like this as much as twice a day 4 or more days out of the week?	1. Yes_____ 2. No_____
33C.	Do you usually bring up phlegm at all on getting up or first thing in the morning?	1. Yes_____ 2. No_____
33D.	Do you usually bring up phlegm at all during the rest of the day or at night?	1. Yes_____ 2. No_____

If YES to any of the above (33A, B, C, or D), answer the following questions. If NO to all, check Does not apply and skip to 34A.

33E.	Do you bring up phlegm like this on most days for 3 consecutive months or more during the year?	1. Yes_____ 2. No_____ 3. Does not apply_____
33F.	For how many years have you had trouble with phlegm?	Number of years_____ Does not apply_____

EPISODES OF COUGH AND PHLEGM

34A.	Have you had periods or episodes of (increased*) cough and phlegm lasting for 3 weeks or more each year? *(For persons who usually have cough and/or phlegm)	1. Yes_____ 2. No_____
	If YES to 34A:	
34B.	For how long have you had at least 1 such episode per year?	Number of years_____ Does not apply_____

WHEEZING

35A.	Does your chest ever sound wheezy or whistling	
	1. When you have a cold?	1. Yes_____ 2. No_____
	2. Occasionally apart from colds?	1. Yes_____ 2. No_____
	3. Most days or nights?	1. Yes_____ 2. No_____
	If YES to 1, 2, or 3 in 35A:	
35B.	For how many years has this been present?	Number of years_____ Does not apply_____
36A.	Have you ever had an attack of wheezing that has made you feel short of breath?	1. Yes_____ 2. No_____
	If YES to 36A:	

Figure 4.3 (*Continued*)

36B.	How old were you when you had your first such attack?	Age in years_____ Does not apply_____
36C.	Have you had 2 or more such episodes?	1. Yes_____ 2. No_____ 3. Does not apply_____
36D.	Have you ever required medicine or treatment for the(se) attack(s)?:	1. Yes_____ 2. No_____ 3. Does not apply_____

BREATHLESSNESS

37.	If disabled from walking by any condition other than heart or lung disease, please describe and proceed to 39A. Nature of conditions(s)_____	
38A.	Are you troubled by shortness of breath when hurrying on the level or walking up a slight hill?	1. Yes_____ 2. No_____
	If YES to 38A:	
38B.	Do you have to walk slower than people of your age on the level because of breathlessness?	1. Yes_____ 2. No_____ 3. Does not apply_____
38C.	Do you ever has to stop for breath when walking at your own pace on the level?	1. Yes_____ 2. No_____ 3. Does not apply_____
38D.	Do you ever have to stop for breath after walking about 100 yards (or after a few minutes) on the level?	1. Yes_____ 2. No_____ 3. Does not apply_____
38E.	Are you too breathless to leave the house or breathless on dressing or climbing one flight of stairs?	1. Yes_____ 2. No_____ 3. Does not apply_____

TOBACCO SMOKING

39A.	Have you ever smoked cigarettes? (No means less than 20 packs of cigarettes or 12 ounces of tobacco in a lifetime or less than 1 cigarette a day for 1 year.)	1. Yes_____ 2. No_____
	If YES to 39A:	
39B.	Do you now smoke cigarettes (as of one month ago)?	1. Yes_____ 2. No_____ 3. Does not apply_____

Figure 4.3 (*Continued*)

39C.	How old were you when you first started regular cigarette smoking?	Age in years _____ Does not apply_____
39D.	If you have stopped smoking cigarettes completely, how old were you when you stopped?	Age stopped_____ Check if still smoking_____ Does not apply_____
39E.	How many cigarettes do you smoke per day now?	Cigarettes per day_____ Does not apply_____
39F.	On the average of the entire time you smoked, how many cigarettes did you smoke per day?	Cigarettes per day_____ Does not apply_____
39G.	Do or did you inhale the cigarette smoke?	1. Does not apply_____ 2. Not at all_____ 3. Slightly_____ 4. Moderately_____ 5. Deeply_____
40A.	Have you ever smoked a pipe regularly? (Yes means more than 12 ounces of tobacco in a lifetime.) If YES to 40A:	1. Yes_____ 2. No_____

FOR PERSONS WHO HAVE EVER SMOKED A PIPE

40B.	1. How old were you when you started to smoke a pipe regularly?	Age_____
	2. If you have stopped smoking a pipe completely, how old were you when stopped?	Age stopped _____ Check if still smoking pipe _____ Does not apply _____
40C.	On the average over the entire time you smoked a pipe, how much pipe tobacco did you smoke per week?	_____ounces per week (a standard pouch of tabacco contains $1\frac{1}{2}$ ounces) _____ Does not apply
40D.	How much pipe tobacco are you smoking now?	Ounces per week _____ Not currently smoking a pipe _____
40E.	Do you or did you inhale the pipe smoke?	1. Never smoked _____ 2. Not at all _____ 3. Slightly _____ 4. Moderately _____ 5. Deeply _____
41A.	Have you ever smoked cigars regularly? (Yes means more than 1 cigar a week for a year.) If YES to 41A:	1. Yes_____ 2. No_____

Figure 4.3 (*Continued*)

FOR PERSONS WHO HAVE EVER SMOKED CIGARS

41B.	1. How old were you when you started smoking cigars regularly?	Age_____	
	2. If you have stopped smoking cigars completely, how old were you when you stopped.	Age stopped	_____
		Check if still smoking cigars	_____
		Does not apply	_____
41C.	On the average over the entire time you smoked cigars, how many cigars did you smoke per week?	Cigars per week	_____
		Does not apply	_____
41D.	How many cigars are you smoking per week now?	Cigars per week	_____
		Check if not smoking cigars currently	_____
41E.	Do or did you inhale the cigar smoke?	1. Never smoked	_____
		2. Not at all	_____
		3. Slightly	_____
		4. Moderately	_____
		5. Deeply	_____

Signature ______________________________ Date ______________________________

Part 2

PERIODIC MEDICAL QUESTIONNAIRE

1. Name______________________________
2. Social Security No. ___ ___ ___ ___ ___ ___ ___ ___ ___ (1 2 3 4 5 6 7 8 9)
3. Present Occupation ____________________
4. Plant ____________________
5. Address ____________________
6. ______________________________ (zip code)
7. Telephone No.______________________________
8. Interviewer______________________________
10. Date____________________
11. What is your marital status? 1. Single _____ 2. Married _____ 3. Widowed _____ 4. Separated/divorced _____

12. **OCCUPATIONAL HISTORY**

12A. In the past year, did you work full (30 hours per week or more) for 6 months or more? 1. Yes_____ 2. No_____

If YES to 12A:

Figure 4.3 (*Continued*)

12B.	In the past year, did you work in a dusty job?		1. Yes_____	2. No_____
			3. Does not apply _____	
12C.	Was dust exposure:	1. Mild_____	2. Moderate_____	3. Severe_____
12D.	In the past year, were you exposed to gas or chemical fumes in your work?		1. Yes_____	2. No_____
12E.	Was exposure:	1. Mild_____	2. Moderate_____	3. Severe_____
12F.	In the past year, what was your:	1. Job/occupation?________________ 2. Position/job title?________________		

13. **RECENT MEDICAL HISTORY**

13A.	Do you consider yourself to be in good health?	1. Yes_____	2. No_____
	If NO, state reason____________________________		
13B.	In the past year have you developed:	Yes	No
	Epilepsy?	_____	_____
	Rheumatic fever?	_____	_____
	Kidney disease?	_____	_____
	Bladder disease?	_____	_____
	Diabetes?	_____	_____
	Jaundice?	_____	_____
	Cancer?	_____	_____

14. **CHEST COLDS AND CHEST ILLNESSES**

14A.	If you get a cold, does it usually go to your chest? (Usually means more than $\frac{1}{2}$ the time.)	1. Yes_____	2. No_____
		3. Don't get colds _____	
15A.	During the past year, have you had any chest illnesses that have kept you off work, indoors at home, or in bed?	1. Yes_____	2. No_____
		3. Does not apply _____	
	If YES to 15A:		
15B.	Did you produce phlegm with any of these chest illnesses?	1. Yes_____ 2. No_____ 3. Does not apply _____	
15C.	In the past year, how many such illnesses with increased phlegm did you have that lasted a week or more?	Number of illnesses _____ No such illnesses _____	

Figure 4.3 (*Continued*)

16. **RESPIRATORY SYSTEM**

In the past year have you had:

	Yes or No	Further Comment on Positive Answers
Asthma	_____	
Bronchitis	_____	
Hay fever	_____	
Other allergies	_____	

	Yes or No	Further Comment on Positive Answers
Pneumonia	_____	
Tuberculosis	_____	
Chest surgery	_____	
Other lung problems	_____	
Heart disease	_____	

Do you have:

	Yes or No	Further Comment on Positive Answers
Frequent colds	_____	
Chronic cough	_____	
Shortness of breath when walking or climbing one flight of stairs	_____	

Do you:

Wheeze	_____	
Cough up phlegm	_____	
Smoke cigarettes	_____	Packs per day _____ How many years_____

Date_____________________ Signature_____________________

Figure 4.3 (*Continued*)

REFERENCES

American Conference of Governmental Industrial Hygienists (1996a). *Threshold Limit Values for Chemical Substances and Physical Agents and Biological Exposure Indices,* ACGIH, Cincinnati, OH.

——— (1996b). *Guide to Occupational Exposure Values: 1996,* ACGIH, Cincinnati, OH.

American Industrial Hygiene Association (1989). *Odor Thresholds for Chemicals with Established Occupational Health Standards,* AIHA Fairfax, VA.

American National Standards Institute (1993). "Hazardous Industrial Chemicals: Material Safety Data Sheets—Preparation," ANSI Z400.1, ANSI, New York.

Code of Federal Regulations, 29 CFR 1910 Subpart Z, U.S. Department of Labor, Occupational Safety and Health Administration, Washington, DC.

Oak Ridge Associated Universities (1988). *A Compendium of Major U.S. Radiation Protection Standards and Guides: Legal and Technical Facts,* prepared by W. A. Mills et al., ORAU, Oak Ridge, TN.

BIBLIOGRAPHY

American Conference of Governmental Industrial Hygienists (1996). *Guide to Occupational Exposure Values: 1996,* ACGIH, Cincinnati, OH.

American Petroleum Institute (1983). *Surveillance of Reproductive Health in the U.S: A Survey of Activity Within and Outside Industry,* monograph prepared by M. Hatch et al., API, Washington, DC.

——— (1990). *A Case-Control Study of Kidney Cancer Among Petroleum Refinery Workers,* API Publication 4504, API, Washington, DC.

Ashford, N. A., and C. S. Miller (1991). *Chemical Exposures: Low Levels and High Stakes,* Van Nostrand Reinhold, New York.

Chemical Manufacturers Association (1991). *Draft ANSI Standards for the Preparation of Material Safety Data Sheets,* CMA, Washington, DC.

——— (1991). *Occupational Epidemiology Resource Manual,* CMA, Washington, DC.

Clayton, G. D., and F. E. Clayton, eds. (1991). *Patty's Industrial Hygiene and Toxicology,* Vols. 1A and 1B, 4th ed., Wiley, New York.

Dillon, H. K., and M. H. Ho, eds. (1991). *Biological Monitoring of Exposure to Chemicals: Metals,* Wiley, New York.

Environmental Protection Agency (1984). *Biological Effects of Radiofrequency Radiation,* EPA/600/8-83/026F, prepared by D. F. Cahill and J. A. Elder, eds., Health Effects Research Laboratory, Office of Research and Development, U.S. EPA, Research Triangle Park, NC.

Ho, M. H., ed. (1987). *Biological Monitoring of Exposure to Chemicals: Organic Compounds,* Wiley, New York.

Hodgson, E. (1988). *Dictionary of Toxicology,* Van Nostrand Reinhold, New York.

Kusnetz, S., and M. K. Hutchinson (1979). *A Guide to the Work-Relatedness of Disease, NIOSH,* Cincinnati, OH.

Lewis, R. J., Sr. (1990). *Rapid Guide to Hazardous Chemicals in the Workplace,* 2nd ed., Van Nostrand Reinhold, New York.

Lipton, S., and J. Lynch (1987). *Health Hazard Control in the Chemical Process Industry,* Wiley, New York.

National Council for Radiation Protection and Measurements (1989). *Radiation Protection for Medical and Allied Health Personnel,* Report 105, NCRP,Bethesda, MD.

National Institute for Occupational Safety and Health (1987). *Pocket Guide to Chemical Hazards,* DHHS (NIOSH) Publication 85-114, U.S. Department of Health and Human Services, Washington, DC.

Nuclear Regulatory Commission (1992). *Occupational Radiation Exposure at Commercial Nuclear Power Reactors and Other Facilities,* 22nd Report, NUREG-0713-V11/XAB, NRC, Washington, DC.

Proctor, N. H., J. P. Hughes, and M. L. Fischman (1990). *Chemical Hazards of the Workplace,* 2nd ed., Van Nostrand Reinhold, New York.

Reichert, R. J. (1990). "Pitfalls and Protocols for Medical Surveillance of Hazardous Waste Workers," pp. 119–129, *Proceedings of the 3rd Annual Hazardous Materials Management Conference/Central,* Chicago.

Sherman, J. (1988). *Chemical Exposure and Disease: Diagnostic and Investigative Techniques,* Van Nostrand Reinhold, New York.

Technology Assessment Task Force (1990). *Reproductive Health Hazards in the Workplace,* Van Nostrand Reinhold, New York.

Williams, P. L., and J. L. Burson (1989). *Industrial Toxicology: Safety and Health Applications in the Workplace,* Van Nostrand Reinhold, New York.

5

WORKPLACE AND PERSONAL MONITORING

Area or general workplace monitoring and personal monitoring are two means of measuring exposures associated with industrial activity. Area monitoring is the process of using real-time air monitoring devices to chart airborne concentration as a function of time at fixed locations. This type of sampling provides qualitative and/or quantitative data about the chemical atmosphere of a work location or area. Typical applications include hazardous waste site characterization, verification of safe entry after evacuation, and location chemical investigations resulting from unusual odors or other problems. Some continuous monitoring systems are used in the workplace to detect leaks and improper operations that cause higher than normal chemical exposures.

Personal monitoring provides quantitative data about employee exposure associated with a specific set of work tasks. The samples are taken in the employee's breathing zone to represent actual inhalation exposures. This type of sampling is used widely throughout industry to ensure that permissible exposure limits are being met. In this chapter we discuss both general workplace and personal monitoring and provide information about standard test methods, types of instruments available for sampling, and other information pertinent to the topic. Sampling of ducts and stacks is described in Chapter 15.

STANDARD TEST METHODS AND PRACTICES

Standard test methods, practices, and guides for workplace and personal monitoring are defined by the American Society of Testing and Materials (ASTM) in the *Annual Book of ASTM Standards* (see the references) and include:

- D1356-73(1991)—Standard Definitions of Terms Relating to Atmospheric Sampling and Analysis

- D1357-82(1989)—Standard Practice for Planning the Sampling of the Ambient Atmosphere
- D1605-60(1990)—Standard Practice for Sampling Atmospheres for Analysis of Gases or Vapors
- D3686-89—Standard Practice for Sampling Atmospheres to Collect Organic Compound Vapors (Activated Charcoal Tube Adsorption Method)
- D3687-89—Standard Practice for Analysis of Organic Compound Vapors Collected by Activated Charcoal Tube Adsorption Method (Using Gas/ Liquid Chromatography)
- D3824-88—Standard Test Method for Continuous Measurement of Oxides of Nitrogen in the Ambient or Workplace Atmosphere by the Chemiluminescent Method
- D4240-83(1989)—Standard Test Method for Airborne Asbestos Concentration in Workplace Atmosphere
- D4490-90—Standard Practice for Measuring the Concentration of Toxic Gases or Vapors using Detector Tubes
- D4532-85(1990)—Standard Test Method for Respirable Dust in Workplace Atmospheres
- D4597-87—Standard Practice for Sampling Workplace Atmospheres to Collect Organic Gases or Vapors with Activated Charcoal Diffusional Samplers
- D4598-87—Standard Practice for Sampling Workplace Atmospheres to Collect Organic Gases or Vapors with Liquid Sorbent Diffusional Samplers
- D4599-86—Standard Practice for Measuring the Concentration of Toxic Gases or Vapors using Length-of-Stain Dosimeter
- D4844-88—Guide for Air Monitoring at Waste Management Facilities for Worker Protection
- D4861-91—Standard Practice for Sampling and Analysis of Pesticides and Polychlorinated Biphenyls in Indoor Atmospheres
- D4947-89—Standard Test Method for Chlordane and Heptachlor Residues in Indoor Air
- E1370-90—Guide to Air Sampling Strategies for Worker and Workplace Protection

As can be seen from the list above, standards exist for workplace monitoring of all types of chemicals, including gases or vapors, particulates, pesticides, PCBs, asbestos, and dusts. Personal sampling methods, including passive or diffusional sampling (D4597 and D4598) and sampling with a personal sampling pump (D3686), are also addressed in the standards.

In addition to ASTM standards, useful air sampling information can be found in ACGIH's handbook entitled *Air Sampling Instruments for Evaluation of Atmospheric Contaminants* (ACGIH 1995). Other useful information is published by NIOSH, OSHA, and EPA. Acceptable analytical methods for workplace applications are documented in *NIOSH Manual of Analytical Methods* (NIOSH 1987), *OSHA Analytical Methods Manual* (OSHA 1990, 1991), and *Compendium of Methods for the Determination of Toxic Organic Compounds in Indoor Air* (EPA 1990). NIOSH, OSHA, the U.S. Coast Guard, and EPA have also published information about air monitoring equipment for use at hazardous waste sites in the manual *Occupational Safety and Health Guidance Manual for Hazardous Waste Site Activities* (NIOSH et al. 1985).

GENERAL SAMPLING CONSIDERATIONS

Typical elements to consider when developing an air sampling strategy include the nature of air contaminants, type of sampling—either area or personal, sampling duration, the need for size-selective sampling, and selection of sampling method. Typically, air contaminants are categorized by physical characteristics. Gases are contaminants that have extreme molecular mobility and are capable of diffusing and expanding rapidly in all directions, thus occupying the entire space of their enclosure. They can be liquefied under pressure or decrease in temperature. Vapors are evaporation products of substances that are liquid at normal temperatures. Although gases and vapors may behave similarly thermodynamically, in many cases they are collected by different sampling collection devices.

Particulate matter, another category of air contaminant, consists of minute, separate particles. This category is further characterized by size and phase (solid or liquid) of the particulates. Dusts are solid particles formed from inorganic or organic materials that originate from processes such as grinding, crushing, blasting, drilling, and pulverizing. Typically, these particles range in size from the visible to the submicroscopic, with the range below 10 microns (μm) being of most concern, since this small size can reach the deepest parts of the lung. Fumes, another type of particulate, are fine particles formed from solid materials by evaporation, condensation, and gas-phase molecular reactions. These particles range in size from 1 μm to less than 0.01 μm. Smoke is a product of incomplete combustion of organic material. The size of particles in smoke is typically less then 0.5 μm. Minute droplets formed from atomization or by condensation from the gaseous state are categorized as liquid particles. Atomized droplets are generally greater than 5 μm in diameter. Condensation of low-volatility organic and inorganic species typically produces submicron aerosols.

The use of area or personal sampling is dependent on the application. Whenever possible, the environment in the worker's breathing zone should be characterized; however, there are some instances when area sampling can be used. Field applications of area and personal sampling techniques are described in the section of this chapter entitled "Field Applications."

Sampling duration can be brief or extended, with brief duration being termed a *grab* or instantaneous sample. This type of sampling is typically taken over a time period of 1 to 5 minutes and is used for determining ceiling or short-term exposures. These samples are best used to characterize processes that are cyclic in nature or that have periodic peak concentrations such as at startup or during cleaning operations. Sampling over an extended period—such as a 1- to 8-hour period—is performed when verifying that permissible exposure limits are not exceeded. Whenever possible, a full 8-hour work shift should be sampled, and in some cases—such as when sampling for coal mine dust exposures—an average over 10 work shifts has been used (ACGIH 1995).

Size-selective sampling is used for sampling particulates of specified aerodynamic equivalent diameter of concern. This type of sampling is used for asbestos and other particulates that can reach the inner compartment of the human lung.

Many types of sampling methods are used to evaluate workplace atmospheres. The following sections of this chapter delineate methods used for area and personal sampling.

AREA MONITORING

Area or general workplace monitoring can be performed using a wide range of analyzers. These include single chemical or total concentration analyzers, which are relatively easy to use and typically provide quantitative information about a single chemical or chemical species. A more sophisticated instrument is needed to evaluate a multichemical environment.

Single Chemical or Total Concentration Analyzers

There are numerous single chemical or total concentration analyzers available for workplace monitoring. In general, these analyzers are uncomplicated and easy to operate. In the next several pages we present a synopsis of several analyzers of this type, with information about the operation of the instrument, chemicals that the instrument can detect, and some brief comments on the limitations of the equipment.

Chemiluminescence Analyzer for Monitoring Oxides of Nitrogen

- *Instrument operation.* Oxides of nitrogen (NO_x) are converted to nitric oxide (NO) andreacted with ozone to generate light emissions that are monitored by a photomultiplier tube, as shown in Figure 5.1. Nitrogen dioxide (NO_2) concentrations are determined by intermittent direct sampling of the stream (without NO_x conversion) and by subtracting the NO concentration from the NO_x concentration.

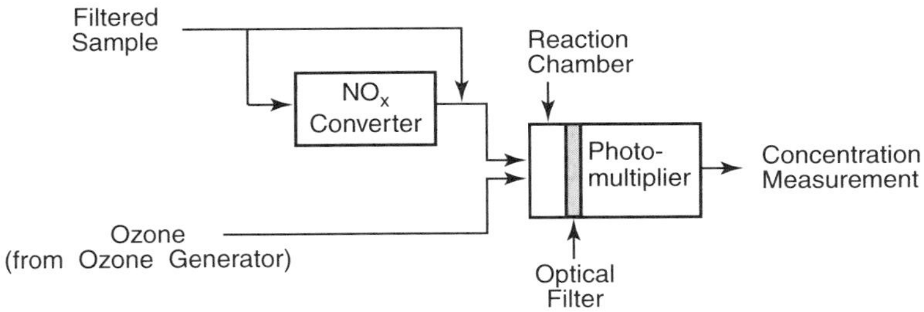

Figure 5.1 Chemiluminescence analyzer. (Adapted from ASTM D3824.)

- *Chemicals detected.* NO_x or NO_2 can be detected in the parts per million (ppm) range.
- *Limitations* Negative interferences may occur at high humidities for instruments calibratedwith dry span gas. Also, olefins and organic sulfur compounds, if present, will positively interfere with NO_x monitoring (ASTM 1997a).

Oxygen Meter with an Electrochemical Sensor

- *Instrument operation.* An oxygen meter typically uses an electrochemical sensor, as shown in Figure 5.2. The electrochemical sensor has a semipermeable membrane to allow air into the cell through diffusion, and uses an electrolytic, current-conducting solution to register current

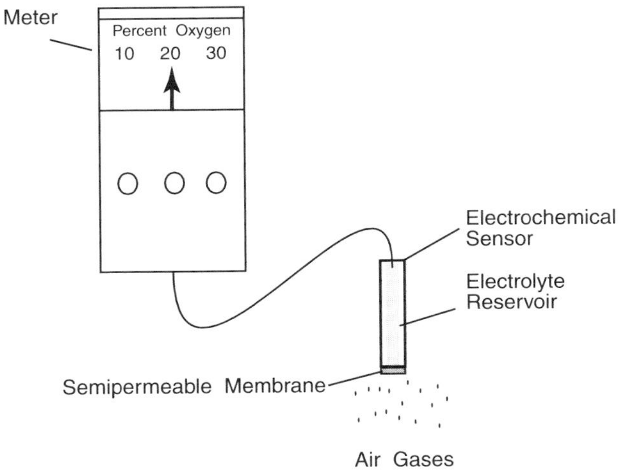

Figure 5.2 Oxygen meter with an electrochemical sensor.

changes directly, proportional to the amount of oxygen in the atmosphere. The current is amplified and displayed on a meter. Alarms can be set to sound if the oxygen concentration drops below a preset percentage.

- *Chemicals detected.* This instrument is programmed to read the percent of oxygen in the atmosphere. Other applications for an electrochemical sensor include a sulfur dioxide sensor and a hydrogen sulfide analyzer.
- *Limitations.* The sensor responds to the partial pressure of oxygen and is therefore altitude sensitive, with reduced readings at higher altitudes. The sensor in an oxygen meter may be affected by oxidants such as ozone. Carbon dioxide will poison the detector cell (NIOSH et al. 1985).

Gas Analyzer Using a Conductivity Meter

- *Instrument operation.* Air is pumped into a detection cell wherein the sample contacts a flowing steam of distilled water or other reagent. The change in conductivity is determined in the detector by two electrode sections, which measure the difference conductivity before and after the introduction of the air sample.
- *Chemicals detected.* Typical chemicals detected include sulfur dioxide, ammonia, hydrogen sulfide, and carbon dioxide in the ppm range. Some models detect mercury.
- *Limitations.* Interferences may require that the sample be pretreated. Sample results are usually temperature dependent.

Combustible Gas Indicator

- *Instrument operation.* A combustible gas indicator (CGI) utilizes a sensor to measure the relative resistance changesproduced by gases burning on hot filaments, one of which is coated with a catalyst. All readings on the combustible gas meter are relative to the calibrant gas, usually hexane, methane, or pentane. A CGI is used to determine whether flammable/combustible material is present in a concentration that could be dangerous. Concentrations between the lower and upper explosive limits (LEL and UEL, respectively) are considered immediately dangerous.
- *Chemicals detected.* Flammable gases can be measured with this instrument. The concentrations are measured in percent LEL and can be converted to ppm using vendor response curves and conversion factors.
- *Limitations.* Oxygen variations will affect combustion and, consequently, proper operation of the instrument. Lean and enriched mixtures will give inaccurately low and high readings, respectively. Temperature differences between calibration and use also can affect the instrument's accuracy. In addition, silicones, halides, and lead compounds will coat the detector unit and render it inoperable (NIOSH et al. 1985). Thus its use in atmospheres containing unknown vapors can be limited. Corrosive atmospheres, over time, can damage internal components of the instrument and

limit its use. Under oxygen-deficient conditions, the instrument will not provide an accurate reading.

Length-of-Stain Detector Tube

- *Instrument operation.* The length-of-stain detector tube is a colorimetric visual indicator. Itis operated by drawing a fixed volume of sample air through a tube with a squeeze bulb or small hand pump. Diffusional models (dosimeters) are also available. A length-of-stain dosimeter operates by allowing an air sample to diffuse through the tube over a 1- to 8-hour period.

  With both types of tubes, the tube contains a length of granulated resin or gel impregnated with a reactive chemical. The granules change color when specific types of air contaminants are introduced. Generally, length-of-stain detector tubes that use a pump allow the chemical concentration to be read directly on the tube, based on the length of color change. In the diffusional models, a color chart is provided by the manufacturer. An example of a hypothetical calibration graph for a length-of-stain dosimeter is shown in Figure 5.3.
- *Chemicals detected.* The detector tube can be used for determining the presence of mosthydrocarbons, acids, bases, organic amines, and alcohols. Accuracy is typically better than ±25% of the actual concentration in the ppm range (ASTM 1991a).
- *Limitations.* Temperature and humidity can affect the length-of-stain color change, and calibration charts must be used properly to make these corrections. In addition, similar chemicals can interfere positively with the detector tube's reading.

Photoionization Detector

- *Instrument operation.* A photoionization detector (PID) uses high-energy ultraviolet (UV) light to ionize volatile organic compounds in an air

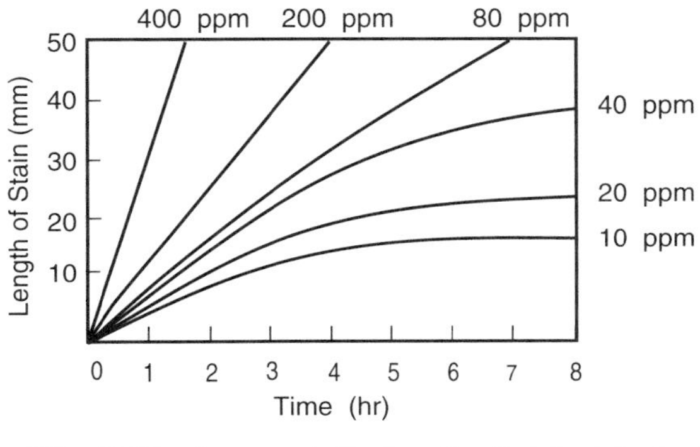

Figure 5.3 Calibration graph for length-of-stain detector tubes.

sample. Ionization of the sample produces a current that is proportional to the number of ions measured.

- *Chemicals detected.* The PID can measure total concentrations of many organic and some inorganic gases. Major air components such as oxygen, nitrogen, and carbon dioxide are not ionized in the process. The detector can quantify chemicals in the parts per billion (ppb) to ppm range, depending on the compound (Daisey 1987).
- *Limitations.* The instrument cannot detect some compounds if the probe used has a lower energy level than the compound's ionization potential. In addition, high humidity can dampen the detector's response. If the instrument does not have a filter, charged particles can damage the internal parts. Similarly, atmospheres containing corrosive gases can cause damage unless a corrosion-resistant instrument is used.

Flame Ionization Detector

- *Instrument operation.* The flame ionization detector (FID) mixes an air sample with hydrogen and combusts the sample in a detector cell, ionizing the gases and vapors. The instrument measures the current flow through the flame electronically as the sample is burned and amplifies this measurement on an analog display.
- *Chemicals detected.* Total hydrocarbons are measured in the ppm range.
- *Limitations.* Ultrahigh-purity hydrogen fuel and high-purity air, free of hydrocarbons, must be used to ensure accurate readings. Also, the combustion chamber is very sensitive and can be damaged by corrosive or reactive gases. The portable version of the instrument is limited in atmospheres that are oxygen deficient, since lack of oxygen can cause flame-outs (Erb et al. 1990).

Infrared Analyzer

- *Instrument operation.* The infrared (IR) analyzer uses a spectrometer to read chemical "fingerprints" or concentration intensities created by passing infrared frequencies through a heated air sample. The basic instrument has a fixed cell path length and is calibrated for one or a few predetermined chemicals.
- *Chemicals detected.* The instrument can detect volatile organic compounds in the spectralrange of 2.5 to 15 μm. In the most sophisticated instruments, numerous compounds can be detected in the ppb to ppm ranges (Daisey 1987).
- *Limitations.* This instrument cannot be used in flammable or explosive atmospheres. In addition, excessively humid or corrosive atmospheres can cause damage to the instrument. Water vapors and carbon dioxide can interfere with the instrument's readings (NIOSH et al. 1985).

Portable Gas Chromatograph

- *Instrument operation.* A portable gas chromatograph (GC) can be used with other detectors, such as a PID or FID, to identify and measure specific compounds. The portable model uses a packed or capillary column to concentrate and separate the compounds according to their vapor pressures. After separation, a detector quantifies the individual compounds (peaks). The identity of the compounds or peaks is qualitatively or quantitatively determined by its retention time in the GC.
- *Chemicals detected.* The GC can separate organic gases and vapors for quantitation with a detector. Quantitation limits are dependent on the detector used.
- *Limitations.* For selective quantitative results, the instrument must be calibrated with the specific analyte. The instrument may not be consistently sensitive to all organic compounds. In addition, mixtures of polar and nonpolar compounds can cause peak superimpositions, which may require changes in column type, length of column, or operating conditions (ASTM 1997a).

Organic Vapor Analyzer

- *Instrument operation.* Portable instrument that measures trace quantities of organic materials in air using a hydrogen flame ionization detection system. The readout is scaled logarithmically. The instrument has an alarm system to provide warning at preset levels.
- *Chemicals detected.* Total organics in the ppm range 1 to 10,000 (ACGIH 1995).
- *Limitations.* Specific organics not quantified.

Thermal Conductivity Analyzer

- *Instrument operation.* Uses a Wheatstone bridge to detect gases by measuring heat loss rate in air sample relative to reference air. Some models are capable of analyzing the concentration of one component in a mixture of gases.
- *Chemicals detected.* Utility gases, toxic gases, and other gases in the range 10 to 100 ppm.
- *Limitations.* Instruments require good temperature and flow control.

Multichemical Analyzers

Multichemical analyzers have become more readily available over the last decade and are starting to be used more frequently in industry. In general, this type of analyzer is more costly than a single chemical or total concentration analyzer since it is more complex. However, many of these analyzers have the capability to monitor more than one port, which gives the user added sampling flexibility. The following paragraphs provide information about several multichemical analyzers.

On-line Mass Spectrometer

- *Instrument operation.* This on-line system utilizes a central mass spectrometer (MS) unit and a vacuum pump system to draw samples and analyze chemicals sequentially from locations as far away as 1500 feet. As many as 50 different workplace locations can be connected with chemical-resistant tubing. Up to 25 chemicals can be analyzed per location. Depending on the number of chemicals analyzed and number of ports attached, the time between samples can range from a few minutes to several hours (Erb et al. 1990).
- *Chemicals detected.* An on-line MS can quantify most chemicals that normally are read on a lab MS, at ppm levels. Metals cannot be quantified. The lower the detection limit required, the longer the dwell time in the analyzer. Lengthened dwell times can substantially increase the time between samples.
- *Limitations.* In certain cases, high chemical concentrations may not be monitored optimally by this system. Corrosive chemicals, dust, and humidity may cause damage to the system. In addition, the system is operationally complex and may require extensive calibration and maintenance.

On-line GC/FID

- *Instrument operation.* The GC/FID is an automatic system designed for continuous monitoring of a wide variety of volatile organic compounds. The automated GC separates chemicals by vapor pressure, and the FID quantifies the chemicals. The system can be configured to sample up to 24 lines sequentially from 50 to 100 feet away (Coleman et al. 1990).
- *Chemicals detected.* Organic vapors can be detected in the low ppm ranges.
- *Limitations.* The system is operationally complex and may require extensive calibration and maintenance.

On-line Fourier Transform Infrared Spectrometer

- *Instrument operation.* This instrument is commonly used for quantitative spectroscopic applications in the mid-infrared range. The instrument operates similarly to a standard IR, but uses variable path lengths and wavelengths to allow for quantitation of numerous chemicals (EPA 1989).
- *Chemicals detected.* Chemicals detected will vary, depending on the setup of the instrument, but the equipment is capable of detecting the same chemicals as those detected by any IR spectrometer.

- *Limitations.* Sample conditioning, which can potentially remove chemicals of interest, is required for sample streams containing moisture or corrosive chemicals. In addition, system complexity requires extensive calibration and may require extensive maintenance.

PERSONAL MONITORING

Personal monitoring is performed to determine actual chemical concentrations that workers are exposed to during the workday. This type of monitoring is performed periodically under typical, representative working conditions. The samples generally are taken over an 8-hour shift if chemical-use operations are continuous. In some cases, sampling may be performed over a short period of time within a shift to quantify peak exposures.

Personal sampling generally is accomplished using two methods: sampling with a personal sampling pump and passive or diffusional sampling. Sampling with a personal sampling pump generally is accepted by OSHA for documenting chemical exposures. Passive sampling is typically used for supplemental sampling and screenings. Sampling verification using OSHA methods, when specified, should be used. The following paragraphs provide information about both of these types of sampling methods.

Sampling with a Personal Sampling Pump

Sampling with a personal sampling pump is standard across the industry and is documented in ASTM standards and by ACGIH. For this type of sampling, a battery-powered personal sampling pump typically is used in conjunction with an air metering device and an adsorbent tube sampler such as charcoal or silica gel tube samplers. Sampling with a personal sampling pump yields accurate results since the volume of air sampled is metered and can be quantified accurately. Factors that must be considered when calculating sampled air volume include time, flow rate, pressure, and workplace temperature and humidity.

Charcoal Tube Adsorption Sampler. This sampler uses a charcoal sampling tube containing two sections of activated charcoal, as shown in Figure 5.4, and a sampling pump that draws a sample at a stable flow rate. The carbon tube is taken to a lab and desorbed into a GC using carbon disulfide (or other recommended desorber) to determine the identity and concentration of the chemical adsorbed. Based on airflow rate and time, a workplace concentration is determined.

Numerous organic chemicals can be detected using this sampling device at detection limits in the ppm range. The method is useful for determining airborne time-weighted average concentrations of many of the organic chemicals listed by OSHA in 29 CFR §910.1000 (ASTM 1997a).

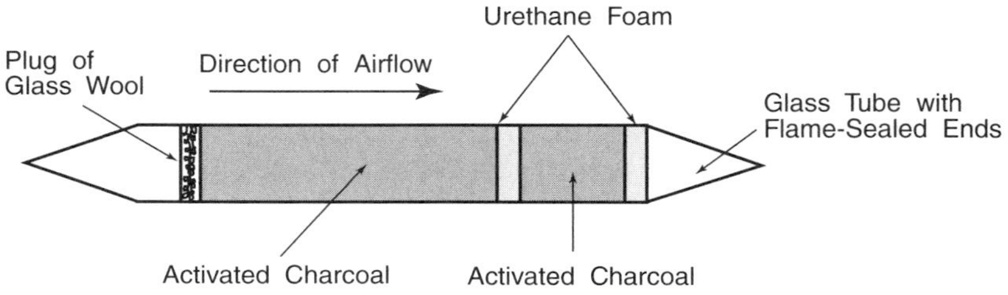

Figure 5.4 Activated charcoal tube. (Adapted from ASTM D3686.)

There are some limitations to sampling with charcoal tube samplers. High humidity can reduce the adsorptive capacity of activated charcoal for some chemicals. Further, mixtures of polar and nonpolar compounds are difficult to recover (desorb) from activated charcoal (ACGIH 1995).

Silica Gel Tube Sampler. This sampler is very similar to the charcoal tube sampler except that silica gel, which is an amorphous form of silica derived from sodium silicate and sulfuric acid, is used instead of charcoal. Unlike charcoal, silica gel is well suited for sampling polar contaminants since they are removed easily from the adsorbent with common solvents. In addition, amines and some inorganic substances not suitable for charcoal sampling can be sampled with a silica gel tube sampler. The major disadvantage of this type of sampler is that it will adsorb water (ACGIH 1995).

Passive or Diffusional Sampling

Passive or diffusional sampling is performed without the use of a personal sampling pump. Sampling devices consisting of badges or dosimeters are attached to the worker near the breathing zone. An example is presented in Figure 5.5. Diffusional monitors can be used to detect numerous contaminants, including acetone, ammonia, carbon monoxide, ethylene oxide, formaldehyde, mercury, nitrogen dioxide, and organic vapors. The dosimeters collect vapors by diffusion onto a medium such as charcoal or other adsorbent to indicate chemical concentrations in the breathing zone. The vapors diffuse through the medium over time at a rate dependent on the cross-sectional area of the diffusion cavity, the diffusion coefficient, and the length of diffusion path. Diffusion factors can be determined using the calibration factors supplied by the manufacturer, or by applying diffusion laws. Samplers can be stored for several weeks to a year, depending on the temperature of the storage area.

For some samplers, the adsorbent will have to be desorbed and analyzed. Other types of passive samplers are colorimetric and use length of stain or color change as the concentration indicator. Like the charcoal tube sampler

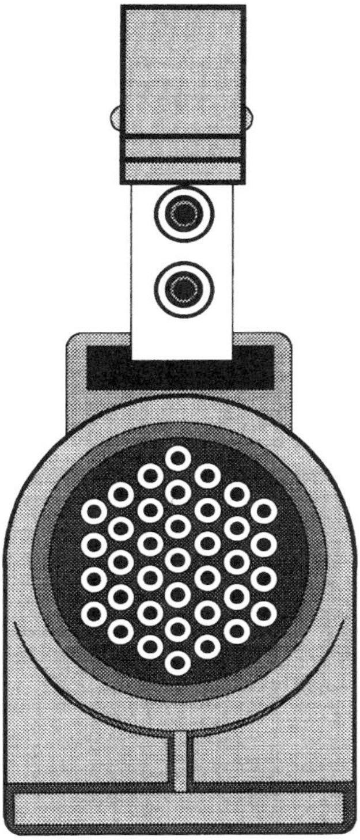

Figure 5.5 Diffusional (passive) monitor.

described previously, passive charcoal adsorbent tube samplers can detect numerous organic chemicals in the ppm range. Limitations are similar.

For colorimetric badges or dosimeters, the manufacturer's color graph must be used. The main limitation of these types of samplers is that they can give only a gross assessment of chemical concentration. In addition, humidity and temperature must be taken into account when reading the calibration curves.

Other Gas and Vapor Samplers

In addition to the samplers described previously, numerous other samplers are available for collecting gases and vapors for analysis from the workplace atmosphere. Examples include gas sample tubes, glass bottles or containers, collapsed bags, and bubblers.

Gas sample tubes use air displacement as a means of collecting an air sample. Generally, an aspirator is used to sweep out the air that is in the tube and replace it with air that is to be sampled. Glass bottles or containers can be filled similarly, using air displacement, or they can be filled using water displacement. During water displacement, the glass container is filled with water and drained at the area to be sampled, allowing air into the container. This method should not be used for sampling water soluble gases. Once the sample tube or container is filled, it is capped with an impermeable cap and transported to a laboratory for analysis.

Collapsed bags provide another method of collecting air samples. The bags are rolled tightly to exclude extraneous air and are opened and filled with air at the monitoring site. For outdoor applications, if the wind velocity is sufficient to fill the bag, the bag can be held open directly into the wind. If not, a blower will be needed to blow air into the bag. For indoor applications, a small pump can be used to fill the bags. The bags are made of polyethylene or other nonreactive material.

Bubblers, including simple bubblers and bubblers with diffusers, can also be used for sampling. A bubbler with a diffuser is shown in Figure 5.6. Both types use pumping devices to pull a sample through the apparatus. Bubblers are generally easy to use and absorb gases and vapors by gas–liquid contact. Bubblers with diffusers allow for better contact than simple bubblers, but they are subject to more frequent clogging. The effectiveness of bubblers is dependent on several factors, including the size of the bubble, the length of the path through the absorbent, the rate of gas flow, transfer coefficients, and the degree of solubility of the contaminant in the absorbent.

Dust Sampler

Dust samplers are used to determine the amount of fines and particulates in the workplace atmosphere that could create a health hazard. ASTM D4532 specifies the use of a personal sampling pump and a sampling head consisting

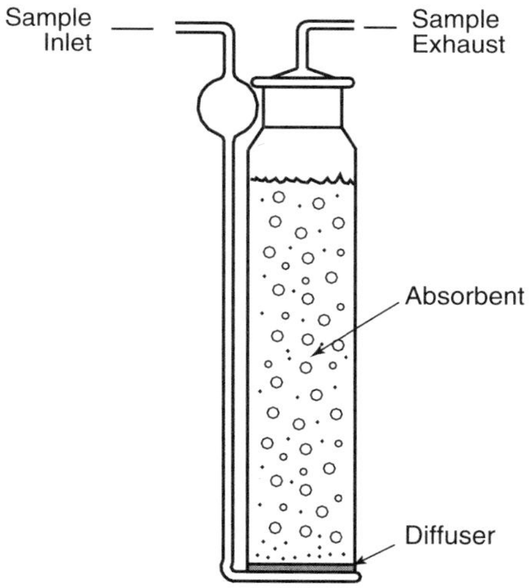

Figure 5.6 Bubbler absorber. (Adapted from ASTM D1605.)

of an aerosol preclassifier (in the form of a cyclone) and a filter assembly for sampling respirable dust. The filter, which must be nonhygroscopic and have a collection efficiency of greater than 95%, is weighed before and after sampling to determine the collected dust weight. The sample volume is calculated and then used to calculate the mass concentration of the respirable dust of interest.

An equivalent sampling method, described by ACGIH, also uses a two-stage dust sampler. Examples of first-stage collectors include cyclones, elutriators, and impactors. Second-stage collectors are generally high-collection-efficiency filters (ACGIH 1995). Once the sample is collected and weighed, light-scattering instruments can provide information on size distribution, if necessary.

Sampling for Asbestos Exposure

Sampling for airborne asbestos is required during any demolition or renovation project that involves any of the OSHA-listed asbestos minerals: chrysotile, crocidolite, amosite, anthophyllite, tremolite, and actinolite. Procedures are defined by OSHA in 29 CFR §910.1001, Appendix A, by ASTM in D4240, by NIOSH (1979), and by ACGIH–AIHA (1975).

The sample is collected by pumping air through an open-faced filter membrane. OSHA specifies a mixed cellulose ester filter membrane designated by the manufacturer as suitable for asbestos counting. A section of the membrane is converted to an optically transparent homogeneous gel, and the asbestos particles are sized and counted by phase-contrast microscopy at a magnification of 400 to 500 times.

All sampling must be conducted so as to be representative of typical working conditions. Since there are OSHA worker exposure limits for times as short as 30 minutes and as long as 8 hours, selection of an appropriate sampling time is an important consideration. For the most reliable results, several samples should be taken over an 8-hour shift to allow for quantitation of peaks as well as an 8-hour time-weighted average. These samples must include personal samples taken in the worker's breathing zone and can include static samples at fixed locations if the dust is distributed uniformly over a large area.

Sampling for Radiation Exposure

Most sampling for worker radiation exposure is performed using a personal dosimeter. Personal sampling must be performed on all workers who have the potential to receive a dose in any calendar quarter excess of 25% of the allowable radiation dose limits. In addition, all workers who enter a high radiation area (an area with potential for a major portion of the body to receive a dose of greater than 100 millirem in 1 hour) must be sampled. Some states may have more stringent or additional requirements.

The personal dosimeter must be processed by processors accredited through the National Voluntary Laboratory Accreditation Program (NVLAP) (DOE 1991). Personal dosimeter performance standards are outlined in ANSI Standard N13.11 (ANSI 1983).

Fixed and portable samplers are also available for sampling indoor and outdoor atmospheres that are potentially radioactive. These samplers include alpha-particle analyzers, beta-particle analyzers, gamma-particle analyzers, scintillation counting systems, radon monitors, and others.

MONITORING INDOOR AIR QUALITY

Indoor air quality of nonmanufacturing areas such as office space or low-chemical-use labs is a key part of environmental health. As a result of negative publicity about Legionnaire's disease, excessive radon exposure in homes and workplaces, and the phenomena known as "sick building syndrome," indoor air quality has gained national focus. Periodic monitoring of indoor air quality can help ensure that the air in a building is fresh and free from contamination. If occupants complain of specific symptoms such as headaches, dizziness, or general malaise, the complaints should be investigated immediately.

Several ASTM standards are published for monitoring indicators of adequate (or poor) indoor air quality. These include D3824 (Standard Test for Continuous Measurement of Oxides of Nitrogen), D4861 (Standard Practice for Sampling and Analysis of Pesticides and PCBs), D4947 (Standard Test Method for Chlordane and Heptachlor Residues), and others.

In addition to ASTM standards, useful information can be obtained from ACGIH's guidance for the sampling of airborne microorganisms and aeroallergens (bioaerosols), which are considered sources of indoor air contamination (ACGIH 1995). Elevated levels of carbon dioxide also contribute to poor indoor air quality and can be monitored using one of several marketed monitors. A checklist for sampling for bioaerosols is presented in Table 5.1.

FIELD APPLICATIONS

With numerous types of analyzers available, it is often difficult to select one that will be appropriate for a given application. ACGIH's handbook pertaining to air sampling instruments has an index of instruments, which makes researching the literature considerably easier. In addition, vendors carry literature about their products, including applications of the instruments and operational information. Each type of analyzer should be investigated carefully before purchase to ensure that it meets all requirements of the application.

TABLE 5.1 Checklist for Bioaerosol Sampling

General

_____ Identify sampling locations.
_____ Determine number of samples and collection times.
_____ Investigate effects of seasonal and temporal variations in bioaerosol concentrations.
_____ Determine assay system limitations.

Determine Material to be Sampled

_____ Bacteria
_____ Fungi
_____ Virus
_____ Antigen
_____ Other microorganisms

Select Sample Collection Method

_____ Agar-based culture media
_____ Slides and/or filters
_____ Airborne sampler
_____ Impactor
_____ Impinger

Evaluate Sampling Efficiency

_____ Efficiency with which particles are collected
_____ Efficiency with which collected particles are transferred to an assay system
_____ Efficiency with which bioaerosol viability and other critical characteristics are preserved during collection and transfer

Check Aspects Associated with Transferring the Sample

_____ Sample preparation
_____ Sample storage
_____ Equipment and media handling
_____ Quality assurance samples

Select Assay Methods (Select, as Appropriate)

_____ Direct examination
_____ Immunoassay (radioimmunoassay staining techniques, enzyme-linked immunosorbent assay, radioimmunoassay, immunochemical assay)
_____ Assessment of infectivity

General Considerations

The accuracy of any sampling data is determined by numerous elements, including instrument sensitivity, instrument calibration, chemical interferences, temperature and humidity factors, sample pump inconsistencies, and other factors. If approached systematically, most of the variables can be eliminated or mathematically adjusted using real-time data.

Similarly, the determination of where to collect the sample must be addressed in an organized manner. Airflow patterns, sources of emissions, vapor density of the chemical, location of workers, and other considerations should be taken into account when selecting sample locations. When prioritizing workplace monitoring efforts, occupancy frequency versus hazard or risk must be assessed. The most frequently occupied areas normally will require more frequent monitoring unless the hazards of the chemicals in the area are low. High-risk areas such as toxic gas storage rooms are equally important and should be included in a monitoring plan. Normally unoccupied areas such as trenches, equipment housings, and support areas might also be candidates for monitoring, since leaks and spills could go unnoticed for long periods of time in these places. When selecting a monitoring instrument, analysis turnaround time should be taken into consideration. Charcoal adsorption tubes, silica gel adsorption tubes, and air and water displacement samplers require laboratory analysis and therefore cannot be used if real-time data are required. In addition, concentrations are integrated over time, so peak concentrations cannot be identified with these samplers if the sample time is an 8-hour shift.

On-line analyzers allow for immediate analytical feedback. These instruments also allow for quantitation of peak concentrations. When installed on line, the system output can be programmed to indicate an alarm at preset concentrations with the use of programmable logic controls or other means.

Leaks, Spills, and Unknown Atmospheres

For situations where a known chemical has leaked or spilled and the concentrations in the atmosphere must be quantified, portable instruments that can monitor and quantify those chemicals are appropriate for use. If the material is flammable/combustible, a combustible gas indicator is useful for evaluating the explosiveness of the atmosphere at the point of entry and, as results allow, at the area of the spill. For pinpointing fugitives and leaks around equipment, colorimetric detector tubes are adequate instruments. Portable FIDs and PIDs can also provide accurate real-time data for known constituents.

Confined-space atmosphere evaluation would require use of instruments to determine oxygen concentration, LEL, and hydrogen sulfide. Concentrations of known chemicals should also be evaluated prior to entry, so that proper personal protection can be specified. Guidelines under 29 CFR §910.146 should be reviewed before initiating entry into a confined space.

For unknown atmospheres such as Superfund and other hazardous waste cleanup sites, abandoned warehouses, or rooms with unlabeled containers, more sophisticated equipment is commonly used. IR analyzers and portable GCs with a detector can be used to read multichemical atmospheres. Setup and analytical readings are more complicated for these instruments, so skilled personnel must be on hand during the evaluation.

For all unknown atmospheres, proper protective clothing, including maximum respiratory and skin protection, must be worn until the atmosphere can be evaluated as safe for entry without such protection. The topic of personal protective clothing and respiratory protection is addressed in Chapter 6.

Routine Workplace Monitoring Applications

Routine or periodic workplace monitoring of manufacturing process areas is not uncommon in industry. A portable IR or other portable on-line instrument is useful for these applications since it can be set up in an area for several days with a strip chart or other type of recorder. Once the data collected are adequate to demonstrate that exposure concentrations are acceptable, the analyzer can be moved to another area. For a less expensive approach, area diffusional samplers or dosimeters can be used instead of a portable IR. For monitoring hazardous chemical and toxic gas storage rooms, continuous monitors or semicontinuous systems that speciate chemicals and toxic gases are used.

Personal Monitoring Applications

For preplanned personal monitoring, charcoal and silica gel adsorbent tubes, both utilized with a personal sampling pump, are the mainstay of the industry. Passive or diffusional samplers in the form of dosimeters or badges are also used for supplemental sampling and screening. The dosimeter is attached to the wearer at the breathing zone during an entire shift. Minimum supervision from the facility's industrial hygienist is necessary for this type of monitoring. Dosimeters can also be used in Superfund or other cleanup activities for general evaluation of chemical concentrations. In these cases, the dosimeter is taped to the outside of a worker's encapsulated suit. If cartridge respirators are approved for use while working in a cleanup area, the dosimeter can be worn in the breathing zone. Dust samplers are used for personal monitoring of workers in operations where respirable dust is generated. This includes exposures to mining dust, cotton dust, asbestos, and other types of dust.

All asbestos removal jobs must be sampled for airborne asbestos during the demolition or renovation process. Contractors or employees who remove asbestos must be trained adequately to do this type of work. Lengthy procedures must be followed to protect the environment from contamination and the workers from exposure. Required personal protective equipment and other requirements that pertain to asbestos demolition and removal activities are discussed more fully in Chapter 6.

REFERENCES

American Conference of Governmental Industrial Hygienists (1995). *Air Sampling Instruments for Evaluation of Atmospheric Contaminants,* 8th ed., ACGIH, Cincinnati, OH.

American Conference of Governmental Industrial Hygienists–American Industrial Hygiene Association (1975). "Aerosol Hazards Evaluation Committee: Recommended Procedures for Sampling and Counting Asbestos Fibers," *American Industrial Hygiene Journal,* vol. 36.

American National Standards Institute (1983). *Personnel Dosimetry Performance, Criteria for Testing,* ANSI N13.11, ANSI, New York.

American Society for Testing and Materials (1997a). *Annual Book of ASTM Standards,* Vol. 11.03, "Atmospheric Analysis; Occupational Health and Safety," ASTM, Philadelphia.

——— (1997b). *Annual Book of ASTM Standards,* Vol. 11.04, "Pesticides; Resource Recovery; Hazardous Substances and Oil Spill Responses; Waste Disposal; Biological Effects," Philadelphia.

Coleman, D. R., et al. (1990). "Automatic Continuous Air Monitoring at Fixed Sites with Minicams," pp. 114–117, *Proceedings of the 3rd Annual Hazardous Materials Management Conference/Central,* Chicago.

Daisey, J. M. (1987). *Real Time Portable Organic Vapor Sampling Systems: Status and Needs,* prepared for the American Conference of Governmental Industrial Hygienists, Lewis Publishers, Boca Raton, FL.

Department of Energy (1991). *The Status of ANSI N13.11: The Dosimeter Performance Test Standard,* DE91 017874, prepared by C. S. Sims, Oak Ridge National Laboratory, Oak Ridge, TN.

Environmental Protection Agency (1989). *Fourier Transform Infrared Spectroscopy as a Continuous Monitoring Method: A Survey of Applications and Prospects,* EPA/600/D-90/003, prepared by Entropy Environmentalists, Inc., U.S. EPA, Research Triangle Park, NC.

——— (1990). Compendium of Methods for the Determination of Toxic Organic Compounds in Indoor Air, EPA/ 600/4-90/010, prepared by Engineering Science, Inc., for Atmospheric Research and Exposure Assessment Laboratory, Office of Research and Development and Quality Assurance Division, Environmental Monitoring Systems Laboratory, Research Triangle Park, NC.

Erb, J. E. Ortiz, and G. Woodside (1990). "On-line Characterization of Stack Emissions," *Chemical Engineering Progress,* vol. 86, no. 5, pp. 40–45.

National Institute of Occupational Safety and Health (1979). *Membrane Filter Method for Evaluating Airborne Asbestos Fibers,* DHEW (NIOSH) Publication 79-127, prepared by N. A. Leidel et al., U.S. Department of Health, Education, and Welfare, Rockville, MD.

——— (1984, rev. 1987). *NIOSH Manual of Analytical Methods,* DHHS (NIOSH) Publication 84-100, 3rd ed., edited by P. M. Eller for U.S. Department of Health and Human Services, NIOSH, Cincinnati, OH.

National Institute of Occupational Safety and Health, Occupational Safety and Health Administration, U.S. Coast Guard, and U.S. Environmental Protection Agency, (1985). *Occupational Safety and Health Guidance Manual for Hazardous Waste Site Activities,* DHHS (NIOSH) Publication 85-115, U.S. Department of Health and Human Services, U.S. Government Printing Office, Washington, DC.

Occupational Safety and Health Administration (1990). *OSHA Analytical Methods Manual,* Part 1, "Organics," OSHA Analytical Laboratories, Salt Lake City, UT.

——— (1991). *OSHA Analytical Methods Manual,* Part 2, "Inorganics," OSHA Analytical Laboratories, Salt Lake City, UT.

BIBLIOGRAPHY

American Conference of Governmental Industrial Hygienists, (1995). *Advances in Air Sampling,* ACGIH, Cincinnati, OH.

American Industrial Hygiene Association, (1985). *Biohazards Reference Manual,* AIHA, Akron, OH.

——— (1987). *Cotton Dust Exposures,* Vol. 2, AIHA, Akron, OH.

——— (1990). "The Practitioner's Approach to Indoor Air Quality Investigations," *Proceedings of the Indoor Air Quality International Symposium,* AIHA, Akron, OH.

American Society for Testing and Materials, (1989). *Design and Protocol for Monitoring Indoor Air Quality,* N. L. Nagda, and J. P. Harper, eds., Special Technical Publication 1002, ASTM, Philadelphia.

——— (1990). *Biological Contaminants in Indoor Environments,* Morey and Feeley, eds., Special Technical Publication 1071, ASTM, Philadelphia.

Department of Energy, (1988). *The Evaluation of Four Models of Personal Air Samplers,* DE88-005563, prepared by J. F. Boyer and B. J. Held, Lawrence Livermore National Laboratory, U.S. Doe, Washington, DC.

Environmental Protection Agency (1985). *Measuring Airborne Asbestos Following an Abatement Action,* EPA/600/4-85/049, Quality Assurance Division, Environmental Monitoring Systems Laboratory, Office of Research and Development, Research Triangle Park, NC and Exposure Evaluation Division, Office of Toxic Substances, Office of Pesticides and Toxic Substances, U.S. EPA, Washington, DC.

Environmental Protection Agency and National Institute for Occupational Safety and Health (1991). *Building Air Quality,* U.S. EPA/NIOSH, Washington, DC.

Hanford Works (1949, declassified 1991). *Manual of Standard Procedures for 100, 200, and 300 Area Survey Work,* DE92-002891, compiled by J.M. Smith, Jr., Health Instrument Operational Division, Hanford Works, Hanford, WA.

Knoll, G. F. (1989). *Radiation Detection and Measurement,* 2nd ed., Wiley, New York.

Lodge, J. P., Jr., ed. (1988). *Methods of Air Sampling and Analysis,* 3rd ed., Lewis Publishers, Boca Raton, FL.

National Council of Radiation Protection and Measurements (1988). *Measurement of Radon and Radon Daughters in Air,* NCRP Report 97, NCRP, Bethesda, MD.

Ness, S. (1991). *Air Monitoring for Toxic Exposures,* Van Nostrand Reinhold, New York.

Sheldon, L. S., C. M. Sparacino, and E. D. Pellizzari (1984). "Review of Analytical Methods for Volatile Organic Compounds in the Indoor Environment," *Indoor Air and Human Health, Proceedings of the 7th Life Sciences Symposium,* Knoxville, TN.

——— (1986). *Standard Operating Procedures Employed in Support of an Exposure Assessment Study,* Vol. 4, prepared by R. W. Handy et al., eds., for Air, Toxics, and Radiation Monitoring Research Division, Office of Monitoring, System and Quality Assurance, Office of Research and Development, U.S. EPA, Washington, DC.

Yocom, J. E., and S. M. McCarthy (1991). *Measuring Indoor Air Quality: A Practical Guide,* Wiley, New York.

6

PERSONAL PROTECTIVE EQUIPMENT

Personal protective equipment (PPE)—including protective equipment for eyes, face, head, and extremities, protective clothing, respiratory devices, and protective shields and barriers—is often needed when working in a hazardous materials or hazardous waste environment. The use of PPE is addressed by OSHA in 29 CFR §§1910.132, 133, 135, and 136. The need for PPE is based on several factors, including types of chemicals and concentrations in the work area, the amount of time the worker is exposed, and activities involved in the work task. In this chapter we discuss types and use of PPE, as well as related emergency equipment such as emergency showers and eyewashes. Special PPE requirements for asbestos removal and hazardous materials emergency response and waste cleanup operations are also addressed.

There is much useful information pertaining to PPE published by agencies and standards organizations. Examples include:

- *Guidelines for Selection of Chemical Protective Clothing* (ACGIH 1987)
- *Respiratory Protection: A Manual and Guideline* (AIHA 1991)
- *Emergency Eyewash and Shower Equipment,* ANSI Standard Z358.1 (ANSI 1990)
- *Practices for Respiratory Protection,* ANSI Standard Z88.2 (ANSI 1980)
- *ASTM Standards on Protective Clothing* (ASTM 1990)
- *Occupational Safety and Health Guidance Manual for Hazardous Waste Site Activities* (NIOSH 1985)

Work tasks involving the use or handling of hazardous materials or hazardous waste must be evaluated for PPE requirements by a knowledgeable person such as an industrial hygienist or chemical.

GENERAL REQUIREMENTS

Employers must ensure that PPE is provided and that the equipment is used and maintained in a sanitary and reliable condition. Where employees provide their own PPE, the employer is responsible to ensure its adequacy. In addition, all PPE must be of safe design and construction for the work to be performed.

It is the employer's responsibility to assess the workplace to determine if hazards are present or likely to be present that necessitate the use of PPE. In addition, the employer must select and have employees use PPE that is adequate for the task. The employer must provide training to each employee, which consists of the following information:

- When PPE is necessary
- What PPE is necessary
- How to don, doff, adjust, and wear PPE properly
- The limitations of the PPE
- The proper care, maintenance, and useful life and disposal of the PPE

The employer must verify that each affected employee has received and understood the required training through written certification that includes the name of the employee, date of training, and subject of certification.

CHEMICAL PROTECTIVE CLOTHING

Material Selection

Chemical protective clothing (CPC) is commonly used in manufacturing operations, since most manufacturing chemicals present some level of risk to workers, with the most common hazard being dermal hazard. Dermal hazards include the following:

- Those that can damage the skin, such as corrosives
- Those that can cause a reaction, such as sensitizers and irritants
- Those that can produce toxicity through permeation through the skin

CPC can range in complexity from simple finger cots or arm shields to fully encapsulating suits. When adequate for the application, CPC can substantially lower the risk of injury or illness.

There are three types of material failures that can render CPC ineffective. These are failures caused from penetration, permeation, and degradation. *Penetration* is the bulk flow of chemical through the protective material. Penetration is typically caused by tears, rips, pinholes, and material defects. These failures are nonchemical related.

Permeation is the diffusion of a chemical through the protective clothing. In essence, the chemical molecularly "passes through" the material. *Degra-*

dation is the change in the physical properties of the material as a result of exposure to a chemical. This can include change in flexibility (or brittleness), elasticity, thickness, texture, and other properties.

Some physical resistance factors of CPC to consider before evaluating chemical compatibility include abrasion resistance, cut resistance, tear strength, tensile strength, flammability, resistance to the effects of heat and cold, closure strength, seam strength, bursting strength, flexibility, weight, and thermal insulation. Contact or exposure potential should also be addressed and categorized. Categories of potential contact include immersion, spraying, splashing, surface contact, mist, and vapors.

When evaluating chemical resistance, permeation test data, manufacturer's recommendations, published CPC guides, and experience should all be taken into account. Typically, CPC must afford protection against varying types of chemical hazards. A single material is not resistant to all chemicals and wastes; thus engineering judgments must be made as to the best materials for face shields, gloves, aprons, boots, and protective suits for use during a specific work task. Once the CPC needed to perform a task is identified, all workers performing these tasks must wear what is specified.

Common materials used in personal protective clothing include butyl, nitrile, neoprene, polyvinyl chloride, chlorinated polyethylene, viton, polycarbonates, polyvinyl alcohols, and others. In the *Hazardous Materials Response Handbook* (NFPA 1989), there is a useful summary of chemical compatibilities for commonly used protective clothing materials. This summary combines six studies and includes compatibility tests for seven materials and over 1000 chemicals. An overview of the tests for each of the seven materials, in terms of resistance to families of chemicals, is presented in Table 6.1.

Vapor-Protective Suits

Requirements for Vapor-Protective Suits. Protective suits are used for many tasks, including normal work activities, chemical emergency response, hazardous waste cleanup activities, fire emergency response, and other activities. Before donning any type of PPE, including protective suits, the user should be fully trained.

Minimum requirements for protective suits used in hazardous chemical emergencies are defined by NFPA in the following standards:

- NFPA 1991 (1990)—Vapor-Protective Suits for Hazardous Chemical Emergencies
- NFPA 1992 (1990)—Liquid Splash-Protective Suits for Hazardous Chemical Emergencies
- NFPA 1993 (1990)—Support Function Protective Garments for Hazardous Chemical Operations

Included in the requirements are:

TABLE 6.1 Chemical Compatibilities for Selected Protective Materials[a]

Material	Compatibilities	Incompatibilities
Butyl	Moderate to strong acids, ammonia solutions, alcohols, inorganic salts, ketones, phenols, and aldehydes	Petroleum distillates, solvents, alkanes; limited use with esters and ethers
Polycarbonates	Weak acids, ammonium solutions, alcohols, inorganic salts, phenols, some bases and aldehydes	Agressive petroleum distillates; limited use with ketones
Polyvinyl chloride	Moderate to strong acids, bases, ammonium solutions, inorganic salts, petroleum distillates, alcohols, alkanes, and some aldehydes	Petroleum distillates, ketones, concentrated solvents, and phenols
Neoprene	Moderate acids, bases, ammonium solutions, alcohols, inorganic salts, some solvents, some phenols, some aldehydes, and ethers	Petroleum distillates, esters; limited use with ketones
Chlorinated polyethylene	Moderate to strong acids, bases, ammonium solutions, some petroleum distillates, alcohols, inorganic salts, phenols, alkanes, and aldehydes	Limited use with ketones and solvents
Nitrile	Moderate to strong acids, bases, most ammonium solutions, some petroleum distillates, some solvents, and some alcohols	Phenols, ketones; limited use with inorganic salts and aldehydes
Viton	Acids, bases, some ammonium solutions, petroleum distillates, alcohols, inorganic salts, most solvents, and phenols	Ketones; limited use with aldehydes

[a]This list provides general information about chemical-resistant materials. The specific chemical or chemical mixture used in the workplace should be evaluated by a trained specialist for compatibility with any material used for protective clothing.

Source: Information from NFPA (1989).

- A product (suit) certification program which ensures that adequate inspection and testing are performed
- Documentation of materials, chemical permeation resistance, and other technical data
- Design and performance criteria for the overall suit and suit components

- Test methods for water penetration, chemical permeation and penetration, flammability resistance, abrasion resistance, cold temperature performance, tear resistance, flexural fatigue, and other tests, as appropriate

Test methods for evaluating adequacy of protective suits should follow ASTM methods for fabric evaluation. In addition to testing information for the chemical protective suit, the manufacturer should provide user information. This information should include items such as cleaning instructions, marking and storage instructions, frequency and details of inspections, maintenance criteria, and other information.

Once a protective suit has been worn, it must be decontaminated thoroughly and inspected for signs of chemical penetration, puncture, tears, or other signs of failure. If any penetrations or signs of failure are noticed, the suit should be disposed of properly unless the damage is limited and repairable. Repair of protective suits, if applicable, should be in accordance with the manufacturer's instructions.

Ergonomics of Vapor-Protective Suits. The ergonomics of wearing vapor-protective suits is important because it affects the worker's ability to be able to perform the job easily and effectively. A vapor-protective suit should not present the worker with additional hazards, such as tripping hazards, safety hazards (i.e., getting caught in equipment), heat stress, or physical loss of senses such as feel. Tripping and safety hazards can be addressed with proper sizing of suits. Heat stress and loss of senses can be addressed by allowing the worker to try different products and materials to minimize these consequences. The employer might want the worker to fill out an environmental symptoms questionnaire to characterize the effects on the worker of wearing a vapor-protective suit for several hours. An example of this type of questionnaire is presented in Table 6.2. This type of questionnaire might be useful for workers who are required to wear the vapor-protective suits for long periods of time, either regularly or periodically.

An example of a fully encapsulating protective suit is shown in Figure 6.1. This type of suit offers the highest level of skin and eye protection and can be worn in an unknown chemical environment or in a chemical environment that is known to be dangerous. Radiation-contamination protective suits are also available for protection against alpha and beta particles (NFPA 1989). These types of suits should be selected and worn under the guidance of a specialist.

RESPIRATORY PROTECTION

Respiratory protection requirements are defined by OSHA in 29 CFR §1910.134. Additional requirements are included in specific sections of 29

TABLE 6.2 Sample Environmental Symptoms Questionnaire

Rating scale:

0 = not at all; 1 = slight; 2 = somewhat; 3 = moderate; 4 = quite a bit; 5 = extreme

Physical Symptoms	Emotional Symptoms
_____ Headaches	_____ Sleeplessness
_____ Lightheadness	_____ Forgetfulness
_____ Dizziness	_____ Listlessness
_____ Feeling of faintness	_____ Feelings of boredom
_____ Feelings of weakness	_____ Nervousness
_____ Dimming of vision	_____ Feelings of worry
_____ Blurring of vision	_____ Restlessness
_____ Shortness of breath	_____ Irritability
_____ Difficulty in breathing	_____ Depression
_____ Pain when breathing	_____ Lack of alertness
_____ Faster-than-normal heartbeat	_____ Loss of appetite
_____ Pounding heartbeat	_____ Feelings of alienation
_____ Chest pains	_____ Disorientation
_____ Chest pressure	
_____ Muscle cramps	
_____ Stomach cramps	
_____ Muscle stiffness	
_____ Aching in legs or feet	
_____ Backache	
_____ Sweaty feet or palms	
_____ Ear aches	
_____ Numbness	
_____ Itchy skin	
_____ Nausea	

Source: Information from Navy Clothing and Textile Research Facility and U.S. Army Research Institute of Environmental Medicine (1992).

CFR Part 1910 that cover OSHA-listed hazardous materials such as acrylonitrile, vinyl chloride, inorganic arsenic, cotton dust, lead, and benzene. NIOSH provides a testing, approval, and certification program for respiratory protection. The Mine Safety and Health Administration (MSHA) also evaluates and approves some respirators in conjunction with NIOSH. OSHA provides standards for respirator use and fit testing.

The need for respiratory protection is determined through sampling a worker's exposure to a specific chemical. If the permissible exposure limit as a time-weighted average is expected to be exceeded during work activities, a respirator is required. Similarly, if the short-term exposure or ceiling limit is expected to be exceeded at any time, a respirator is required.

Typical respiratory protective devices include the chemical cartridge, particulate respirator, gas mask, supplied air respirator, and self-contained

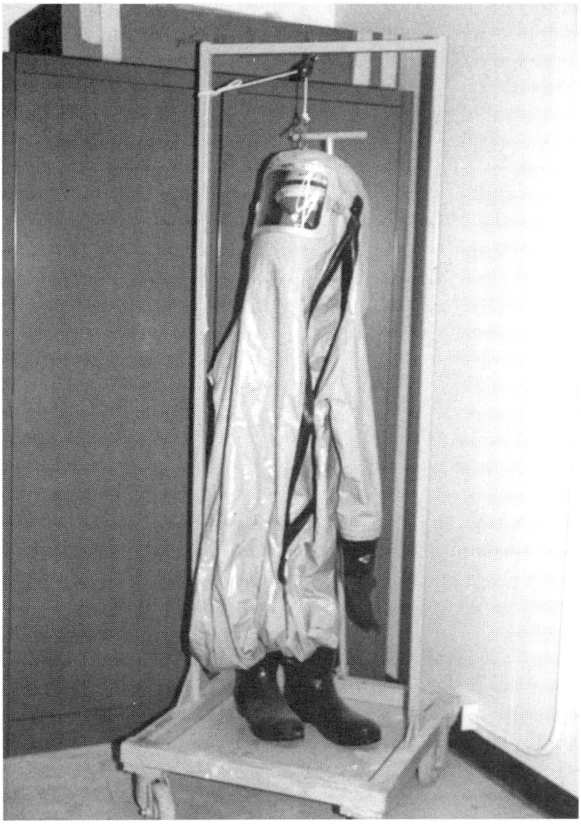

Figure 6.1 Fully encapsulating protective suit.

breathing apparatus (SCBA). Examples of SCBAs which are stored after use are shown in Figure 6.2. Selection of a respirator is based on chemicals used in the area, concentrations, expected exposure time, and oxygen present in the atmosphere. Training is essential before use of any type of respirator. For industrial applications, OSHA requires that respirator models and applications be approved by NIOSH and by NIOSH/MSHA for substances such as asbestos, cotton dust, and other dusts and mists.

Chemical Cartridges

Chemical cartridges are air-purifying respirators used for respiratory protection against specific chemicals of relatively low concentrations. Since they have no air-supplying capability, the cartridges cannot be used in atmospheres containing less than 19.5% oxygen or in atmospheres immediately dangerous to life or health (IDLH). The respirators come in half-face (orinasal) or full-face-piece models, with canister holders attached on the face piece or belt.

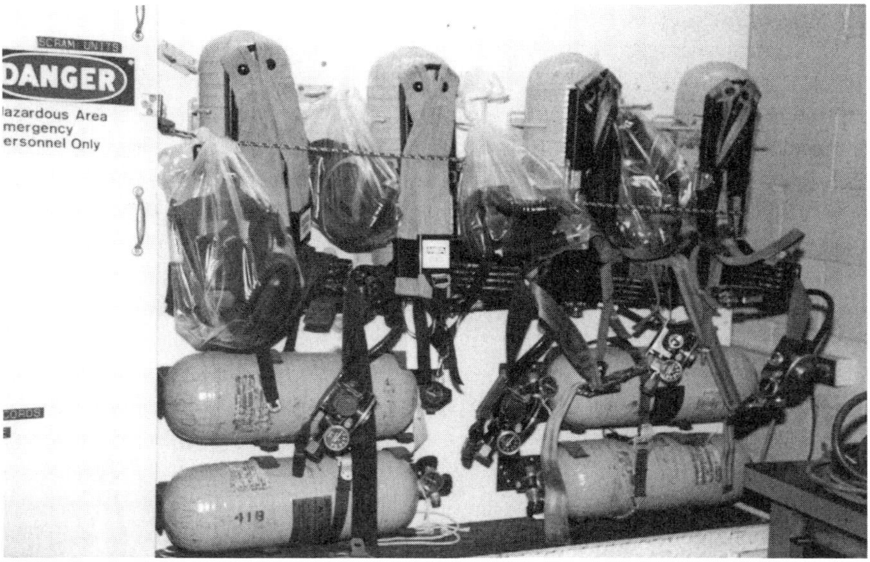

Figure 6.2 SCBAs which are stored after use.

The length of time before chemical breakthrough of the canister depends on the concentration of the chemical in the work area. All models are designed to permit canister replacement. Table 6.3 summarizes the types of chemical cartridges and applications approved by NIOSH.

Particulate Respirators

Particulate respirators are designed for protection against dusts, mists, fumes, and other particulates. Before using this type of respirator, the atmosphere should be tested and must contain at least 19.5% oxygen and cannot be IDLH. Designs of particulate respirators include:

- Single-use particulate respirators
- Particulate respirators for protection against dusts and mists
- Particulate respirators for protection against dust, fumes, and mists
- Particulate respirators with a high-efficiency filters

The first three designs listed above are approved for use in atmospheres where the permissible exposure limit as a time-weighted average is not less than 0.05 milligram per cubic meter (mg/m^3) or 2 million particles per cubic foot. The high-efficiency particulate respirator has the greatest versatility and can be used for protection against dusts, mists, asbestos-containing dusts and mists (if approved by NIOSH/MSHA), and radionuclides. These respirators

TABLE 6.3 Chemical Cartridge Types and Applications Approved by NIOSH

Cartridge Type	Effective Concentrations	Other Protection Included on Some Models
Ammonia	<300 ppm	Methyl amine, dusts, mists
Methyl amine	<100 ppm	Ammonia, dusts, mists, fumes of various metals, radionuclides, radon
Chlorine	<10 ppm	Hydrogen chloride, hydrogen sulfide, organic vapors, dusts, fumes, mists, radionuclides
Hydrogen chloride	<50 ppm	In general, most models provide no other protection
Chlorine, sulfur dioxide, hydrogen chloride	<10 ppm, <50 ppm, <50 ppm	Organic vapor, hydrogen sulfide, formaldehyde, chlorine dioxide, dusts, mists, radionuclides
Organic vapor	<1000 ppm by volume	Sulfur dioxide, hydrogen chloride, chlorine, pesticides, dusts, mists, mists of paints/lacquers/enamels, chlorine dioxide, formaldehyde
Mists of paints/lacquers/enamels	<1000 ppm by volume	Organic vapors, pesticides
Pesticides	Specific to model	Organic vapor, mosts of paints/lacquers/enamels
Vinyl chloride	<10 ppm	Model provides no other protection
Other gases and vapors	Specific to model	Hydrogen sulfide, chlorine dioxide, formaldehyde, organic vapors, chlorine, dusts, mists, fumes of various metals, radon, radionuclides

Source: Information from NIOSH (1990).

provide protection against dusts, fumes, and mists having permissible exposure limits as a time-weighted average of less than 0.05 mg/m^3.

Gas Masks

Gas masks (nonpowered) are similar to chemical cartridge respirators, except they are all full-face masks and are for a one-time application in an emergency situation. Like chemical cartridge respirators, they are not effective in atmospheres of less than 19.5% oxygen or in IDLH atmospheres (NIOSH 1990).

Gas masks afford protection against a specific chemical or chemical type such as ammonia, chlorine, acids gases, organic vapors, carbon monoxide, sulfur dioxide, and pesticides. Many models have approval for protection against additional vapors, gases, and dusts or mists. OSHA has defined colors for identification of types of canisters in §1910.135, Table 1-1. These defined colors are presented in Table 6.4.

Supplied-Air Respirators

Supplied-air respirators are used for the same applications as those of cartridge respirators. As with the other respirators described previously, they are not approved for use in IDLH atmospheres or atmospheres containing less than 19.5% oxygen. This limitation is based on safety considerations, since the air supply or air line could fail. However, there is an exception to this standard if an auxiliary tank of air permitting escape is incorporated into the respirator system (AIHA 1991). Approved types of supplied-air respirators include continuous flow, pressure demand, and continuous-flow abrasive blasting.

When using a supplied-air respirator, a line is attached to the user from a supply of respirable air. Full face pieces or hoods must be worn with these devices. Hose lengths and air supply pressure ranges specified by the manufacturer (and approved by NIOSH/MSHA) must be used. The maximum length of hose allowed is 300 feet and the maximum inlet pressure is 125 pounds per square inch gage (psig).

Self-Contained Breathing Apparatus

The self-contained breathing apparatus (SCBA) is the only respirator approved by NIOSH for use in oxygen-deficient and IDLH atmospheres. SCBAs also are worn in emergency situations and when unknown atmospheres must be entered. A widely used SCBA is the open-circuit pressure demand system, which maintains a positive pressure inside the face piece at all times, which provides the highest level of respiratory protection. The system has a full face piece and a face piece- or belt-mounted regulator that regulates airflow from a bottle worn on the worker's back. Service life is generally 30 to 60 minutes.

Other types of SCBAs include open-circuit demand system and closed-circuit demand and pressure demand systems. The open-circuit demand system maintains a positive pressure in the face piece during exhalation but allows for negative pressure during inhalation. Closed-circuit systems allow the worker's exhalation to be filtered (chemically scrubbed) and rebreathed, while providing supplemental oxygen from a supply source. This type of system has a longer service life but does not provide protection as adequately as the open-circuit systems.

Open-circuit pressure demand systems for "escape only" typically have a service life of 5 to 15 minutes. These devices are full face pieces or hoods

TABLE 6.4 Colors for Canisters Assigned by OSHA

Atmospheric Contaminants to Be Protected Against	Colors Assigned[a]
Acid gases	White
Hydrocyanic acid gas	White with $\frac{1}{2}$-inch green stripe completely around the canister near the bottom
Chlorine gas	White with $\frac{1}{2}$-inch yellow stripe completely around the canister near the bottom
Organic vapors	Black
Ammonia gas	Green
Acid gases and ammonia gas	Green with $\frac{1}{2}$-inch blue stripe completely around the canister near the bottom
Carbon monoxide	Blue
Acid gases and organic vapors	Yellow
Hydrocyanic acid gas and chloropicrin vapor	Yellow with $\frac{1}{2}$-inch blue stripe completely around the canister near the bottom
Acid gases, organic vapors, and ammonia gases	Brown
Radioactive materials, excepting tritium and noble gases	Purple (magenta)
Particulates (dusts, fumes, mists, fogs, or smokes) in combination with any of the gases or vapors above	Canister color for contaminant, as designated above, with $\frac{1}{2}$-inch gray stripe completely around the canister near the top
All of the atmospheric contaminants above	Red with $\frac{1}{2}$-inch gray stripe completely around the canister near the top

[a]Gray shall not be assigned as the main color for a canister designed to remove acids or vapors. Orange shall be used as a complete body or stripe color to represent gases not included in this table. The user will need to refer to the canister label to determine the degree of protection the canister will afford.

and do not have a large air bottle associated with them. Closed-circuit escape-only systems usually have only a mouthpiece instead of a face piece or hood. These systems have a service life of up to 60 minutes.

EYE AND FACE PROTECTION

Protective eye and face devices must comply with ANSI Z87.1-1989 (ANSI 1989). Although not a complete list, the following occupations should consider eye and face protection for routine use: carpenters, electricians, machinists, mechanics and repairers, millwrights, plumbers and pipe fitters, sheet metal workers and tinsmiths, assemblers, sanders, grinding machine operators, lathe and milling machine operators, sawyers, welders, laborers, chemical process operators and handlers, and timber cutting and logging workers. General guidance for the proper selection of eye and face protection is presented in Table 6.5.

EYEWASHES AND EMERGENCY SHOWERS

ANSI Standard Z358.1 (ANSI 1990) provides recommended practices for eyewashes and emergency showers. Eyewashes and emergency showers must be available in all chemical work areas and must be inspected periodically to ensure proper flow and, in the case of eyewashes, flow direction. Additionally, the area around eyewashes and emergency showers must be kept clear for easy access. Easily readable signs that mark the location of the chemical safety devices should be posted.

In most chemical-use work locations, eyewashes and emergency showers are permanently installed in the building. For field applications such as hazardous waste cleanup operations or when water service is not available due to scheduled maintenance of the process and potable water system, portable eyewash and shower units can be used. An example of these portable units is shown in Figure 6.3.

Other emergency equipment that should be available in the workplace includes a portable fire extinguisher and first-aid equipment. An evacuation route should be posted clearly in the work area, and all personnel should be trained on how to respond in emergencies.

PERSONAL PROTECTION DURING ASBESTOS REMOVAL

An asbestos fiber is a mineral fiber longer than 5 micrometers (μm) with a length/diameter ratio of at least 3:1, which consists of chrysotile, crocidolite, amosite, anthophyllite, tremolite, or actinolite asbestos. The permissible exposure limit set by OSHA is 0.1 fiber per cubic centimeter for an 8-hour time weighted average and 1.0 fiber/cm^3 for a 30-minute excursion limit (short-term exposure limit).

Requirements for PPE when removing asbestos are defined in OSHA 29 CFR §§1910.1001 and 1926.58 and under TSCA in §763.121. Guidance for workplace safety and health as it relates to asbestos exposure is detailed in ASTM E849-86. Demolition and renovation workers are required to wear appropriate NIOSH/MSHA-approved respiratory protection, as specified in the regulations, which includes:

- *Not in excess of 1 fiber/cm^3 (10 times the PEL):* half-mask air-purifying respirator, other than a disposable respirator, equipped with high-

TABLE 6.5 General Guidance for Selection of Eye and Face Protection

Source	Assessment of Hazard	Protection
Impact: chopping, grinding, machining, masonry work, woodworking, sawing, drilling, chiseling, powered fastening, riveting, and sanding.	Flying fragments, objects, large chips, particles, sand, dirt, etc.	Spectacles with side protection, goggles, face shields. See notes (1), (3), (5), (6), (10). For severe exposure, use a face shield
Heat: furnace operations, pouring, casting, hot dipping, and welding	Hot sparks	Face shields, goggles, spectacles with side protection. For severe exposure, use a face shield See notes (1), (2), (3).
	Splash from molten metals	Face shields worn over goggles. See notes (1), (2), (3).
	High-temperature exposure	Screen face shields, reflective face shields. See notes (1), (2), (3).
Chemicals: acid and chemicals handling, degreasing, plating	Splash	Goggles, eyecup, and cover types. For severe exposure, use a face shield. See notes (3), (11).
	Irritating mists	Special-purpose goggles.
Dust: woodworking, buffing, general duty conditions	Nuisance dust	Goggles, eyecup, and cover types. See note (8).
Light and/or radiation		
Welding		
Electric arc	Optical radiation	Welding helmets or welding shields. Typical shades: 10–14. See notes (9), (12).
Gas	Optical radiation	Welding goggles or welding face shield. Typical shades: gas welding 4–8, cutting 3–6, brazing 3–4. See note (9).
Cutting, torch brazing, torch soldering	Optical radiation	Spectacles or welding face shield. Typical shades, 1.5–3. See notes (3), (9).
Glare	Poor vision	Spectacles with shaded or special-purpose lenses, as suitable. See notes (9), (10).

TABLE 6.5 (*Continued*)

Notes:
(1) Care should be taken to recognize the possibility of multiple and simultaneous exposure to a variety of hazards. Adequate protection against the highest level of each of the hazards should be provided. Protective devices do not provide unlimited protection.
(2) Operations involving heat may also involve light radiation. As required by the standard, protection from both hazards must be provided.
(3) Face shields should only be worn over primary eye protection (spectacles or goggles).
(4) As required by the standard, filter lenses must meet the requirement for shade designations in §1910.133(a)(5). Tinted or shaded lenses *are not* filter lenses unless they are marked or identified as such.
(5) As required by the standard, persons whose vision requires the use of prescription lenses must wear either protective devices fitted with prescription lenses or protective devices designed to be worn over regular prescription eyewear.
(6) Wearers of contact lenses must also wear appropriate eye and face protection devices in a hazardous environment. It should be recognized that dusty and/or chemical environments may represent an additional hazard to contact lens wearers.
(7) Caution should be exercised in the use of metal frame protective devices in electrical hazard areas.
(8) Atmospheric conditions and the restricted ventilation of the protector can cause lenses to fog. Frequent cleaning may be necessary.
(9) Welding helmets or face shields should be used only over primary eye protection (spectacles or goggles).
(10) Non-sideshield spectacles are available for frontal protection only but are not acceptable eye protection for the sources and operations listed for "impact."
(11) Ventilation should be adequate, but well protected from splash entry. Eye and face protection should be designed and used so that it provides both adequate ventilation and protects the wearer from splash entry.
(12) Protection from light radiation is directly related to filter lens density. See note (4). Select the darkest shade that allows task performance.
Source: 29 CFR §1910, Subpart I, Appendix B.

efficiency filters (filters that are at least 99.97% efficient against monodispersed particles of 0.3 μm or larger).

- *Not in excess of 5 fibers/cm^3 (50 times the PEL):* full-face-piece air-purifying respirator equipped with high-efficiency filters.
- *Not in excess of 10 fibers/cm^3 (100 times the PEL):* any powered air-purifying respirator equipped with high-efficiency filters or any supplied-air respirator operated in continuous-flow mode.
- *Not in excess of 100 fibers/cm^3 (1000 times the PEL):* full-face-piece supplied-air respirator operated in pressure demand mode.
- *Greater than 100 fibers/cm^3 (1000 times the PEL) or unknown concentration:* full-face-piece supplied-air respirator operated in pressure demand mode equipped with an auxiliary positive-pressure SCBA.

At any time, respirators assigned for higher environmental concentrations may be used at lower concentrations.

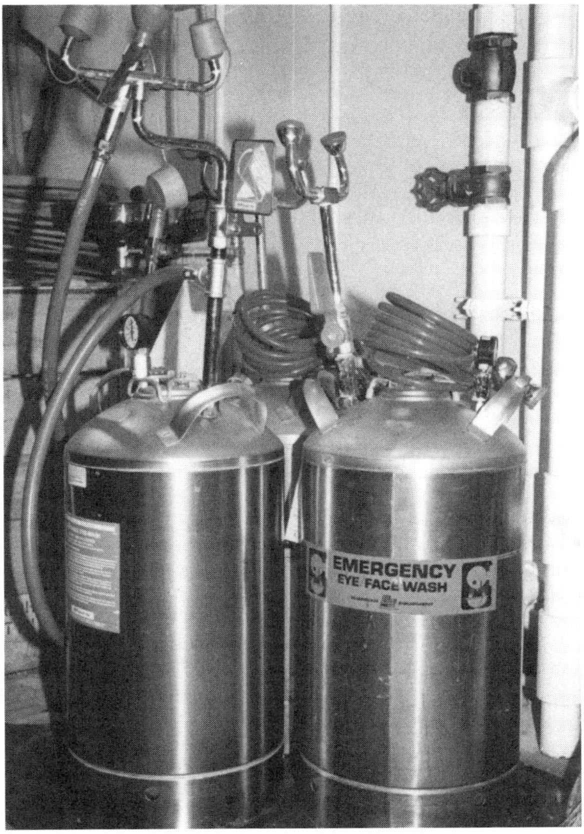

Figure 6.3 Portable eyewash and shower units.

During the respirator selection process, the worker should be allowed to pick a comfortable respirator from a selection of various sizes from different manufacturers. A qualitative respirator fit test using saccharin nebulizer, irritant smoke, or odor threshold screening should be performed before initial use. Additionally, quantitative fit tests can be performed. Quantitative fit tests generally use a particle counting method. Field checks by positive and negative pressure methods should be performed before each entry into the contaminated area.

Full-body work clothing is required with hoods, boot covers, and gloves. Arm cuffs of coveralls should be taped to seal them tightly against the gloves. Overboots also should be taped tightly to the legs of the coveralls. Face shields, goggles, or other appropriate protective equipment also might be warranted, depending on the respiratory protection required. The employer is responsible for cleaning, laundering, and replacement of protective clothing.

Engineering controls are needed during demolition to control the airborne particles. These include, but are not limited to:

- Complete enclosure of the area
- Wetting of the area and particularly the asbestos-containing material that is to be removed
- Exhausting the area through an air pollution control device such as a baghouse, high-efficiency particulate air (HEPA) filter, furnace exhaust filter, or wet collector.
- Controlled-access change rooms with clothes lockers and showers for decontamination after work.
- Controlled sampling that minimizes worker and public exposure to airborne asbestos fibers.

Determinations of airborne concentrations must be made following OSHA's sampling and analytical procedure specified in Appendix A to 29 CFR §1910.1001. This method is described in Chapter 5. Additionally, personal sampling is required at least once every six months for employees whose exposure to asbestos is expected to exceed the exposure limits (discounting wearing a respirator). Samples must be collected from within the breathing zones of employees and in areas representative of such breathing zones. Affected employees must have the opportunity to observe monitoring and must have access to the monitoring records.

Asbestos waste includes all friable asbestos, contaminated clothing discarded for disposal, and other contaminated material, such as plastic bags and enclosure material. The waste must be sealed in leaktight bags or drums prior to proper disposal.

PERSONAL PROTECTION DURING HAZMAT RESPONSE AND WASTE CLEANUP

Types of PPE acceptable for hazardous materials (HAZMAT) emergency response and hazardous waste cleanup activities have been established EPA in guidance (EPA 1984) and by OSHA in 29 CFR §1910.120, Appendix B, and have been incorporated into training manuals (International Fire Service 1988, Noll et al. 1988). During a HAZMAT response or waste cleanup activity, there are four levels of clothing that can be worn by responders, depending on the hazard potential associated with the job performed. These include:

- *Level A.* This type of protection should be worn when the highest level of protection is needed for respiratory, skin, and eye protection. In general, this is used in unknown atmospheres and known high hazard atmospheres. Level A protection includes a positive-pressure SCBA, a fully encapsulating chemical resistant suit, chemical-resistant inner and outer gloves, and chemical-resistant safety boots. A two-way communications device such as a two-way radio or other device is also necessary. Depending on the situation, other safety clothing might be added, such as a hard hat, coveralls, long cotton underwear, or a cooling unit.

- *Level B.* This type of protection should be selected when the highest level of respiratory protection is needed but a lower level of skin and eye protection is required. In general, this type of protection is used in oxygen-deficient and IDLH atmospheres, where the substances do not represent a severe skin hazard. Level B protection includes a positive-pressure SCBA, chemical-resistant clothing (hooded one- or two-piece chemical splash suit, disposable chemical-resistant coveralls, overalls and long-sleeved jacket, or equivalent clothing), inner and outer chemical-resistant gloves, chemical-resistant safety boots, and a two-way communication device. Optional clothing could include a hard hat, long cotton underwear, and disposable boot covers.
- *Level C.* This type of protection should be selected when the atmosphere contaminants are known and the concentrations have been measured and meet the criteria for use of an air-purifying respirator. Additionally, skin and eye exposure must be unlikely. Level C protection includes use of a full-face-piece air-purifying respirator, chemical-resistant clothing, chemical-resistant inner and outer gloves, chemical-resistant safety boots, and a two-way communication device. Optional clothing could include a hard hat, disposable boot covers, long cotton underwear, a face shield, and an escape mask.
- *Level D.* This is type of protection should be selected when there is no respiratory hazard and when minimum skin and eye protection are required. This is primarily a work uniform, and the atmosphere must have no known hazards. Level D protection includes coveralls or other work uniform, safety glasses, and safety boots.

The type of PPE used during a HAZMAT response or waste cleanup activity should be reevaluated periodically as new information becomes available. Any change in level of PPE used should be reviewed and approved by an industrial hygienist, safety engineer, or other specialist in charge of specifying PPE. Other aspects of HAZMAT response, including incident handling techniques, are described in Chapter 19.

REFERENCES

American Conference of Governmental Industrial Hygienists (1987). *Guidelines for Selection of Chemical Protective Clothing,* 3rd ed., A. D. Schwope, P. P. Costas, J. O. Jackson, and D. J. Weitzman, eds., ACGIH, Cincinnati, OH.

American Industrial Hygiene Association (1991). *Respiratory Protection: A Manual and Guideline,* 2nd ed., C. E. Colton, L. R. Birkner, and L. M. Brosseau, eds., AIHA, Akron, OH.

American National Standards Institute (1980). *Practices for Respiratory Protection,* ANSI Z88.2, ANSI, New York.

——— (1989). *Practice for Occupational and Educational Eye and Face Protection,* ANSI Z87-1, ANSI, New York.

——— (1990). *Emergency Eyewash and Shower Equipment,* ANSI Z358.1, ANSI, New York.

American Society for Testing and Materials (1990). *ASTM Standards on Protective Clothing,* ASTM, Philadelphia.

Environmental Protection Agency (1984). *Standard Operating Safety Guides,* Office of Emergency and Remedial Response, Hazardous Response Support Division, U.S. EPA, Edison, NJ.

International Fire Service Training Association (1988). *HAZMAT for First Responders,* Fire Protection Publications, Oklahoma State University, Stillwater, OK.

National Fire Protection Association (1989). *Hazardous Materials Response Handbook,* H. F. Martin, ed., NFPA, Quincy, MA.

——— (1990). *Certified Equipment List,* DHHS (NIOSH) Publication 90-102, U.S. Department of Health and Human Services, Washington, DC.

National Institute for Occupational Safety and Health, Occupational Safety and Health Administration, U.S. Coast Guard, and U.S. Environmental Protection Agency (1985). *Occupational Safety and Health Guidance Manual for Hazardous Waste Site Activities,* DHHS (NIOSH) Publication 85-115, U.S. Department of Health and Human Services, Washington, DC.

Navy Clothing and Textile Research Facility and U.S. Army Research Institute of Environmental Medicine (1992). *Heat Stress Induced by the Navy Fire Fighter's Ensemble Worn in Various Configurations.* Technical Report NCTFR 192, NCTFR/IEM, Natick, MA.

Noll, G. G., M. S. Hildebrand, and J. G. Yvorra (1988). *Hazardous Materials: Managing the Incident,* Fire Protection Publications, Oklahoma State University, Stillwater, OK.

BIBLIOGRAPHY

American Society for Testing and Materials (1989). *Chemical Protective Clothing Performance in Chemical Emergency Response,* Special Technical Publication 1037, Perkins and Stull, eds., ASTM, Philadelphia.

——— (1989). *Performance of Protective Clothing: Second Symposium,* Special Technical Publication 989, Mansdorf, Sager, and Nielsen, eds., ASTM, Philadelphia.

Center for Labor Education and Research (1990). *Worker Protection During Hazardous Waste Remediation,* Hazardous Waste Training Program, University of Alabama, Birmingham, AL.

Code of Federal Register, 40 CFR Part 61 Subpart M, "National Emission Standard for Asbestos," U.S. EPA, Washington, DC.

Defence Research Establishment Ottawa (1991). *Heat Stress Caused by Wearing Different Types of CW Protective Garment,* DREO Technical Note 91-14, prepared by S. D. Livingstone and R. W. Nolan, DREO, National Defence, Ottawa, Ontario, Canada.

Department of Energy (1990). *Recent Investigation in Personal Protective Equipment (Including Respirators),* DE90 005785, Lawrence Livermore National Laboratory, Livermore, CA.

——— (1990). *Respirator Standards, Regulations, and Approvals in the U.S.A.,* prepared by B. J. Held, Lawrence Livermore National Laboratory, Livermore, CA.

Department of the Navy (1991). *Survey of Hazardous Chemical Protective Suit Materials,* NCTRF Report 186, Navy Clothing and Textile Research Facility, Natick, MA.

Environmental Protection Agency (1984). *NESHAPS Asbestos Demolition and Renovation Inspection Workshop Manual,* EPA/340/1-85/008, prepared by GCA Corporation for U.S. EPA, Washington, DC.

——— (1987). *Development and Assessment of Methods for Estimating Protective Clothing Performance,* EPA/600/2-87/104, prepared by R. Goydan et al., U.S. EPA, Washington, DC.

——— (1990). *Environmental Asbestos Assessment Manual, Superfund Method for Determination of Asbestos in Ambient Air,* U.S. EPA, Washington, DC.

——— (1992). *Limited-Use Chemical Protective Clothing for EPA Superfund Activities,* EPA/600/R-92/014, prepared by Arthur D. Little, Inc., for Risk Reduction Engineering Laboratory, Office of Research and Development, U.S. EPA, Cincinnati, OH.

Forsberg, K., and L. H. Keith (1989). *Chemical Protective Clothing Performance,* Wiley, New York.

Forsberg, K., and S. Z. Mansdorf (1993). *Quick Selection Guide to Chemical Protective Clothing,* 2nd ed., Van Nostrand Reinhold, New York.

Government Institutes (1987). *Asbestos in Buildings, Facilities and Industry,* GI, Rockville, MD.

Keith, L. H., and D. B. Walters (1992). *The National Toxicology Program's Chemical Data Compendium,* Vol. 6, "Personal Protective Equipment," Lewis Publishers, Boca Raton, FL.

National Institute of Building Sciences (1988). *Asbestos Abatement and Management in Buildings: Model Guide Specification,* 2nd ed., NIBS, Washington, DC.

National Institute for Occupational Safety and Health (1989). *Federal Research on Chemical Protective Clothing and Equipment: A Summary of Federal Programs for Fiscal Year 1988,* DHHS (NIOSH) Publication 89-119, U.S. Department of Health and Human Services, Washington, DC.

7

WORKPLACE AND BUILDING SAFETY

Workplace and building safety are described in this chapter in terms of fire protection systems, life safety management, and ventilation systems. Fire protection systems are the main safety systems used at all facilities to mitigate fires, explosions, and other chemical emergencies. Life safety management, which is related to fire protection, includes means of egress, building occupancy management, and building construction. Ventilation systems are critical in maintaining adequate air exchange throughout the workplace for proper control of workplace contamination.

FIRE PROTECTION

An integral part of loss prevention and worker safety is fire protection. Fire protection equipment can range from portable extinguishers to elaborate dry chemical, foam, or sprinkler systems. Design and installation of a fire protection system is based on chemicals used in the area, room layout, and occupancy. Two useful references on the subject of fire protection are the *Fire Protection Handbook* (NFPA 1997) and the *Uniform Fire Code* (ICBO 1991a). In the following paragraphs the detail some basic fire protection information described in these references and in other NFPA documents that can be used in industrial applications.

Common Terms Related to Flammability

The terms *flammable* and *combustible* were defined in Chapter 3. There are other terms related to flammable and combustible materials (ASSE 1988, NFPA 1997), including:

- *Flammable or explosive limits:* the range of concentration of a combustible material in air at which propagation of flame occurs on contact with an ignition source. Concentrations are expressed in percent vapor or gas in the air by volume. If the concentration is below the lower flammable limit (LFL) or lower explosive limit (LEL), the air mixture is considered too lean and flame propagation will not occur. Conversely, if the concentration is above the upper flammable limit (UFL) or upper explosive limit (UEL), the mixture is too rich and flame propagation likewise will not occur. The lower explosive limits of some commonly used solvents are presented in Table 7.1.

- *Flash point:* the lowest temperature at which a liquid gives off enough vapor to form an ignitable mixture with air near the surface of the liquid or within a test vessel, which is capable of flame propagation away from the source of ignition.
- *Explosion:* a rapid increase of pressure in a confined space followed by its sudden release due to rupture of the container, vessel, or structure; the increase in pressure is generally caused by an exothermic chemical reaction or overpressurization of a system. Bursting or rupture of a building or a container due to the development of internal pressure.
- *BLEVE (boiling liquid expanding vapor explosion):* an explosion caused when fire impinges on the shell of a bulk liquid container, tank, or vessel above the liquid level, causing loss of strength of the metal and consequently, an explosive rupture of the structure from internal pressure.
- *Deflagration:* an exothermic reaction that expands rapidly from the burning gases to the unreacted material by conduction, convection, and radiation and where the combustion zone progresses through the material at a rate that is less than the velocity of sound in the unreacted material. Propagation of a combustion zone at a velocity that is less than the speed of sound in the unreacted medium.
- *Detonation:* an exothermic reaction characterized by the presence of a shock wave in the material that establishes and maintains the reaction, usually causing an explosion, and where the reaction zone expands at a rate greater than the speed of sound in the unreacted material. Propagation of a combustion zone at a velocity that is greater than the speed of sound in the unreacted medium.

Many facilities that manufacture, use, or store hazardous materials use the NFPA fire diamond to communicate fire hazards. Indeed, numerous cities have local regulations or ordinances requiring the use of fire diamonds or equivalent communication/warning labels. The fire diamond hazards and related colors are presented in Figure 7.1. Examples of special instructions that might be included on a fire diamond are presented in Figure 7.2.

In addition to colors representing hazards, a numbering system has been established to communicate degree of hazard. This numbering system is described in Table 7.2.

The *Uniform Fire Code* provides classifications of materials according to hazard. These are presented in Table 7.3.

TABLE 7.1 Communication of Fire Hazards Through the NFPA Fire Diamond and Other Means

Solvent	Cubic Feet per Gallon of Vapor of Liquid at 70°F	Lower Explosive Limit in Percent by Volume of Air at 70°F
Acetone	44.0	2.6
Amyl acetate (iso)	21.6	1.0[a]
Amyl alcohol		
n	29.6	1.2
iso	29.6	1.2
Benzene	36.8	1.4[a]
Butyl acetate (*n*)	24.8	1.7
Butyl alcohol (*n*)	35.2	1.4
Butyl Cellosolve	24.8	1.1
Cellosolve	33.6	1.8
Cellosolve Acetate	23.2	1.7
Cyclohexanone	31.2	1.1[a]
1, 1-Dichloroethylene	42.4	5.9
1, 2-Dichloroethylene	42.4	9.7
Ethyl acetate	32.8	2.5[a]
Ethyl alcohol	55.2	4.3
Ethyl lactate	28.0	1.5[a]
Methyl acetate	40.0	3.1
Methyl alcohol	80.8	7.3
Methyl Cellosolve	40.8	2.5
Methyl ethyl ketone	36.0	1.8
Methyl *n*-propyl ketone	30.4	1.5
Methyl (VM&P) (76° naphtha)	22.4	0.9
Naphtha (100° flash) safety solvent—Stoddard solvent	23.2	1.0
Propyl acetate		
n	27.2	2.8
iso	28.0	1.1
Propyl alcohol		
n	44.8	2.1
iso	44.0	2.0
Toluene	30.4	1.4
Turpentine	20.8	0.8
Xylene (*o*)	26.4	1.0

[a]At 212°F.

Portable Extinguishers

NFPA has published a standard—NFPA 10—that outlines criteria for the selection, installation, maintenance, and testing of portable fire extinguishers. According to the standard, fires can be classified into four categories:

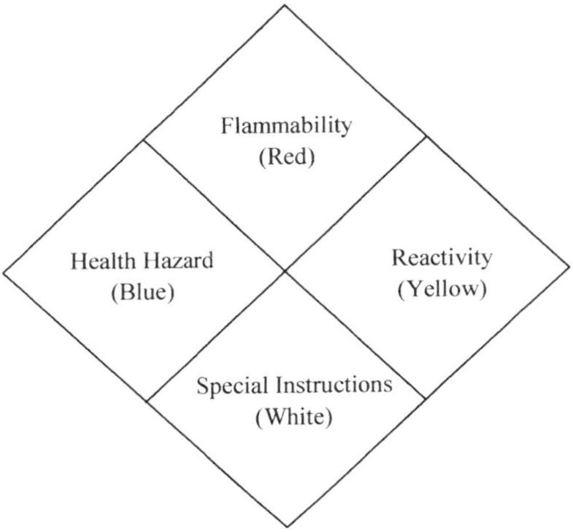

Figure 7.1 NFPA diamond.

- *Class A.* These are fires in ordinary combustible materials such as paper, wood, and plastics. These fires can be quenched with water or water solutions that absorb heat or with the use of certain dry chemicals that interrupt the combustion chain reaction. Class A fire-extinguishing agents include water, saturated steam, and multipurpose dry chemicals such as ammonium phosphate.
- *Class B.* These are fires in flammable or combustible liquids and flammable gases. These fires require the use of dry chemicals that interrupt the combustion chain reaction, inhibit vapor release, or displace oxygen. Class B extinguishing agents include carbon dioxide, aqueous film-

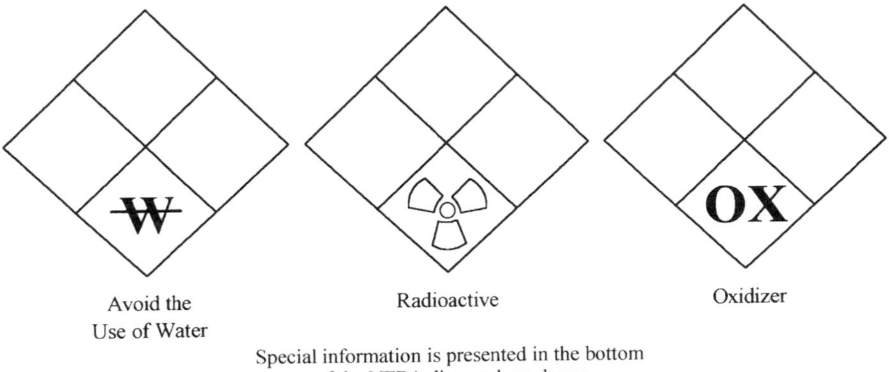

Figure 7.2 Special information symbols.

TABLE 7.2 NFPA Numbering System Associated with the Fire Diamond

Flammability Ratings		*Health Hazard Ratings*	
4	Extremely flammable	4	Too dangerous to enter
3	Ignites at normal temperatures	3	Extremely dangerous
2	Ignites when moderately heated	2	Hazardous
1	Must be preheated to burn	1	Slightly hazardous
0	Will not burn	0	Normal material
Reactivity Ratings		*Special Instructions*	
4	May detonate	~~W~~	Avoid use of water
3	Shock or heat may cause detonation	OX	Oxidizer
		COR	Corrosive
2	Violent chemical change possible	ACID	Acid
1	Unstable if heated	ALK	Alkali
0	Normally stable	Other	Symbols for radioactive materials, biohazard, and other hazards may be used

forming foam, Halons, and dry chemical agents such as potassium bicarbonate, potassium chloride, monoammonium phosphate, and potassium carbamate.

- *Class C.* These are fires in electrical equipment. These fires require nonconductive extinguishing agents such as carbon dioxide and dry chemical agents.
- *Class D.* These are fires in combustible metals such as magnesium, sodium, lithium, and potassium. These fires require nonreactive, heat-absorbing extinguishing agents such as graphite, dry soda ash, dry sand, and dry powder such as sodium chloride.

There are several types and sizes of portable fire extinguishers suitable for fighting the various classes of fires that can be stored in the workplace. Water-based portable extinguishers with a capacity of $2\frac{1}{2}$ gallons are commonly used for class A fires. This extinguisher weighs approximately 30 pounds. Bigger units that supply 25 to 60 gallons of water are available but must be mounted on wheels to make them portable.

Dry chemical portable extinguishers for class A, B, and C fires weigh from 2 to 50 pounds. As was the case for water-based systems, larger and heavier units can be mounted on wheels. Dry sand is often stored in barrels on pallets and can be moved using a fork truck or dolly.

Basic operational designs of water, dry chemical, and carbon dioxide extinguishers (NFPA 1997) include:

- *Stored-pressure unit.* This unit is used for water-based agents and dry chemicals. The unit contains the extinguishing agent and the propellant

TABLE 7.3 Uniform Fire Code's Classifications of Hazards

Liquid and Solid Oxidizer

- *Class 4:* can undergo an explosive reaction as a result of contamination or exposure to thermal or physical shock, will enhance the burning rate and may cause spontaneous ignition of combustibles; examples include ammonium permanganate and guanidine nitrate.
- *Class 3:* will cause a severe increase in the burning rate of combustible materials with which it comes in contact or will undergo vigorous self-sustained decomposition resulting from contamination or exposure to heat; examples include ammonium dichromate and chloric acid.
- *Class 2:* will cause a moderate increase in the burning rate or may cause spontaneous ignition of combustible materials with which it comes in contact; examples include copper chlorate and zinc permanganate.
- *Class 1:* will slightly increase the burning rate but does not cause spontaneous ignition when it comes in contact with combustible materials; examples include ammonium persulfate and zinc peroxide.

Organic Peroxide

- *Unclassified:* capable of detonation and presents an extremely high explosion hazard through rapid explosive decomposition
- *Class I:* capable of deflagration, but not detonation, and presents a high explosion hazard through rapid decomposition; examples include fulfonyl peroxide and diisopropyl peroxydicarbonate.
- *Class II:* burns rapidly and presents a severe reactivity hazard; examples include acetyl peroxide (25%) and peroxyacetic acid (43%).
- *Class III:* burns rapidly and presents a moderate reactivity hazard; examples include benzoyl peroxide (78%) and cumene hydroperoxide (86%).
- *Class IV:* burns in the same manner as ordinary combustibles and presents a minimum reactivity hazard; examples include benzoyl peroxide (70%) and laurel peroxide (98%).
- *Class V:* does not burn or present a decomposition hazard; examples include benzoyl peroxide (35%) and 2,4-pentanedione peroxide (4% active oxygen).

Unstable (Reactive) Material

- *Class 4:* readily capable of detonation or of explosive decomposition or explosive reaction at normal temperatures and pressures due to mechanical or localized thermal shock; examples include acetyl peroxide and ethyl nitrate.
- *Class 3:* capable for detonation or of explosive decomposition or explosive reaction, but requires a strong initiating source, such as thermal or mechanical shock at elevated temperatures and pressures, before initiation; examples include hydrogen peroxide (52%) and perchloric acid.
- *Class 2:* normally unstable and readily undergoes rapid release of energy at normal temperatures and pressures, and can undergo violent chemical change at elevated temperatures and pressures, but does not detonate; examples include acrylic acid and styrene.
- *Class 1:* normally stable but can become unstable at elevated temperatures and pressures; examples include acetic acid and paraldehyde.

TABLE 7.3 (*Continued*)

Water Reactive Material

- *Class 3:* reacts explosively with water without heat or confinement; examples include isobutylaluminum and diethylzinc.
- *Class 2:* may form potentially explosive mixtures with water; examples include calcium carbide and potassium peroxide.
- *Class 3:* may react with water with some release of energy, but not violently; examples include acetic anhydride and sodium hydroxide.

Source: ICBO (1994a).

gas, which are mixed in a sealed cylinder. Once activated, the extinguishing agent can be discharged intermittently using a release lever and hose attachment.

- *Pump tank.* This type of design is used only for plain water and water-based agents. The unit uses a hand-operated, vertical piston pump to force water from the tank. Force, range, and duration of the discharge are operator dependent.
- *Cartridge-operated extinguisher.* This extinguisher is used for dry chemicals. The unit utilizes a small cartridge of propellant gas attached to the shell and threaded into the nonpressurized chamber containing the dry chemical. When activated, the propellant gas is released into the chamber to pressurize the agent in the cylinder.
- *Self-expelling extinguisher.* This extinguisher contains carbon dioxide. The carbon dioxide is retained as a liquid under high pressure and is self-expelling when released.

In addition to water-based, dry chemical, and carbon dioxide portable fire extinguishers, there are portable aqueous film-forming extinguishers that are designed much like the stored-pressure water units. A portable dry powder extinguisher is available that is cartridge operated.

Fire Protection Systems

Common fire protection systems found in industrial applications include automatic sprinkler systems, carbon dioxide systems, Halon systems, foam systems, and dry chemical systems. NFPA standards pertaining to these different types of systems are presented in Appendix D. Pertinent information about the various fire suppressant systems, including industrial applications and toxicity of the fire-extinguishing agent, is summarized in Table 7.4.

Fire Protection in Storage Rooms

NFPA Requirements. In large chemical storage areas, segregation by chemical type or category provides the best fire protection. Separate rooms or buildings

TABLE 7.4 Industrial Fire Protection Systems

Type	Applications	Toxicity	Considerations
Automatic water sprinkler	Combustibles such as fuel oils; water-soluble liquids such as acetone	Nontoxic	Can slow down rate of combustion in class B fires; special pipe and fitting materials needed in corrosive atmospheres; severe dust explosions can damage system
Carbon dioxide	Flammable liquids such as gasoline; flammable gases such as propane; electrical fires	Asphyxiant in closed rooms	Extinguishing atmosphere short-lived; special storage requirements in extreme climates
Halons-1301 and 1211	Flammable liquids such as methyl ethyl ketone; flammable gases such as methane; flammable solids such as thermoplastics	Generally nontoxic	Clean, noncorrosive; ozone-depleting potential (are being replaced)
Dry chemical agents	Flammable or combustible liquids in enclosed areas; electrical fires	Slightly corrosive	Can coat surfaces; ineffective if reignition occurs
Foam agents	Flammable or combustible liquids such as gasoline and alcohols	Nontoxic; biodegradable	Used on spills and tank fires; not suitable for water-reactive materials; conductive, so not suitable for use on electrical fires

Source: Information from NFPA (1991a, 1997).

should be used to segregate oxidizing chemicals, flammable or combustible chemicals, unstable chemicals, corrosives, chemicals reactive to water or air, radioactive materials, and materials subject to self-heating. In addition, chemicals and other hazardous materials should be stored separately from non-chemical parts and other manufacturing supplies. Examples of fire extinguishing methods for the various chemical categories, as defined by NFPA, are detailed in Table 7.5.

Drums or packages of chemicals may be stored on pallets and lifted into place on storage racks by forklifts or placed in racks by an automated system. The design of storage rack systems is presented in NFPA 231, Appendix A, for general storage and in NFPA 231C for rack storage of materials. Although the latter standard does not address chemical storage, per se, it is a good guide for sprinkler design for materials of varying flammability.

In all cases, the storage areas must be clean and orderly, and fire protection equipment must be well maintained. Access to the racks and the load/unload areas must be roomy, with a clear path for egress. Aisle widths generally range from 4 to 8 feet but should provide enough space for egress when fork trucks and other equipment is in use.

Design of rack storage systems inside buildings is based generally on storage height and commodity class. Storage height is measured from floor level to the top of the storage container resting on the top rack. In design schemes, the most flammable products and packaging require in-rack sprinklers as well as ceiling sprinklers of adequate water supply. Horizontal barriers that cover the entire rack, including the flue spaces, are also required in many cases. Other considerations for storage areas include total water demand (gallons per minute per square foot), rack configuration, storage to ceiling clearance, and aisle width. Specific design criteria should be researched thoroughly before sprinkler installation.

Uniform Fire Code Requirements. Requirements for fire protection in storage rooms that contain hazardous materials are also outlined in the *Uniform Fire Code* (ICBO 1994a). Requirements are defined for storage of compressed gases (Article 74), storage of cryogenic fluids (Article 75), storage of explosive materials (Article 77), and storage of flammable and combustible liquids (Article 79).

Hazardous materials are included as a discrete category in Article 80. This category includes additional requirements for the above-mentioned hazardous materials, as well as storage requirements for other hazardous materials, such as flammable solids, liquid and solid oxidizers, organic peroxides, pyrophoric materials, unstable and water-reactive materials, highly toxic solids and liquids, radioactive materials, corrosives, and carcinogens, irritants, sensitizers, and other health hazard solids, liquids, and gases. For each category there is a defined exempt storage amount. Exemptions typically are based on type of storage and type of fire protection provided (i.e., unprotected by sprinkler or cabinet, stored within cabinet in unsprinklered building, stored in sprinklered building but not in cabinet, or stored in cabinet inside sprinklered building).

TABLE 7.5 Fire Extinguishing Methods by Chemical Category

Category	Examples	Fire Extinguishing Methods
Oxidizing chemicals	Potassium permanganate, potassium perchlorate, and sodium nitrate	Large quantities of water should be used as the fire extinguishing agent; fire fighters should wear self-contained breathing apparatus; area should be ventilated.
Flammable/combustible chemicals	Gasoline, heavy oils, and sulfur/sulfides of sodium, potassium, and phosphorus	Large quantities of water or foam should be used as the fire-extinguishing agent; sulfur and sulfide compounds could cause irritating gases.
Unstable chemicals	Picric acid, TNT, and organic peroxides	Large quantities of water should be used as the fire-extinguishing agent, preferably through automatic sprinklers since explosion potential exists.
Corrosive chemicals	Sodium hydroxide, hydrochloric acid, sulfuric acid, and perchloric acid	Water spray is recommended in acid and alkali storage areas; toxic fumes can result in some cases; an explosion potential may also exist in some cases.
Water-reactive chemicals	Calcium carbide, chromyl chloride, and alkali metals	Graphite base powder or inert materials can be used for some water-reactive chemicals.
Air-reactive chemicals	Elemental sodium and yellow phosphorus	Water, followed by application of dirt or sand.
Radioactive materials	Plutonium, radium, and uranium	Fire and explosion characteristics are not affected by radioactivity of material; automatic fire protection is necessary to avoid human exposure to radioactivity; preplanning for disposal of contaminated water is necessary
Self-heating materials	Elemental sulfur and activated charcoal	Water or steam should be used only on fire area; wet material is more susceptible than dry material to self-heating.

Source: Information from NFPA (1997).

The next several paragraphs summarize some of the key requirements specified in the *Uniform Fire Code*. The summary is not meant to be all inclusive but should provide a general idea of the types of requirements defined in the code. For inclusive information, the code and its companion publication, *Uniform Fire Code Standards* (ICBO 1994b), should be reviewed.

General Storage of Hazardous Materials. General storage requirements address the topics of spill control, drainage, containment, ventilation, separation of incompatibles, construction of hazardous materials storage cabinets, fire-extinguishing systems, and other relevant topics. Any special requirements related to a particular material are addressed separately in discussions of that material.

Storage of Compressed Gases. Compressed gases are required to be adequately secured to prevent falling or being knocked over. Additionally, legible operating instructions must be maintained at the operating location. "No Smoking" and other warning signs must be posted around enclosures that hold the compressed gas cylinders. Additional items pertain to toxic gases and highly toxic gases. These include ventilation, use of gas cabinets and gas detection systems, acceptable rates of release, treatment systems for exhausted gas cabinets, and other topics.

Storage of Cryogenic Fluids. Cryogenic fluids are classified in the code as flammable, nonflammable, corrosive/highly toxic, and oxidizer. These materials generally are not stored in storage rooms, but rather, in aboveground, belowground, or in-ground containers. Requirements for each type of installation is addressed in the standard. Additionally, minimum distances from the tank are specified for the different classes of cryogenics.

Storage of Explosive Materials. Explosive materials must be handled and stored carefully. Smoking, matches, open flames, or other spark-producing devices are prohibited within 50 feet of magazines. Also, dried grass, leaves, trash, and debris must be cleared within 25 feet of magazines. Explosives cannot be unpacked or repacked within 50 feet of a magazine or in close proximity to other explosives. Any deteriorated explosive must be reported to the fire chief immediately and destroyed by an expert. Numerous other requirements pertain to explosive and blasting agents, and these requirements should be investigated thoroughly before these materials are brought on site for storage.

Storage of Flammable and Combustible Liquids. Flammable and combustible liquids are addressed fully in the code. Storage room requirements include the use of in-rack sprinkler systems. Drainage systems must be designed to carry off to a safe location any anticipated spill plus a minimum calculated flow of the sprinkler system. Aisle widths must be 4 feet between racks and

8 feet for main aisles. Class I flammable liquids (which have a flash point below 100°F) cannot be stored in basements. Additional requirements for individual classes of flammable and combustible liquids and types of storage are also addressed.

Storage of Flammable Solids. Indoor storage of these materials sometimes requires explosion venting or suppression. If the material stored is over 1000 cubic feet, the material must be separated into piles of 1000 cubic feet. Aisle widths between piles must equal the height of the pile or 4 feet, whichever is greater. For exterior storage, flammable solids must be stored in piles no greater than 5000 cubic feet. Aisle widths between piles must be one-half the height of the piles or 10 feet, whichever is greater. Additionally, for exterior storage applications, distance from storage to any building, property line, or public way must be at least 20 feet unless the storage area has an unpierced 2-hour fire-resistive surrounding barrier, defined as a wall extending not less than 30 inches above and to the sides of the storage area. Other requirements pertaining to ventilation, spill control, and other topics are defined.

Storage of Liquid and Solid Oxidizers. For indoor storage, some amounts of these materials require detached storage, as specified in the code. Detached storage buildings must be single story without a basement or crawl space. Requirements for distance from storage to any building, property line, or public way depends on the class of oxidizer. Class 4 oxidizers must be separated from other hazardous materials by not less than 1-hour fire-resistive construction. Detached buildings for class 4 oxidizers must be located a minimum of 50 feet from other hazardous materials storage. For exterior storage, maximum quantities and arrangements are specified in the code. Other requirements pertaining to safe distance to property boundary or public way, smoke detection systems, ventilating systems, and other items are addressed.

Storage of Organic Peroxides. For indoor storage, some amounts of organic peroxides require detached storage, depending on class. Separation distance of detached storage from any building, property line, or public way is also dependent on class. Additionally, special storage requirements pertain to this class of material. For instance, 55-gallon containers cannot be stored more than one container high, and a minimum 2-foot clear space must be maintained between storage and uninsulated metal walls. Containers and packages in storage areas must be closed, and other requirements apply.

Storage of Pyrophorics. Special storage requirements apply to pyrophorics that exceed exempt amounts. These include a limit of palletized storage of 10 feet by 10 feet by 5 feet high. Individual containers cannot be stacked. Also, aisle space between storage piles must be at least 10 feet. Tanks or containers inside a building cannot exceed 500 gallons. Exterior storage and other requirements are also outlined.

Storage of Unstable (Reactive) Materials. These materials must be stored in a detached building if the storage amounts exceed 2000 pounds of a class 3 material or 50,000 pounds of a class 2 material. Most class 4 materials also require detached storage. The floor must be liquidtight and there must be a means to vent smoke and heat in a fire or other emergency. The material should not be stored in piles greater than 500 cubic feet, and aisle width must be equal to the height of the piles or 4 feet, whichever is greater. Exterior storage allows for larger piles of 1000 cubic feet, with aisle widths of one-half the height of the pile or 10 feet, whichever is greater. Additionally, exterior storage of unstable materials should not be within 20 feet of any building, property line, or public way unless the storage area has an unpierced 2-hour fire-resistive barrier. If the material may deflagrate, the safe storage distance is 75 feet. Other requirements apply.

Storage of Water-Reactive Materials. Like unstable materials, these materials must be stored in a detached building if the storage amounts exceed 2000 pounds of a class 3 material or 50,000 pounds of a Class 2 material. The floor must be liquid tight and the room must be waterproof. External storage requirements include a safe storage distance of 20 feet or an unpierced 2-hour fire-resistive barrier, except for class 3, which requires 75 feet between the storage area and any building, property line, or public way. Other requirements include a means of venting smoke and heat, fire-extinguishing system requirements, secondary containment requirements, and others.

Storage of Highly Toxic Solids and Liquids. In some cases, exhaust scrubbers are required in areas that store these materials to ensure that vapors from an accidental release or spill are mitigated. Additionally, these materials should be isolated from other materials by 1-hour fire-resistive construction or stored in approved hazardous materials storage cabinets. Exterior storage piles of these materials cannot exceed 2500 cubic feet. Aisle widths between piles must be at least one-half the height of the pile or 10 feet, whichever is greater. Highly toxic liquids that liberate highly toxic vapors cannot be stored outside a building unless effective collection and treatment systems are provided. Additional requirements include specification of fire-extinguishing systems, safe distances from storage to exposures, and others.

Storage of Radioactive Materials. Areas used for the storage of radioactive materials must be provided with detection equipment suitable for determining surface-level contamination at levels that would pose a short-term hazard condition. All storage areas must be in compliance with the requirements of the Nuclear Regulatory Commission and state or local requirements. Additional requirements are defined including requirements for fire-extinguishing systems, safe distances from storage to exposures, and others. NFPA requirements for fire protection at these types of facilities are addressed in the next section.

Storage of Corrosives. General requirements for storage of corrosives include a liquidtight floor and adequate containment. Exterior storage requirements include a distance of 20 feet from storage to any building, property line, or public way or an unpierced 2-hour fire-resistive barrier. Additionally, secondary containment is required for outdoor storage.

Storage of Carcinogens, Irritants, Sensitizers, and Other Health Hazard Solids, Liquids, and Gases. A liquidtight floor and secondary containment is required for indoor storage of these materials. For exterior storage, requirements for safe distance from storage to exposure is 20 feet or a 2-hour fire-resistive barrier. Outdoor piles of this type of material cannot be larger than 2500 cubic feet. Aisle widths between piles must be one-half of the height of the piles or 10 feet, whichever is greater.

Fire Protection for Facilities Handling Radioactive Materials

Fire protection for facilities handling radioactive materials should be designed to minimize the spread of radioactive contamination during a fire incident. Design considerations include items such as ventilation, fire-water drainage, and other considerations. In addition, good administrative controls, which can aid in minimizing effects of fire at any facility, are especially important for facilities that handle radioactive materials.

NFPA Standards. Fire protection practices for facilities handling radioactive materials are documented by NFPA in the following standards:

- NFPA 801 (1998)—Recommended Fire Protection Practice for Facilities Handling Radioactive Materials
- NFPA 803 (1998)—Standard for Fire Protection for Light Water Nuclear Power Plants

These standards address key aspects of fire prevention and fire response, including fire protection systems and equipment, general facility design, administrative controls, and other topics.

Design Considerations. Several design considerations are particularly important to facilities that handle radioactive materials. These include:

- *Ventilation systems.* Ventilation systems in a facility that handles radioactive materials must include capabilities for heat removal, fire isolation, and filtration of radioactive gases and particles. Fresh-air inlets should be located to reduce the possibility of radioactive contaminants being introduced.

- *Duct systems.* Fire-resistant or fire-retardant materials should be used in ducts. Fire dampers should be provided (unless shutdown of ventilation system is not allowed) to resist spread of contaminated smoke.
- *Drainage systems.* Drainage systems should be designed to remove liquids to a safe area for testing and proper disposal. The liquid handling area should be designed to include the volume of a spill from the largest container, a 20-minute discharge from fire hoses or other suppressant system, and for outdoor applications, a typical amount of rain or snow. The drains should be designed to minimize fire hazard and radioactive contamination of clean areas (NFPA, 1998).
- *Fire detection system.* A fire detection system, such as a system that alarms at a fixed temperature or if there is an accelerated rate of rise in temperature, is important. This detection system will alert building occupants of fire conditions and allow for immediate evacuation. Additionally, the alarm can be used by management as a signal to implement the fire emergency plan rapidly.
- *Electrical power.* Electrical operations equipment including transformers, control panels, and main switches should be located well away from areas that handle radioactive and ignitible materials.
- *Other design considerations.* Other design considerations include emergency lighting for a means of egress, lightning protections, interior finish, and other considerations addressed as part of life safety management.

Administrative Controls. Administrative controls are a necessary part of the fire safety program. Included in a well-managed fire prevention program are items such as a documented fire prevention program, a fire hazard analysis, a fire emergency plan, and an adequate testing, inspection, and maintenance program.

If an emergency does occur that requires response and mitigation, proper personal protective equipment must be worn at all times, including face masks, clothing that prevents the entry of radioactive materials into the body, and a self-contained breathing apparatus. This protective equipment should be specified by a trained industrial hygienist or safety specialist.

LIFE SAFETY

In addition to fire protection, there are other safety standards that generally pertain to the category of life safety. These include standards for methods of egress, structural features of the building such as handrails and stairwells, floor loads, use of building materials, fire-resistant walls, and other life safety standards. These standards are based on occupancy classification. Two useful references on this subject include the *Life Safety Code* (NFPA 1997b) and the *Uniform Building Code* (ICBO 1994c).

Life Safety Code

The *Life Safety Code* (NFPA 1997b) and accompanying handbook, the *Life Safety Code Handbook* (NFPA 1997c), address requirements for life safety. Included is information pertaining to:

- *Classification of occupancy and hazard of contents.* Industrial, storage, and health care classifications are addressed along with other classifications that would not pertain to hazardous materials and hazardous waste management such as assembly, educational, residential, and other classifications.
- *Means of egress.* Means of egress components, arrangement of means of egress, discharge from exits, emergency lighting, marking of means of egress, and other topics are addressed.
- *Features of fire protection.* Included is information on construction and compartmentation, smoke barriers, special hazard protection, and interior finish.
- *Building service and fire protection equipment.* Information pertaining to utilities, ventilation, heating and air conditioning, and smoke control is provided. Other topics covered include elevators, escalators, conveyors, rubbish chutes, incinerators, and laundry chutes. Additionally, fire detection, alarm and communications systems, automatic sprinklers, and other extinguishing equipment is discussed.
- *New and existing occupancies (by occupancy classification).* Each occupancy classification is addressed in terms of new and existing occupancies. Included is information about general requirements, means of egress requirements, protection, special provisions, building services, and special requirements within the occupancy classification.

Divisions of Group H, Hazardous Occupancy, are presented in Table 7.6. The *Life Safety Code* has been in existence for more than 60 years and supplements NFPA's fire protection standards.

Uniform Building Code

The *Uniform Building Code* (ICBO 1994c) and its companion publication, the *Uniform Building Code Standards* (ICBO 1994d) also provide standards for building and life safety in terms of occupancy. These requirements include items such as lighting, ventilation, sanitation, sprinkler and standpipe systems, fire alarms, construction type, construction height, allowable area, and special hazards.

In addition to requirements for occupancy classifications, numerous other building standards are detailed. These include items such as fire-resistive standards for interior walls and ceilings, engineering regulations pertaining to

TABLE 7.6 Divisions of Hazardous Occupancies

- *Division 1:* occupancies with materials exceeding exempt limits, which present a high explosion hazard
- *Division 2:* occupancies where combustible dust is manufactured, used, or generated in such a manner that concentrations and conditions create a fire or explosion potential; occupancies with materials exceeding exempt limits, which present a moderate explosion hazard or a hazard from accelerated burning
- *Division 3:* occupancies where flammable solids, other than combustible dust, are manufactured, used, or generated; includes materials exceeding exempt limits that present a high physical hazard
- *Division 4:* repair garages not classified as Group S, Division 3 Occupancies
- *Division 5:* aircraft repair hangars not classified as Group S, Division 5 Occupancies and heliports
- *Division 6:* semiconductor fabrication facilities and comparable research and development areas in which hazardous production materials are used and the aggregate quantity exceeds exempt limits
- *Division 7:* occupancies with materials exceeding exempt limits that are health hazards

quality and design of the materials of construction, and requirements based on types of construction.

Legal Requirements for Fire Protection and Life Safety

NFPA fire standards, the *Uniform Fire Code,* the *Uniform Building Code,* the *Life Safety Code,* and equivalent standards are considered model codes. Some municipalities, however, have adopted these codes or parts of these or other codes into city ordinances or statutes. In this manner, these codes can become enforceable standards.

VENTILATION SYSTEMS

Ventilation is important in preventing buildup of toxic or flammable concentrations of chemicals stored or used in a process. Ventilation removes contaminants or flammable vapors from an area as well as providing fresh dilution air. The following paragraphs discuss the ventilation process in industrial and other applications.

Types of Air Contaminants

Air contaminants can be hazardous to health or life safety or cause nuisance conditions. Air contaminants can be in the form of gas, vapor, dust, fume, smoke, or mist. These terms have distinct meanings in the environmental, health, and safety field. Definitions published by the American Society of Safety Engineers (1988) are presented below:

- *Gas:* a state of matter in which the material has a very low density and viscosity, can expand and contract greatly in response to changes in temperature and pressure, easily diffuses into other gases, and readily and uniformly distributes itself throughout any container. A gas can be changed to the liquid or solid state only by the combined effect of increased pressure and decreased temperature (below the critical temperature).
- *Vapor:* the gaseous phase of a substance that is a liquid at normal temperature and pressure.
- *Dust:* suspended particles of solid matter (such as pollen or soot) in such a fine state of subdivision that they may be inhaled, swallowed, or absorbed in the body. Dusts do not diffuse in air but settle under the influence of gravity. Dust is a descriptive term for airborne solid particles that range in size from 0.1 to 25 μm. Dusts above 5 μm in size usually will not remain airborne long enough to present an inhalation problem.
- *Fume:* a gaslike emanation containing minute solid particles arising from the heating of a solid body such as lead, as distinct from a gas or vapor. This physical change is often accompanied by a chemical reaction, such as oxidation. Fumes flocculate and sometimes coalesce. Odorous gases and vapors should not be called fumes. As distinguished from dusts, fumes are finely divided solids produced by other methods of subdividing, such as chemical processing, combustion, explosion, or distillation. Some solids when heated to a liquid produce a vapor which, while it arises from the molten mass, immediately condenses to a solid without returning to its liquid state. Fumes are much finer than dusts, containing particles from 0.1 to 1 μm in size.
- *Smoke:* carbon or soot particles less than 0.1 μm in size that result from the incomplete combustion of carbonaceous materials such as coal or oil. An air suspension (aerosol) of particles, often originating from combustion or sublimation.
- *Mist:* fine liquid droplets or particles, measuring from 40 to 500 μm, suspended in or falling through the air, or a thin film of moisture condensed on a surface in droplets. Mist is generated by condensation from the gaseous to the liquid state, or by breaking up a liquid into a dispersed state by splashing, foaming, or atomizing. Examples in industrial operations are the oil mist produced during cutting and grinding, acid mists from electroplating, acid or alkali mists from pickling operations, and paint spray mists.

Ventilation Methods

There are several methods used to ventilate industrial buildings. Common methods include natural ventilation, dilution ventilation, and local exhaust systems.

Natural Ventilation. Natural ventilation is generally sufficient for warehouse buildings that house inert materials, nontoxic gases, and chemical handling articles such as sealed empty drums, spare pallets, and maintenance tools. In addition, small-volume, off-the-shelf chemicals such as lubricating oils and epoxies often are stored in cabinets inside a building of this design. Design considerations for a naturally ventilated building include building structure, surrounding structures, prevailing winds, and other climatological factors.

Dilution Ventilation. Dilution ventilation combines general exhaust with a supply air system and normally includes heat control, moisture control, and air conditioning. This type of ventilation is adequate for industrial applications that have low-level concentrations of chemical contaminants that are uniformly and widely dispersed and do not occur close to the worker's breathing zone. Design of this type of ventilation system takes into account:

- Rate of contaminant generation and concentration of vapors to be removed
- Room volume
- Effective volumetric airflow rate
- Incomplete mixing as dictated by placement of exhaust outlet and supply air inlet

Dilution ventilation requirements vary considerably from one chemical to the next, depending on permissible exposure limits as well as physical properties of a chemical. Examples of air replacement requirements for selected chemicals under uniform conditions are presented in Table 7.7.

Since requirements are varied from chemical to chemical, there is no general guideline for the number of air changes needed to keep chemical concentrations within acceptable limits. Typically, dilution ventilation systems are not adequate for exhausting some chemicals because of their low permissible exposure limits or because of high flammability. Examples include chloroform, carbon disulfide, and gasoline. In addition, dilution ventilation systems normally are not adequate for exhausting particulates or large point source emissions. For those chemicals or emissions that require an excessive number of air changes per hour to meet regulatory or other standards, local exhaust at the process or tool will be necessary.

The number of air changes required for offices, warehouses, and other nonchemical industrial applications will vary depending on the application. Ventilation requirements for areas such as general manufacturing space, machine shops, engine rooms, boiler rooms, and laundries should be evaluated by a ventilation specialist to ensure that persons working in the area have a healthful and comfortable environment.

For boiler rooms and laundries, which are particularly hot environments, ventilation systems that use exhaust and supply air systems typically are used

TABLE 7.7 Examples of Air Replacement Requirements for Dilution Ventilation Systems and Selected Contaminants

Contaminant	OSHA PEL (ppm)[a]	Minimum Air Changes/Hour[b]
n-Butyl acetate	150	11
Cyclohexane	300	7
Ethyl benzene	100	18
Ethyl ether	400	5
Isoamyl alcohol	100	19
Isopropyl alcohol	400	7
Methyl chloroform	350	6
n-Propyl acetate	200	9
Tetrahydrofuran	200	9
Turpentine	100	14

[a]The PEL is the permissible exposure limit based on a time-weighted average (TWA), which is the employee's average airborne exposure in an 8-hour work shift of a 40-hour workweek that must not be exceeded.

[b]Based on standard temperature and pressure, perfect dilution, an evaporation rate of 2 gal/hr, and a room 50 ft long by 50 ft wide by 12 ft high.

Source: PELs from 29 CFR §1910.1000, Table Z-1-A.

for heat control. The supply air is cooled and dehumidified to bring the inside air to an acceptable temperature and relative humidity. When possible, heat sources are enclosed and exhausted separately. Design considerations for these applications include temperature, humidity, airflow path, and potential convection and evaporative cooling effects. Other industries requiring specially designed ventilation systems for heat control include foundries, ceramic manufacturing, and coke processing.

Local Exhaust. Local exhaust is used to capture a chemical, particulate, mist, or other emission at or near its source before the contaminant disperses into the workplace. Local exhaust systems are composed of hoods, air cleaners, and fans. Hood design and hood placement are two of the most important aspects of local exhaust design. Several commonly used hood types are presented in Figure 7.3. Applications for these selected hood type are summarized in Table 7.8. ACGIH guidelines for hood capture velocities are presented in Table 7.9.

The hood should always be located as close to the source as possible, as required capture velocity increases exponentially with distance. Also, the exhaust system should not pull air across the employee's breathing zone, since this would negate the purpose of the system. To ensure proper operation of a local exhaust system, alarm systems can be installed at locally exhausted tools or operations that have specific ventilation requirements for safe working conditions. These systems generally use audible alarms and a flashing light

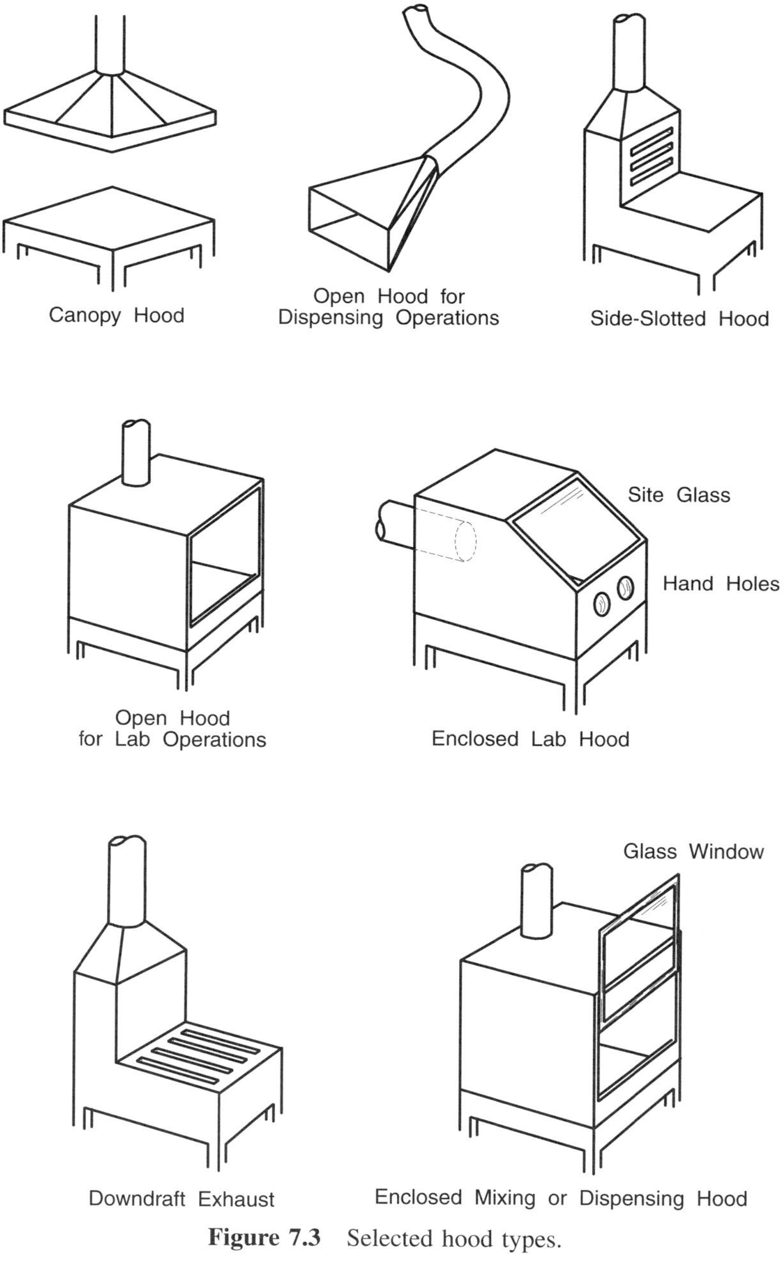

Figure 7.3 Selected hood types.

TABLE 7.8 Examples of Selected Hood Types and Industrial Applications

Hood Type	Typical Industrial Applications[a]
Canopy hood	Automated plating operations, degreasing operations, and hot processes
Enclosed hood or booth	Highly toxic or reactive chemical dispensing, lab operations involving infectious materials, radioactive material applications, extremely dusty operations, furnance or oven applications, spraying operations, high-agitation mixing operations
Slotted hood (side mounted)	Floor applications for operations involving fumes heavier than air, low-hazard bench operations
Open hood	Low- to medium-toxicity chemical dispensing operations, flammable chemical dispensing operations, lab operations
Slotted or open hood (downdraft)	Felting or brushing, operations, grinding operations, low-level dust producing operations

[a]Some of these applications may require respiratory protection.

(for particularly noisy areas) to indicate that the exhaust has fallen below the required velocity. In some cases the equipment can be interlocked to shut down at critical set points. In other cases, evacuation from the area is required until the problem can be resolved.

In addition to exhaust hood design and size, other design considerations (ACGIH 1992, Alden and Kane 1982) include:

TABLE 7.9 ACGIH Guidelines for Hood Capture Velocities

Dispersal Conditions	Capture Velocity (ft/min)	Application Examples
Contaminants released with practically no velocity into quiet air	50–100	Evaporation from tanks and open vessels
Contaminants released at low velocity into moderately still air	100–200	Spray booths, welding, plating, pickling
Contaminants released at considerable velocity or into zone of rapid air movement	200–500	Spray painting, barrel filling
Contaminants release at high initial velocity into zone of very rapid air movement	500–2000	Grinding, abrasive blasting

- Minimum duct velocity based on transport velocity needed for specific contaminants
- Branch duct size
- Friction losses throughout the system
- Airflow balancing
- Fan type and pressure ratings
- Variations in temperature and humidity
- Air abatement devices or contaminant collectors
- Type and location of stacks

Functional considerations include:

- Location of workers with respect to replacement air vents
- Location of workers with respect to exhaust systems
- Heat stress associated with the operation
- Clothing required to perform the operation safely (i.e., long sleeves, arm guards, aprons) and temperature needed to maintain worker comfort
- Local lighting requirements that could add additional heat to the operation

INDOOR AIR QUALITY

Since virtually all buildings utilize mechanical ventilation systems for temperature and humidity control of inside air, the maintenance of these systems for acceptable air quality in buildings is mandatory. As use of building space varies, the air supply system must meet needs and conditions specific to each application. Changes in building use from small-scale to large-scale production or from low occupancy to high occupancy may require ventilation system upgrades. If the ventilation system is inadequate for the building application, it might be evidenced in several ways, including dampened inside or outside walls, high-velocity drafts through windows or doors, stagnant air in rooms or hallways, unpleasant odors, or continually unpleasant temperatures.

In addition to ventilation system inadequacies for specific building use, indoor air quality problems can be caused by other factors. These factors include environmental tobacco smoke, carpet fabrics, carpet and floor tile glues, interior paints and varnishes, and use of janitorial supplies (DOE and EPA 1991). Examples of chemical constituents potentially found in a building and the sources of contamination are shown in Table 7.10.

Odors and chemical fumes from these sources can be picked up in the ventilation system and spread to all parts of the building. In cases where fume buildup becomes extreme, sick building syndrome can occur. Occupants of such a building might experience symptoms such as headaches, dizziness or light headedness, watery eyes, wheezing, or other allergic reactions.

TABLE 7.10 Selected Examples of Chemicals Potentially Found in an Office Building and the Source of Contamination

Contamination Source	Chemicals Potentially Found in an Office Building
Tobacco smoke	Carbon monoxide, carbon dioxide, oxides of nitrogen, ammonia, volatile N-nitrosamine, hydrogen cyanide, volatile hydrocarbons, volatile alcohols, volatile aldehydes, and ketones
Adhesives	Alcohols, ketones, and halogenated and aromatic hydrocarbons
Paints	Aromatic hydrocarbons, halogenated hydrocarbons, and aliphatic hydrocarbons
Varnishes and lacquers	Alcohols, toluene, and aromatic hydrocarbons
Fungicides, germicides, and disinfectants	Formaldehyde, other aldehydes, and ammonia
Carpet	Halogenated hydrocarbons, aliphatic hydrocarbons, and aromatic hydrocarbons
Caulk	Volatile organic compounds, halogenated organics, and aliphatic and aromatic hydrocarbons
Insulation	Volatile organic compounds and aromatic hydrocarbons
Cleaning products	Ammonia and aldehydes

Sources: Information from DOE and EPA (1991).

In addition to chemical fumes emitted from building materials, environmental tobacco smoke, and janitorial supplies, indoor air quality can be negatively affected by biological contaminants, including living organisms and the by-products of living organisms. Common biological contaminants include bacteria, fungi, pollen, molds, and algae. These contaminants can be toxic, pathogenic, or allergenic. Sources of these contaminants include stagnant drains and drip pans, uncleaned filters, and wet or humid surfaces.

Radon may also be a cause of indoor air quality problems. Radon is an inert radioactive gas that has been found to be present in numerous houses and buildings all over the nation. A maximum level of radon for indoor air has been established by the EPA at 4 picocuries per liter (pCi/l). Exposures over this amount are considered hazardous to the occupants of the building or house.

Radon can emanate from several different sources. These include soil and building products and materials. Radon exposure typically will vary within a house or building, with the largest exposure concentrations generally found in basements and first floors. Closed house conditions (i.e., during winter months for most states) also yield higher exposures (Belanger 1990).

Air in buildings that are occupied by a large number of people should be monitored on a periodic basis. Complaints by the occupants should be investigated to assure that proper air quality is maintained. The fresh air intake to the circulating system should be inspected visually on a regular basis, and any change of condition should be addressed immediately. Aspects that might negatively affect the intake air include proximity of load/unload areas that can disperse excessive vehicular exhaust, proximity to cooling towers that might spray treated water into the intake system, and proximity of shrubs and trees that are sprayed periodically with pesticides or fertilizers.

REFERENCES

Alden, J. L., and J. M. Kane (1982). *Design of Industrial Ventilating Systems,* 5th ed., Industrial Press, New York.

American Conference of Governmental Industrial Hygienists (1992) *Industrial Ventilation: A Manual of Recommended Practice,* 21st ed., ACGIH, Cincinnati, OH.

American Society of Safety Engineers (1988). *Dictionary of Terms Used in the Safety Profession,* ASSE, Des Plaines, IL.

Belanger, W. E. (1990). "Prediction of Long-Term Average Radon Concentrations in Houses Based on Short-Term Measurements," *Proceedings of the 1990 EPA/A&WMA International Symposium, Measurement of Toxic and Related Air Pollutants,* U.S. EPA and Air and Waste Management Association, Raleigh, NC.

Department of Energy and Environmental Protection Agency (1991). *Sick Building Syndrome: Sources, Health Effects, Mitigation,* M. C. Baechler, et al., eds., Noyes Data Corporation, Park Ridge, NJ.

International Conference of Building Officials (1994a). *Uniform Fire Code,* 1994 ed., ICBO, Whittier, CA.

——— (1994b). *Uniform Fire Code Standards,* 1994 ed., ICBO, Whittier, CA.

——— (1994c). *Uniform Building Code,* 1994 ed., ICBO, Whittier, CA.

——— (1994d). *Uniform Building Code Standards,* 1994 ed., ICBO, Whittier, CA.

National Fire Protection Association (1997a). *Fire Protection Guide to Hazardous Materials,* 11th ed., NFPA, Quincy, MA.

——— (1997b). *Life Safety Code,* NFPA 101, NFPA, Quincy, MA.

——— (1997c). *Life Safety Code Handbook,* NFPA, Quincy, MA.

——— (1997d). *Explosion Prevention Systems,* NFPA 69, NFPA, Quincy, MA.

——— (1997e). *Fire Protection Handbook,* 18th ed., NFPA, Quincy, MA.

——— (1998a). *Recommended Fire Protection Practice for Facilities Handling Radioactive Materials,* (1998a) NFPA 801, NFPA, Quincy, MA.

——— (1998b). *Standard for Portable Fire Extinguishers,* NFPA 10, NFPA, Quincy, MA.

BIBLIOGRAPHY

American Society for Testing and Materials (1989). *Design and Protocol for Monitoring Indoor Air Quality,* N. L. Nagda and J. P. Harper, eds., ASTM, Philadelphia.

——— (1990). *Biological Contaminants in Indoor Environments,* Morley and Feeley, eds., Special Technical Publication 1071, American Society for Testing and Materials, Philadelphia.

American Society of Heating, Refrigerating, and Air-Conditioning Engineers (1989). *Building Systems: Room Air and Air Contaminant Distribution*, L. L. Christianson, ed., ASHRAE, Atlanta, GA.

——— (1989). *Ventilation for Acceptable Indoor Air Quality,* ASHRE Standard 62, ASHRAE, Atlanta, GA.

Bond (1991). *Sources of Ignition,* Butterworth-Heinemann, Newton, MA.

Bonneville Power Administration (1984). *Ventilation in Commercial Buildings,* prepared by Seton, Johnson, & Odell, Inc., for Office of Conservation, BPA, Portland, Oregon.

Brooks, B. O., and W. F. Davis (1992). *Understanding Indoor Air Quality,* Lewis Publishers, Boca Raton, FL.

Burgess, W. A., M. J. Ellenbecker, and R. T. Treitman (1989). *Ventilation for Control of the Work Environment,* Wiley, New York.

Cherry (1988). *Asbestos Engineering, Management and Control,* Lewis Publishers, Boca Raton, FL.

Cothern, C. R., and J. E. Smith, Jr., eds. (1987). *Environmental Radon,* Plenum Press, New York.

Department of Energy (1987). *Indoor Air Quality Environmental Information Handbook: Building System Characteristics,* prepared by Mueller Associates, Inc., U.S. DOE, Washington, DC.

——— (1989). *Radon Research Program,* DE89 007284, Office of Health and Environmental Research, U.S. DOE, Washington, DC.

——— (1990). *Indoor Air Quality Issues Related to the Acquisition of Conservation in Commercial Building,* prepared by M. C. Baechler, D. L. Hadley, and T. J. Marseille of Pacific Northwest Laboratory, U.S. DOE, Washington, DC.

——— (1990). *Health Effects Associated with Energy Conservation Measures in Commercial Buildings,* Vol. 2, "Review of the Literature," prepared by R. D. Stenner and M. C. Baechler of Pacific Northwest Laboratory, U.S. DOE, Washington, DC.

Environmental Protection Agency (1988). *Indoor Air Quality in Public Buildings,* Vol. 1, prepared by L. S. Sheldon et al., U.S. EPA, Washington, DC.

——— (1988). *Indoor Air Quality in Public Buildings,* Vol. 2, EPA-600/6-88-009b, prepared by L. S. Sheldon et al., U.S. EPA, Washington, DC.

——— (1990). *Indoor Air-Assessment Methods of Analysis for Environmental Carcinogens,* EPA 600/8-90/041, Environmental Criteria and Assessment Office, Office of Health and Environmental Assessment, Office of Research and Development, U.S. EPA, Research Triangle Park, NC.

Environmental Protection Agency and National Institute for Occupational Safety and Health, (1991). *Building Air Quality,* ACGIH, Cincinnati, OH.

Fisk, W. J., et al. (1987). *Indoor Air Quality Control Techniques: Radon, Formaldehyde, Combustion Products,* Noyes Data Corporation, Park Ridge, NJ.

Godish, T. (1989). *Indoor Air Pollution Control,* Lewis Publishers, Boca Raton, FL.

Government Institutes (1991). *Fire Protection Management for Hazardous Materials: An Industrial Guide,* GI, Rockville, MD.

Heinsohn, R. J. (1991). *Industrial Ventilation: Engineering Principles,* Wiley, New York.

Kay, J. G., G. E. Keller, and J. F. Miller (1991). *Indoor Air Pollution: Radon, Bioaerosols, and VOCs,* Lewis Publishers, Boca Raton, FL.

Lao, K. Q. (1990). *Controlling Indoor Radon,* Van Nostrand Reinhold, New York.

National Academy of Science (1985). *Building Diagnostics: A Conceptual Framework,* NAS, Washington, DC.

National Council on Radiation Protection and Measurements (1989). *Control of Radon in Houses,* NCRP Report 103, NCRP, Bethesda, MD.

National Research Council (1986). *Environmental Tobacco Smoke: Measuring Exposures and Assessing Health Effects,* Committee on Passive Smoking, Board on Environmental Studies and Toxicology, NRC, National Academy Press, Washington, DC.

Nazaroff, W. W., and A. V. Nero, eds. (1987). *Radon and Its Decay Products in Indoor Air,* Wiley-Interscience, New York.

Stamper, E., and R. Koral (1979). *Handbook of Air Conditioning, Heating, and Ventilating,* 3rd ed., Industrial Press, New York.

8

WORKPLACE MANAGEMENT OF RADIATION EXPOSURE

As a result of the proliferation of electronic devices, there has been increased public attention on the effects of ionizing and nonionizing radiation. The Radiation Control for Health and Safety Act of 1968 covers both ionizing and nonionizing radiations emitted from any electrical product. The Occupational Safety and Health Administration's requirements pertaining to radiation are promulgated under 29 CFR §§1910.96 and 1910.97 for ionizing and nonionizing radiation, respectively. Additionally, radioactive materials and wastes are regulated by the Nuclear Regulatory Commission and the U.S. Environmental Protection Agency.

Nonionizing radiation is radiation within the electromagnetic spectrum, which includes ultraviolet and infrared radiation, radio frequencies, microwaves, and lasers. Ionizing radiation includes alpha, beta, and gamma rays; X-rays; neutrons; high-speed electrons and protons; and other atomic particles. In this chapter we discuss radiation hazards and methods of protection against these hazards.

NONIONIZING RADIATION

Nonionizing radiation—radiation within the electromagnetic spectrum—can be defined as radiation with sufficient energy to excite atoms or electrons, but with insufficient energy to remove electrons from an atom or to cause the formation of ions. Even though nonionizing radiation does not cause the formation of ions, there is still a potential health hazard to employees who are exposed to this type of radiation in the work area. Types of nonionizing radiation that are typically found in the workplace include ultraviolet and infrared radiation, high-intensity visible light, radio frequencies, microwaves, and lasers.

Ultraviolet Radiation

Sources of Exposure. Although the primary source of ultraviolet (UV) radiation is the sun, this type of radiation is also found in the workplace. Sources include fluorescent lighting, welding operations, plasma torches, and laser operations. In water treatment operations and other sterilization applications, UV light is used as a germicide for bacteria and molds. It is also used in wastewater treatment operations—typically in combination with other treatment methods such as ozonation—for destruction of chlorinated compounds and pesticides.

Hazards. The most critical range of UV radiation is from 240 to 320 nanometers (nm). This is the range at which UV radiation has the highest biological impact, the primary biological effect being on the skin and eyes. Symptoms are dependent on skin type, dose, and time of exposure, and include reddening of skin, the formation of blisters, and peeling of skin. Once the source is removed, the symptoms subside. Exposure to continuous ultraviolet radiation results in increased pigmentation in the upper layer of the skin. The American Conference of Governmental Industrial Hygienists (ACGIH) has published threshold limit values (TLVs) for ultraviolet radiation in the spectral region between 180 and 400 nm (ACGIH 1997). It is presumed that continuous exposure over a long period of time (i.e., many years) has the potential to cause skin cancer. Additionally, skin aging may be a consequence of continuous exposure to UV radiation over a long period of time.

Exposure of ultraviolet radiation to the eyes can occur during welding operations. If the exposure is above the TLV, the worker may experience inflammation of the conjunctiva or the cornea if proper protective equipment is not used. The eye does not build up a protective layer, so symptoms may recur with each overexposure. In addition to inflammation of the eye, exposure to ultraviolet radiation may cause blurred or decreased visual acuity. These symptoms are temporary and will disappear when exposure ceases.

Worker Protection. When protecting against ultraviolet radiation, exposure time and shielding are the main considerations. The shorter the exposure, the less effect the ultraviolet radiation will have on the worker. When exposure is such that shielding is required, personal protective clothing for skin protection might include gloves, a long-sleeved shirt, coveralls, and a face shield. Eye protection will include filter lenses of varying shade numbers, depending on the operation. For example, industry standards recommend that filter lenses of shade number 3 or 4 be used for torch brazing and soldering; filter lenses of shade numbers 5 or 6 be used for oxygen cutting, medium gas welding, and arc welding up to 30 amperes (A); and filter lenses of shade number 6 or 8 be used for heavy gas welding and arc welding and cutting from 30 to 75 A. Additionally, flash goggles should be worn under all arc-welding helmets (Talty 1988).

Infrared Radiation

Sources of Exposure. Exposure to infrared (IR) radiation can occur from any surface that is of a higher temperature than the receiver. Sources of IR radiation include heat-producing operations such as drying/baking operations, heating of metal parts, dehydration of materials, and other heating operations that use a furnace or oven.

Hazards. IR radiation is perceptible to the worker in the form of heat to the skin. Exposure in the spectral range of 750 nm to 1.5 μm can cause acute skin burns. Additionally, IR radiation in the shorter-wavelength region can cause damage to the eye, particularly to the cornea, iris, retina, and lens. In particular, IR radiation in the range 400 to 700 nm emits a bluish light, and chronic exposure to this blue light can cause retinal injury. ACGIH has published TLVs for IR radiation in the range 760 nm to 1.4 μm.

Worker Protection. IR radiation is best controlled by administrative controls and personal protective equipment. Exposure to IR radiation should be limited and eye protection should be worn to reduce IR radiation levels.

High-Intensity Visible Light

Sources of Exposure. Sources include welding, carbon arc lamps, and some lasers.

Hazards. High-intensity visible light can cause injury to the eye if energy levels associated with the light are high enough.

Worker Protection. Typical controls for this hazard include shielding or filtering the source from the eye and enclosure of the source.

Radio Frequencies

Sources of Exposure. Radio frequencies can induce electrical currents in conductors, and they can induce the displacement of current in semiconductors. The first application, which results in the transfer of patterned energy, is used for radio and television. The second application allows radio frequency to be used as a heat source. Radio-frequency heating is used in industry for hardening of metals, bonding and laminating, rubber vulcanization, plastics molding, and other applications.

Hazards. The most common hazard associated with exposure to radio frequencies is the thermal effect on the body. A temperature increase can harm the eye lens, male reproductive capability, and central nervous system. Nonthermal effects from exposure to radio frequencies include the potential for demodulation of heart and central nervous system activity and the potential for enhanced reactions to certain types of chemicals.

Worker Protection. TLVs for this type of radiation are established and exposure should be monitored in areas where the TLV could be reached or exceeded. Warning signs might also be needed.

Microwaves

Sources of Exposure. Microwave frequencies have been established by the Federal Communications Commission. Typical uses (sources of exposure) include television, satellite communication, radar, and radio astronomy. In the workplace, microwave ovens are typical sources of exposure.

Hazards. Biological effects attributed to microwaves consist of thermal heating.

Worker Protection. TLVs for this type of radiation are established and exposure should be monitored in areas where the TLV could be reached or exceeded.

Lasers

Sources of Exposure. The word *laser* is an acronym for "light amplification by stimulated emission of radiation." Lasers are used for numerous purposes, including alignment, welding, trimming, fiber optics communication systems, surgical applications, and other purposes. A laser can use ultraviolet, infrared, visible, or microwave radiation, and produces a concentrated light beam that is coherent, monochromatic, and typically of high power density. The three elements of a laser include:

- An optical cavity consisting of at least two mirrors, one of which is partially transmissive
- An active laser medium that can be excited from an unenergized ground state to a relatively long-lived excited state
- A means of "pumping" or supplying the excitation energy to the active laser medium

Examples of types of lasers are presented in Table 8.1.

Hazards. The primary hazard from laser radiation is exposure to the eye and, secondarily, the skin. If radiation levels are kept below those that damage the eye, there will be no harm to other body tissues. The hazards associated with a laser depend on several factors, including wavelength and intensity of beam,

TABLE 8.1 Types of Lasers

Solid Host Lasers

- *Ruby* (*chromium*): pulsed operation, with up to 1000 megawatts (MW) per pulse and 1 watt (W) of power (continuous wave)
- *Neodymium YAG:* rapid pulsed operation with 10 MW per pulse and 100 W of power (continuous wave)
- *neodymium glass:* rapid pulse operations with 10 MW per pulse and 100 W of power (continuous wave)

Gas Lasers

- *Helium neon* (*neutral atom*): continuous wave of up to 100 milliwatts (MW)
- *Argon* (*ion gas*): continuous wave of 1 to 20 W
- *Carbon dioxide* (*molecular gas*): continuous wave of 10 to 5000 W
- *nitrogen* (*molecular gas*): pulsed or rapid pulsed operation with 250 mW of power (continuous wave)

Other Lasers

- *Gallium* (*semiconductor diode*): pulsed or continuous wave operation with 1 to 20 W power per pulse
- *Hydrogen chloride* (*chemical laser*): continuous wave or pulsed operation
- *Organic dye laser* (*liquid laser*): continuous wave or pulsed operation with power per pulse of 1 MW

Source: Information from Talty (1988).

duration of exposure, body part exposed, and means of exposure. NIOSH has provided information about laser class, potential hazard, and requirements by class, as detailed in Table 8.2.

Worker Protection. Special requirements, by class, are presented in Table 8.2. In addition, there are several controls, both engineering and administrative, that can further protect the employee from hazards of lasers. These include:

- Provide interlocks on equipment, as appropriate.
- Ensure proper employee training.
- Ensure proper signing of area and equipment.
- Ensure that surfaces surrounding lasers are nonreflective.
- Ensure proper shielding from beam, as appropriate.
- Ensure that beam direction is controlled.
- Ensure flammable solvents and combustible materials are stored away from lasers.
- Ensure protective eyewear is prescribed and used.

TABLE 8.2 Laser Classifications and Requirements

Class I

- *Potential hazard:* incapable of creating biological damage.
- *Special requirements:*none.

Class II

- *Potential hazard:* low power; beam may be viewed directly under carefully controlled conditions.
- *Special requirements:* posting of signs in area; control of beam direction.

Class III

- *Potential hazard:* medium power; beam cannot be viewed.
- *Special requirements:* well-controlled area; no specular surfaces; terminate beam with diffuse material and minimum reflection; eye protection for direct beam viewing.

Class IV

- *Potential hazard:* high power; direct and diffusely reflected beam cannot be viewed or touch the skin.
- *Special requirements:* restricted entry to facility; interlock; fail-safe system; alarm system; panic button; good illumination of at least 150 footcandles; light-colored diffuse room surfaces; operated by remote control; designed to reduce fire hazard, buildup of fumes, etc.

Class V

Classes II, III, and IV which are completely enclosed so that no radiation can leak out.

Source: Information from Talty (1988).

- Perform periodic medical examination—in particular eye exams—on personnel working on or near lasers, as appropriate.

Exposure limits developed by the American National Standards Institute (ANSI) are published in the ANSI Z136. They include exposure limits for direct ocular exposure intrabeam viewing, viewing a diffuse reflection of a laser beam or an extended-source laser, and exposure of skin to a laser beam.

Extremely Low Frequency Radiation

Sources of Exposure. Sources of exposure to extremely low frequency (ELF) radiation include household appliances, wiring in the home, and power transmission sources.

Hazards. Studies have indicated that exposure to ELF radiation can cause changes in heartbeat and respiration, and that chronic exposure to ELF magnetic fields may increase the risk of cancers such as leukemia and central nervous system tumors (Clayton and Clayton 1991). More research on the effects of ELF radiation is warranted and is being conducted.

Worker Protection. There are no national standards pertaining to exposure to ELF radiation, but guidelines have been generated for medical exposures to magnetic fields (including exposure from MRI devices) and for exposures to static magnetic fields.

IONIZING RADIATION

Ionizing radiation can be defined as any electromagnetic or particulate radiation with enough energy to produce ions when it interacts with atoms and molecules. The main types of ionizing radiation include X-rays, gamma rays, alpha particles, beta particles, neutrons, high-speed electrons and protons, and other atomic particles. Uses include radiography, tracing of flow of materials in pipes, sterilization of food and medical supplies, use in nuclear reactors, and use as cross-linking agents to improve the properties of plastics (National Safety Council 1988.)

Alpha Particles

Definition. Alpha particles originate in the nuclei of radioactive atoms during the process of disintegration. An alpha particle is a high-energy particle composed of two protons and two neutrons. It has a mass of four atomic mass units (amu) and a charge of +2. The energy level of an alpha particle varies, but can be as high as 10 million electron volts (MeV).

The alpha particle causes more ionization than a does a beta particle or gamma radiation in the absorbing material; however, because of its large mass and positive charge, the alpha particle can travel only short distances.

Hazards. Once in the body, alpha particles can concentrate in bone, lungs, kidney, and liver, and can cause tissue and other internal damage.

Worker Protection. Alpha particles can be easily shielded with paper-thin materials. As a result, the danger of exposure from alpha-particle radiation is through respiration and ingestion rather than through penetration of the skin.

Beta Particles

Definition. A beta particle is an electron emitted by the nucleus during radioactive decay. The electrons can be positively or negatively charged. The mass of these particles is insignificant. The energy level of a beta particle typically ranges between 0.017 and 4 MeV.

Hazards. These particles have the potential to form secondary gamma and x-radiation, which make them more hazardous than the alpha particle.

Worker Protection. As a result of its smaller mass, the penetration capability of the beta particle is greater than that of the alpha particle but less than that of gamma rays and X-rays. Shielding materials for beta particles must be more substantial than those used for alpha particles, and typically include aluminum and other materials of low atomic weight

Gamma Radiation

Definition. Gamma radiation is not a particle; instead, it is energy waves in the electromagnetic spectrum. Gamma rays are emitted by the nucleus of certain radionuclides during their decay scheme. The energy levels of gamma radiation can range between 0.008 and 10 MeV. There are approximately 240 radioactive isotopes, or radionuclides, that can undergo radioactive decay to reach a more stable energy level. The rate of this decay is measured in terms of a *half-life,* the amount of time needed for half of the active atoms of any given quantity to decay. A pictorial view of the decay of a radioactive material at 100 millirems per hour (mR/h) is shown in Figure 8.1. As can be seen from the figure, after 7 half-lives, the material has decayed to less than 1%.

Hazards. The process of ionization in the tissue may cause irreparable damage to living cells. Since gamma radiation has no mass or charge and is a wave, it has the capability for deep penetration and presents a greater hazard in the workplace than either alpha or beta particles.

Worker Protection. Shielding to protect against gamma radiation requires dense material such as lead and iron. Shield thickness is calculated based on shield material, radiographic source (e.g., cobalt-60 or radium-226), and distance. Standard tables provide information about half-value thicknesses for typical materials and sources of radiation.

X-Radiation

Definition. X-radiation is also electromagnetic radiation. It is similar in properties to gamma radiation but does not have as high an energy level as gamma radiation. X-radiation is formed with high-speed electrons are slowed down or stopped. In the process of slowing down, the electrons give up energy in the form of X-radiation. The quantity of energy released depends on the speed of the electron and the characteristics of the medium or striking target. Typ-

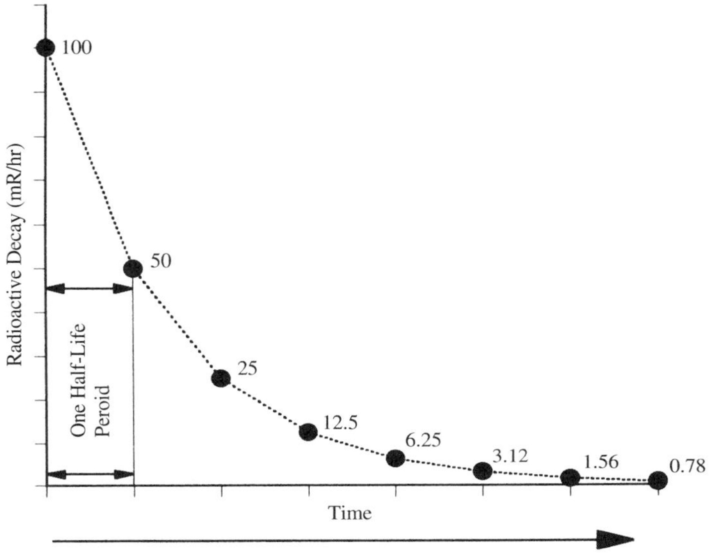

Figure 8.1 Graph of decay of radioactive material.

ically, X-radiation is produced by a machine, although some X-radiation can occur from radioactive decay.

Hazards. X-radiation can cause severe tissue damage.

Worker Protection. Like gamma radiation, shielding to protect against X-radiation requires dense material and should be calculated carefully by an expert.

Radiation Protection Program

Elements to include in a radiation protection program will vary from company to company, depending on size and number and type of radiation sources. Typical elements to include in a program are:

- Program documentation and administration
- Training on safe work practices
- Training on emergency procedures
- Training on warning symbols and signs
- Control measures
- Surveys and monitoring
- Medical surveillance

- Record keeping
- Incident reporting

An excellent source of information that is useful to the safety engineer when setting up a radiation protection program is the National Council on Radiation Protection and Measurements (NCRP). The organization has published over 100 reports on all topics of radiation, including radiation exposure, radiation measurement and assessment, radiation protection, engineering and administrative controls, and radiation in the medical and health fields. A listing of selected publications from NCRP is presented in Appendix C.

Warning Symbol. OSHA requires a standard symbol to depict radiation hazards. That symbol is shown in Figure 8.2.

Record Keeping. The need for accurate records related to radiation safety cannot be overstated. Examples of records that might be maintained as part of a radiation protection program are presented in Table 8.3.

Surveys and Monitoring. An effective radiation protection program must include assessment of radiation exposure in the area. Typical monitoring instruments include film badges, thermoluminescence detectors, dosimeters,

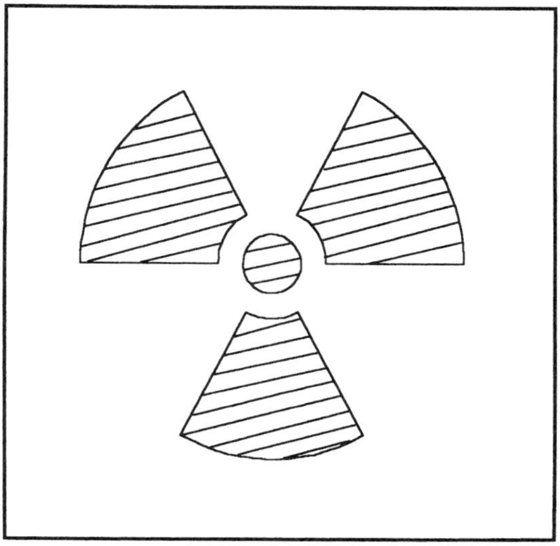

Note: Crosshatched area is to be magenta or purple; background is to be yellow. Size and angle requirements are given in 29 CFR 1910.96.

Figure 8.2 Warning label for radioactive materials.

TABLE 8.3 Examples of Records Related to Radiation Safety That Might Be Maintained as Part of a Radiation Protection Program

Radiation Protection Program Records

- Authorizing documents
- Accreditations and certifications, such as accreditation of dosimetry, hospital accreditations, laboratory accreditation, and accreditation of nuclear plant training program
- Guidance documents
- Records detailing qualifications and positions and training of employee in the radiation protection program
- Written radiation protection program documents, such as radiological training program, instrument calibration program, personal protective equipment program, exposure and assessment programs, ALARA programs, and engineering control programs

Individual Records

- Exposure categories for individuals, such as occupational exposure, occasional exposure, management visitors, and other categories
- Personal data
- External dosimetry assessment records
- Internal dosimetry records
- Occupational exposure history
- Abnormal exposures
- Training records

Workplace Records

- Facility description, including design features relevant to radiation protection, potential affected populations, emission and release point, and other information
- Records of controlled areas
- Documents detailing ventilation and exhaust systems
- Access control points
- Radiation work permits or other work authorizations
- Area radiation and contamination records
- List and description of personal protective equipment
- ALARA documentation
- Records of radioactive material shipments
- Inventory of radioactive material
- Records of accidents and incidents

TABLE 8.3 (*Continued*)

Environmental Records

- Preoperating monitoring program, including climatic, topographic, land use, demographic studies, and radiological surveillance records
- Operation of environmental monitoring program and records
- Reports of radioactive releases and dose assessment
- Reports from off-site investigations and special studies

Radiation Protection Instrumentation Records

- Equipment specifications, such as type of detector, energy, range, sensitivity, and other information
- Records pertaining to calibration facility and source certification
- Calibration records for each instrument
- Maintenance records
- Instrument inventory

Source: NCRP (1992).

Geiger–Müller counters, radon monitors, and ionization chambers. Monitoring should be performed on a specified schedule and equipment should be calibrated according to a schedule recommended by the manufacturer.

Incident Reporting. Employers must notify the proper authorities when there is any incident involving radiation exposure that exceeds specified guidelines. An effective way to ensure that notifications are made promptly is for the safety engineer to develop a matrix that provides information about:

- Exposure levels that require notification to proper authorities
- Statutory time requirements for reporting the incident
- Name of agency, contact person, phone number, and any other information needed to make the incident notification
- Name and phone number of facility security contact
- Name and phone number of communications contract
- Sample information to report, such as date, time, facility name and location, description of incident, threat to surrounding areas, incident response, and other information
- Other pertinent information

Once an incident is reported, there is usually a requirement for a follow-up report, which then becomes part of the public record.

REFERENCES

American Conference of Governmental Industrial Hygienists (1997). *1997–1998 Threshold Limit Values for Chemical Substances and Physical Agents and Biological Exposure Indices,* ACGIH, Cincinnati, OH.

Clayton, G. D., and F. E. Clayton, eds. (1991). *Patty's Industrial Hygiene and Toxicology: General Principles,* Vol. 1, Part B, Wiley, New York.

National Council on Radiation Protection and Measurements, (1992). "Maintaining Radiation Protection Records," NCRP Report 114, NCRP, Bethesda, MD.

National Safety Council (1988). *Fundamentals of Industrial Hygiene,* 3rd ed., NSC, Chicago.

Talty, J. T., ed. (1988). *Industrial Hygiene Engineering: Recognition, Measurement, Evaluation and Control,* prepared for NIOSH, Noyes Data Corporation, Park Ridge, NJ.

BIBLIOGRAPHY

American National Standards Institute (1983). "Personnel Dosimetry Performance Criteria for Testing," ANSI N13.11, ANSI, New York.

American Petroleum Institute, (1988). *Naturally Occurring Radioactive Material (NORM) in Oil and Gas Production Operations,* API, Washington, DC, videotape.

——— (1991). *The Status of ANSI N13.11: The Dosimeter Performance Test Standard,* DE91 017874, prepared by C. S. Sims, Oak Ridge National Laboratory, Oak Ridge, TN.

Department of Energy, (1988). *Review of EPA, DOE, and NRC Regulations on Establishing Solid Waste Performance Criteria,* DE88 015331, ORNL/TM-9322, prepared by A. J. Mattus, T. M. Gilliam, and L. R. Dole, Oak Ridge National Laboratory, Oak Ridge, TN.

——— (1993). *Electric Power Lines: Questions and Answers on Research into Health Effects,* DE94 005637, Bonneville Power Administration, Portland, OR.

Knoll, G. F. (1989). *Radiation Detection and Measurement*, 2nd ed., Wiley, New York.

Oak Ridge Associated Universities, (1988). *A Compendium of Major U.S. Radiation Protection Standards and Guides: Legal and Technical Files,* prepared by W. A. Mills et al., ORAU, Oak Ridge, TN.

9

ADMINISTRATIVE REQUIREMENTS FOR PROPER MANAGEMENT OF HAZARDOUS MATERIALS AND HAZARDOUS WASTE

There are numerous administrative requirements pertaining to proper management of hazardous materials and hazardous waste specified by governmental agencies. These requirements include proper container labeling, periodic training, container and containment inspections, plans and controls, material/waste tracking and scheduled reporting, and release or other noncompliance reporting. In this chapter we discuss administrative requirements that are defined in the regulations. In addition, management of these requirements through computer applications is addressed.

REGULATORY STANDARDS

Proper management of hazardous materials and hazardous waste has been defined through numerous administrative requirements that are applicable to manufacturing and waste management facilities. These requirements have been published in regulations promulgated by various agencies including OSHA, the Nuclear Regulatory Commission (NRC), the Department of Transportation (DOT), and EPA. EPA requirements have been promulgated under several acts, including the Clean Water Act (CWA), Clean Air Act (CAA), Toxic Substance and Control Act (TSCA), Comprehensive Environmental Response, Compensation, and Liability Act (CERCLA), and Superfund Amendments and Reauthorization Act (SARA). In addition, hazardous waste management requirements for treatment, storage, and disposal facilities (TSD) have been promulgated under the Resource Conservation and Recovery Act (RCRA). A summary of key regulatory standards that address administrative requirements is presented in Table 9.1.

TABLE 9.1 Regulatory Standards Applicable to Manufacturing and Waste Management Facilities

Labeling

10 CFR §20.203—Caution Signs, Labeling, Signals, and Controls for Radiation Areas (NRC)
10 CFR §61.57—Labeling of Low-Level Radioactive Waste (NRC)
29 CFR §1910.96—Warning Labels for Radioactive Materials
29 CFR §1910.134—Labeling (and Color Assignment) for Gas Mask Canisters
29 CFR §§1910.1001(j)(1-2) and 1926.58—Labeling of Asbestos Abatement and Project Areas (OSHA)
29 CFR §1910.1200(f)—Chemical Labeling (OSHA)
40 CFR §§262.31–262.34—Waste Labeling (RCRA)
40 CFR §761.40—Marking of PCBs and PCB Items (TSCA)
40 CFR §763.121(k)—Labeling of Asbestos Abatement Project Areas (TSCA)
49 CFR Part 172, Subpart D—Marking of Hazardous Materials Packages (DOT)
49 CFR Part 172, Subpart E—Package or Container Labeling (DOT)
49 CFR Part 172, Subpart F—Placarding (DOT)

Training

10 CFR §19.11—Posting of Notices to Workers (NRC)
10 CFR §19.12—Instructions to Workers (NRC)
10 CFR Part 72, Subpart I—Training of Personnel Working in Spent Fuel and High-Level Radioactive Waste Areas (NRC)
29 CFR §1910.119(g)—Training for Workers Involved with Processes Storing Highly Hazardous Chemicals
29 CFR §§1910.120(e) and 1910.120(p)—Training for Hazardous Waste Operations and Emergency Response (OSHA)
29 CFR §1910.1200(h)—Training Requirements in the Hazard Communication Standard (OSHA)
40 CFR §61.145—Asbestos NESHAP Training for Handling of Regulated Asbestos-Containing Material (CAA)
40 CFR §264.16—Hazardous Waste Handling Training for Permitted TSD Facilities (RCRA)
40 CFR §763.92—Training and Periodic Surveillance for Asbestos-Containing Materials in Schools (TSCA)
49 CFR 172, Subpart H—Training of Individuals Involved in the Transportation of Hazardous Materials (DOT)

Inspections

10 CFR §19.14—Presence of Representatives of Licensees and Workers During Inspections (NRC)
10 CFR §61.82—Commission Inspections of Low-Level Waste Land Disposal Facilities (NRC)
40 CFR §264.15—General Facility Inspections for Permitted TSD Facilities (RCRA)

TABLE 9.1 ***(Continued)***

40 CFR §264.174—Hazardous Waste Container Inspections for Permitted TSD Facilities (RCRA)
40 CFR §264.191—Tank Assessments/Certifications for Permitted TSD Facilities (RCRA)
40 CFR §264.195—Hazardous Waste Tank Inspections for Permitted TSD Facilities (RCRA)
40 CFR §264.226—Hazardous Waste Surface Impoundment Inspections for Permitted TSD Facilities (RCRA)
40 CFR §264.254—Hazardous Waste Pile Inspections for Permitted TSD Facilities (RCRA)
40 CFR §264.347—Hazardous Waste Incinerator and Associated Equipment Inspections for Permitted TSD Facilities (RCRA)
40 CFR §264.574—Hazardous Waste Drip Pad Inspections for Permitted TSD Facilities (RCRA)
40 CFR §264.602—Hazardous Waste Miscellaneous Unit Inspections for Permitted TSD Facilities (RCRA)
40 CFR §280.43-44—Release Detection Requirements for Underground Storage Tank Systems (RCRA)
40 CFR §720.122—Inspections Conducted by EPA for TSCA Information Verification (TSCA)
40 CFR §761.65—PCB Inspections (TSCA)

Plans and Controls

10 CFR §19.13—Notification and Reports to Individuals (NRC)
10 CFR Part 20, Subpart I—Storage and Control of Licensed Material (NRC)
10 CFR Part 20, Subpart K—Waste Disposal (NRC)
10 CFR §61.12(j)—Quality Control Program for Low-Level Radioactive Waste Disposal Facilities (NRC)
10 CFR §61.53—Environmental Monitoring (NRC)
10 CFR Part 61, Subpart E—Financial assurances (NRC)
29 CFR §1910.120(a)–(q) Plans and Controls for Emergency Response and Hazardous Waste Operations (OSHA)
29 CFR §1910.1200(e)—Hazard Communication Plan (OSHA)
40 CFR Part 112—Spill Prevention, Control, and Countermeasure Plan (CWA)
40 CFR §264.13—Waste Analysis Plan for Permitted TSD Facilities (RCRA)
40 CFR §264.99—Compliance Monitoring Program for Permitted TSD Facilities (RCRA)
40 CFR §264.100—Corrective Action Program for Permitted TSD Facilities (RCRA)
40 CFR §264.112—Facility Closure Plan for Permitted TSD Facilities (RCRA)
40 CFR §264.118—Facility Post-closure Plan for Permitted TSD Facilities (RCRA)
40 CFR Part 264, Subpart D—Emergency and Contingency Plan for Permitted TSD Facilities (RCRA)
40 CFR Part 264, Subpart H—Proof of Financial Assurance for Permitted TSD Facilities (RCRA)
40 CFR Part 280, Subpart H—Proof of Financial Assurance for Underground Storage Tank Systems (RCRA)

TABLE 9.1 ***(Continued)***

Material/Waste Tracking, Reporting, and Record Keeping

10 CFR Part 20, Subpart L—Nuclear Waste Tracking/Recordkeeping (NRC)
10 CFR §60.71—Records of Receipt, Handling, and Disposition of Radioactive Waste at a Geologic Repository (NRC)
10 CFR §61.80—Reports of Radioactive Waste Acceptance at Land Disposal Facilities (NRC)
10 CFR Part 74—Material Control and Accounting of Special Nuclear Material (NRC)
40 CFR §61.145—Asbestos Abatement Project Notification (CAA)
40 CFR §259.72—Transporter Notification for Medical Waste Handling in Covered States (RCRA)
40 CFR §259.78—Transporter Report of Medical Wastes Handled in Covered States (RCRA)
40 CFR §§264.71 and 264.72—Manifest Requirements for Permitted TSD Facilities (RCRA)
40 CFR §264.73 and Appendix I—Operating Record for Permitted TSD Facilities (RCRA)
40 CFR §264.75—Biennial Reporting Including Waste Minimization Reporting for Permitted TSD Facilities (RCRA)
40 CFR §264.76—Unmanifested Waste Report for Permitted TSD Facilities (RCRA)
40 CFR §268.7—Waste Analysis and Recordkeeping Associated with Land Disposal Restrictions (RCRA)
40 CFR §280.34—Release Detection Recordkeeping Requirements for Underground Storage Tank Systems (RCRA)
40 CFR §370.25—Inventory Reporting—Tiers I and II (SARA)
40 CFR §372.30—Toxic Chemical Release Reporting (SARA)
40 CFR Part 716—Health and Safety Data Reporting (TSCA)
40 CFR Part 717—Records and Reports of Allegations That Chemical Substances Cause Significant Adverse Reactions to Health or the Environment (TSCA)
40 CFR Part 720—Premanufacturing Notification (TSCA)
40 CFR Part 761, Subpart K—PCB Waste Disposal Records and Reports (TSCA)
40 CFR §761.208—Use of Manifest for PCB Waste (TSCA)
40 CFR Part 763, Subpart G—Asbestos Abatement Project Notification (TSCA)
49 CFR §172.205—Hazardous Waste Manifest Requirements (DOT)

Release and Other Noncompliance Reporting

10 CFR §20.2201—Reports of Theft or Loss of Licensed Material (NRC)
10 CFR §20.2202—Notification of Incidents (NRC)
10 CFR §20.2203—Reports of Exposures, Radiation Levels, and Concentrations of Radioactive Material Exceeding the Limits (NRC)
10 CFR §21.21—Reporting of Defects and Nnoncompliance (NRC)
40 CFR Part 61—Reporting of Noncompliance with NESHAP Standards (CAA)
40 CFR §117.3—Noncompliance Reporting (CWA)
40 CFR §264.56—Release Reporting (RCRA)
40 CFR §270.5—Noncompliance Reporting by the Director (RCRA)

TABLE 9.1 ***(Continued)***

40 CFR §280.50—Reporting of Suspected Releases from Underground Storage Tank Systems (RCRA)
40 CFR §280.53—Reporting and Cleanup of Spills and Overfills Associated with Underground Storage Tank Systems (RCRA)
40 CFR §302.6—Release Reporting (CERCLA)
40 CFR §355.40—Release Reporting (SARA)
49 CFR §171.15—Reporting Hazardous Materials Incidents (DOT)
Standards Applicable to Nonpermitted Waste Management Facilities or Facilities Under Interim Status (RCRA)
40 CFR §262.34—General Requirements for Facilities That Accumulate Waste for Less Than 90 Days
40 CFR Part 265—RCRA Standards for Nonpermitted Facilities

LABELING

Chemical Labeling

Chemical labeling requirements under OSHA specify that the chemical manufacturer, importer, or distributer must ensure that each container of hazardous chemicals is tagged or marked with the following information:

- Identity of the hazardous chemical
- Appropriate hazard warnings
- Name and address of the chemical manufacturer, importer, or other responsible party

Manufacturers may put additional information on the label, such as the DOT shipping label, as appropriate.

It is the responsibility of the manufacturer to ensure that all chemical containers are labeled properly before shipment. The employer has the responsibility to ensure that all chemical containers in the workplace identify the hazardous chemicals therein and display appropriate hazard warnings, as specified under OSHA. Bulk chemical tanks, process tanks, and other containers must all be labeled to meet OSHA requirements.

OSHA allows alternatives to affixing labels on stationery containers. Acceptable alternative labeling methods include signs, placards, process sheets, batch tickets, and equivalent methods. An alternative labeling method must identify the containers to which it is applicable, convey the required information, and be easily accessible to employees in their work area throughout each shift. Additionally, if hazardous chemicals are transferred from a labeled container into a portable container that is intended only for the immediate use of the employee who performs the transfer, labeling of the portable container is not required.

Employer's labels typically are used for processes and auxiliary equipment that contain chemicals such as process baths and holding tanks. Additionally, chemicals that are dispensed from large containers (i.e., 55-gallon drums) into smaller, more easily transported containers (i.e., 1- to 2-gallon containers) require an employer's label if the person dispensing the chemical is not the user. These labels are often similar to a manufacturing label, but also may include internal information such as process center and department numbers, part number, and the phone number of the chemical distribution center. The labels can be designed and printed professionally or can be developed using a PC-based software program and printed on a PC-compatible printer. The using department can be given an array of labels for use with chemicals specific to the area.

Hazardous Waste Labeling

Hazardous waste labeling required under RCRA is the responsibility of the hazardous waste generator. Once a container is put in use for hazardous waste storage, it must be labeled with the words "Hazardous Waste," and the hazard(s) must be specified. Another important part of hazardous waste labeling is the accumulation start date. Waste generators are not allowed to store hazardous waste for over 90 days without a permit; thus the accumulation start date is used to track 90-day storage requirements for generators that do not have a RCRA Part B storage permit. In low-volume waste areas, the accumulation start date and 90-day storage requirements apply when the quantity of waste stored reaches 55 gallons of hazardous waste or 1 quart of acutely hazardous waste. In order to ship a container, the generating facility must comply with DOT regulations specified in 49 CFR §172.304. This includes adding the facility's name, address, manifest number, and EPA waste number to the label.

As was the case with chemical labels, internal waste labels can be prepared using a PC-based software program. Waste labels can be printed as needed, with information such as accumulation start date, internal contact numbers, and other information being filled in at the time the label is printed. Or if a printer is not available in the area, the label can be printed without the accumulation start date and contact numbers, and those items can be filled in manually when the label is put in use. Identification as to whether the waste is hazardous or nonhazardous can be preprogrammed so that labeling mistakes are minimized. Although RCRA labeling requirements do not apply to nonhazardous waste, as a good management practice this waste type should be labeled in a similar manner as hazardous waste, including the name of the waste along with the term "Nonhazardous Waste."

DOT Labeling

Requirements for DOT labeling are discussed in Chapter 10.

Other Labeling Requirements

During any demolition or renovation project involving asbestos the area must be closed to passing pedestrian traffic via signs, cones, tape, or other method, as required by OSHA and TSCA. In addition, the area must be enclosed and labeled. Labels are also required for PCBs, radiation sources, biohazards, and other hazards. Selected examples of these types of labels are presented in Figures 9.1–9.4.

TRAINING

Training is required for employees who work with hazardous materials and hazardous waste. OSHA has three key standards that have training requirements. These are the "Hazard Communication Standard," the "Standard for Process Safety Management of Highly Hazardous Chemicals," and the "Standard for Hazardous Waste Operations and Emergency Response." RCRA and DOT also define training requirements for persons working with hazardous waste and persons involved in transporting hazardous materials, respectively.

Training Required by the Hazard Communication Standard

A widespread training requirement in the manufacturing industry is defined in OSHA's "Hazard Communication Standard." This standard places requirements on chemical manufacturers, chemical distributors, and facilities that use chemicals. This standard is not applicable to workplaces that only use chemicals in the form of consumer products and where the use and frequency of exposure in the workplace is the same as in normal consumer use.

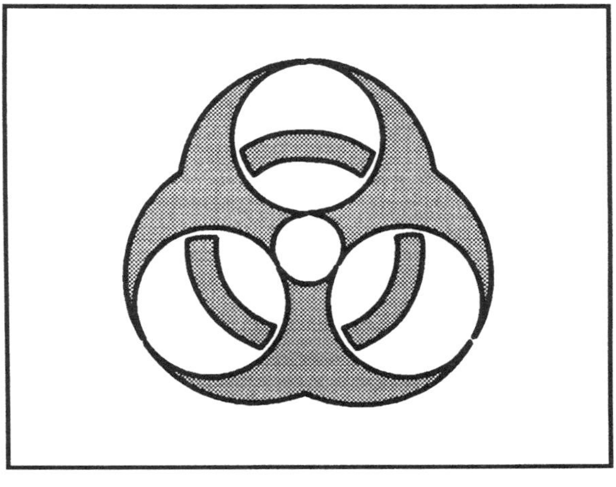

Figure 9.1 Biohazard warning label.

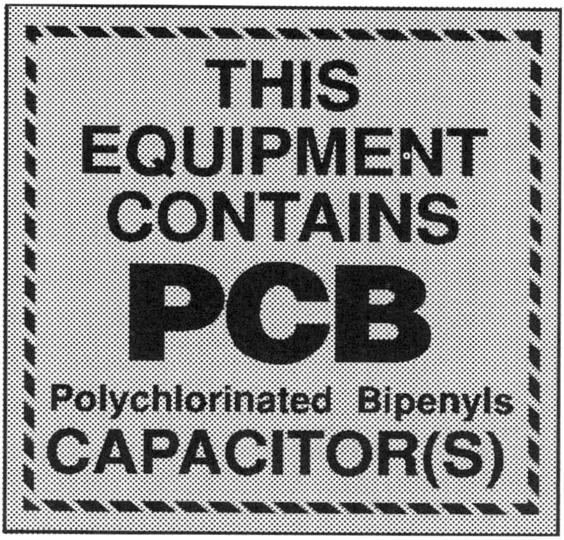

Figure 9.2 Label for PCB-containing capacitors.

The Hazard Communication Standard requires employers to give workers adequate information about hazardous materials that they handle or use in the workplace. Specifically, employers are required to provide employees with information and training on hazardous chemicals and chemical operations at the time of their initial assignment and as new hazards are introduced into the workplace. This training includes information such as:

- Methods and observations to detect the presence or release of a hazardous chemical in the workplace, such as monitoring conducted by the employer, continuous monitoring devices, and visual appearance or odor of hazardous chemicals when being released
- Physical and health hazards of the chemicals in the workplace
- Protective measures required when being exposed to the chemical, such as use of protective equipment, specified work practices, and emergency procedures
- Explanation of the labeling system, information and location of material safety data sheets (MSDS) or equivalent, and other information about the hazard communication program

Hazardous Waste Operations Training

Hazardous waste operations training is required under RCRA and OSHA for those employees who work with hazardous waste. RCRA regulations require

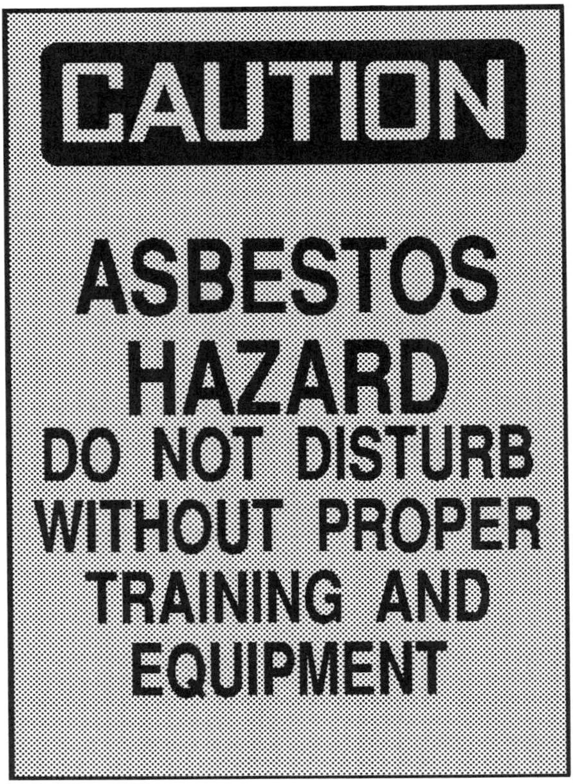

Figure 9.3 Label for asbestos-containing materials.

that employees who work in a hazardous waste area must be trained on the hazards of the job and applicable regulations. In addition, the training program must be be designed to ensure that facility personnel are able to respond effectively to emergencies. Training should include information about emergency procedures, emergency equipment, and emergency systems. RCRA also requires that employers list all job positions that involve hazardous waste management and maintain a list of persons filling those positions. Employers must specify training requirements for each position and maintain personnel training records for each employee filling those positions until facility closure or for three years after the employee leaves that position.

The requirements under OSHA for training of workers in hazardous waste operations and emergency response vary, depending on the job of the worker. Training is required for the following groups of workers:

- *General site workers (such as equipment operators, general laborers, and supervisory personnel) engaged in hazardous substance removal or*

Figure 9.4 Warning label for corrosive material.

other activities that expose or potentially expose workers to hazardous substances and health hazardous. This group must receive a minimum of 40 hours of instruction off site and a minimum of three days of actual field experience under the direct supervision of a trained, experienced supervisor.

- *Workers on site only occasionally for a specific limited task (such as but not limited to groundwater monitoring, land surveying, and other tasks) and who are unlikely to be exposed over the permissible exposure limits or other published exposure limits.* This group must receive a minimum of 24 hours of instruction off the site and a minimum of one day of actual field experience under the direction of a trained, experienced supervisor.
- *Workers regularly on site who work in areas that have been monitored and fully characterized, indicating that exposures are under permissible exposure limits or other published exposure limits, and the characterization indicates there are no health hazards or the possibility of an emergency developing.* This group must receive a minimum of 24 hours of instruction off the site and a minimum of one day of actual field experience under the direct supervision of a trained, experienced supervisor.
- *On-site management and supervisors directly responsible for, or who supervise employees engaged in, hazardous waste operations.* This group must receive a minimum of 40 hours of initial training and three days of supervised field experience plus an additional 8 hours of specialized training. This requirement may be reduced to 24 hours and one day for

supervisors who are only responsible for employees who are not exposed to hazards over the permissible exposure limits or other published exposure limits.

- *Employees who are engaged in emergency response at hazardous waste cleanup sites that may expose them to hazardous substances.* This group must be trained in how to respond to such expected emergencies.
- *Employees conducting operations at hazardous waste TSD facilities.* This group must have 24 hours of initial training to enable them to perform their assigned duties and functions in a safe and healthful manner so as to not endanger themselves or other employees.
- *Employees who are engaged in emergency response at hazardous waste TSD sites.* This group must be trained to a level of competence in recognition of health and safety hazards to protect themselves and other employees.

All workers and supervisors must be certified by the instructor as having completed the training successfully. Annual refresher training of at least 8 hours is part of the training program for most groups. A summary of elements that might be covered during training of workers engaged in hazardous waste operations is presented in Table 9.2. Information about the incident command system used in emergency response and training required for different levels of responders (i.e., first responder, hazardous waste technician, hazardous materials specialist, and on-scene commander) is presented in Chapter 19.

Training for Process Safety Management of Highly Hazardous Chemicals

All employees who are involved in operating a process that is covered in 29 CFR §1910.119 must be trained before operating the process. The training should documented and include an overview of the process and applicable operating procedures. Emphasis must be placed on specific safety and health hazards, emergency operations including shutdown, and safe work practices applicable to the employee's job tasks. Refresher training must be provided at least once every three years to assure that the employee understands and adheres to the current operating procedures of the process.

DOT Training

DOT has outlined training for persons involved in the transportation of hazardous materials. The training requirements include general awareness training, function-specific training, safety training, and OSHA or EPA training, as applicable. Certification of training is necessary, and refresher training is required at least once every two years. If an employee changes hazardous ma-

TABLE 9.2 Summary of Elements That Might Be Covered During Hazardous Waste Operations Training

General Overview

- Requirements under RCRA and OSHA
- Regulations pertaining to specific wastes, such as asbestos, PCBs, benzene, and others (as applicable)
- Names of general supervisor, emergency coordinator, persons and alternates responsible for site safety and health, and other organizational information
- Lines of authority, responsibility, and communication

Hazards Review

- Comprehensive work plan
- Overview of hazards, such as chemical exposures, biological exposures, radiological exposures, heat or cold, other
- MSDS information
- General safety hazards
- Medical symptoms that could indicate overexposure

Safety and Health Review

- Site-specific safety and health plan
- Site control plan
- Work practices that could exacerbate or minimize risk
- Safe use of engineering controls and equipment on site
- Air monitoring methods and equipment
- Confined-space entry procedures and training
- Hazardous waste handling procedures, material handling procedures, and spill containment procedures
- Required personal protective equipment, such as respirators and protective clothing and training on equipment use and inspection
- Required medical surveillance program
- Chemical data and other reference manuals available in the work area
- Hands-on review of work activities

Overview of Emergency Procedures

- Emergency and contingency plan review
- Location of and use of emergency equipment, automatic shutoffs, and emergency communication devices
- Hands-on exercises invoking plan and using emergency equipment

terials job functions, that employee must be trained in the new job function within 90 days. The employee must be supervised by a properly trained employee until the training is completed. A record of training, inclusive of the last two years, must be created and maintained for each employee for as long as the employee works in a hazmat job and for 90 days thereafter.

Specialized Training

Specialized training is required for workers who handle radioactive materials and waste and for workers who remove asbestos from equipment and buildings. Information about the contaminant, permissible exposure limits, personal protective equipment required, monitoring information, and other items are covered in the training.

INSPECTIONS

All TSD facilities must perform inspections as required under RCRA. These inspections must include documented inspection of all equipment on a scheduled basis, as written in the facility inspection plan and as required in various sections of the regulations.

Container Inspections

RCRA regulations require all hazardous waste management facilities to perform inspections of stored hazardous waste containers at least weekly. The containers must be inspected visually for leaks or corrosion.

Tank Inspections

Hazardous waste tank systems and associated spill containment must be inspected daily for system integrity. Inspection items should include:

- Inspection of the aboveground portions of the tank system for cracks, leaks, or corrosion
- Review of data gathered from monitoring and leak detection equipment, such as pressure or temperature gauges and monitoring wells, to ensure that the tank system is being operated according to design
- Inspection of secondary containment and surrounding area for signs of hazardous waste releases, such as wet spots and dead vegetation

Other inspection items include the inspection of the cathodic protection system six months after installation and annually thereafter and bimonthly inspection of sources of impressed currents.

General operating requirements for hazardous waste tank systems include the use of appropriate controls and practices to prevent spills and overflows from the tank and containment system. These devices include:

- Spill prevention controls such as check valves and dry disconnect couplings

- Overfill prevention controls such as level-sensing devices, high-level alarms, automatic feed cutoff, or bypass to a standby tank
- Maintenance of sufficient freeboard in uncovered tanks to prevent overtopping by wave or wind action or by precipitation

Integrity testing and tank assessment by an independent, qualified, registered professional engineer is required for hazardous waste tank systems. The tank assessment should include information pertaining to structural integrity, corrosion potential, overfill and spill prevention controls, and freeboard, as applicable.

Release detection using one of several methods is required for underground storage tank systems containing petroleum products or CERCLA hazardous substances. Acceptable methods include inventory control methods, manual or automatic tank gauging, tank tightness testing, vapor monitoring, groundwater monitoring, and interstitial monitoring.

Inspection of Surface Impoundments

Surface impoundment liners must be inspected during and immediately after construction and installation. The surface impoundment itself, including dikes and vegetation surrounding the dike, must be inspected at least once a week to detect any leaks, deterioration, or failures in the impoundment. Structural integrity certification also is required.

Inspection of Incinerators

Hazardous waste incinerators have several monitoring and inspection requirements. These include:

- Continuous monitoring of combustion temperature, waste feed rate, and the indicator of combustion gas velocity specified in the permit while incinerating hazardous waste
- Monitoring of carbon monoxide on a continuous basis at a point in the incinerator downstream of the combustion zone and prior to release to the atmosphere while incinerating hazardous waste
- Upon request by the regional administrator, sampling and analysis of the waste and exhaust emissions to verify that the operating requirements established in the permit achieve the required performance standards
- Daily inspection of the incinerator and associated equipment, such as pumps, valves, conveyors, and pipes, for leaks, spills, fugitive emissions, and signs of tampering
- Weekly inspection of the emergency waste feed cutoff system and associated alarms to verify operability, unless the applicant demonstrates

to the regional administrator that weekly inspections will unduly restrict or upset operations and that a less frequent inspection will be adequate

- Monthly operational testing

Inspection of Waste Piles, Drip Pads, and Miscellaneous Units

The amount of liquids removed from each leak detection system sump must be recorded at least weekly for waste piles, and other weekly system integrity inspections are required. Drip pads must be inspected weekly and after storms to detect evidence of deterioration, malfunctions or improper operation run-on and runoff control systems, the presence of leakage in and proper functioning of the leak detection system, and deterioration or cracking of the drip pad surface. Hazardous waste miscellaneous units must be inspected at a frequency that protects human health and the environment.

Inspection of Other Materials

Stored PCB waste containers are to be inspected at least once every 30 days for leaks or deterioration. PCB waste must be destroyed at a permitted disposal facility within one year of the waste's initial storage. All radioactive waste packaging and other materials associated with disposal of nuclear waste must be inspected according to an approved schedule specified in the facility's application for a radioactive waste disposal license.

Tanks and containers used for storage of production or process chemicals are presently not required by OSHA to be inspected on a specific schedule, although some state or local governments might require inspection of these storage units. It is good management practice to inspect all types of tanks and containers that hold chemicals, even those not specifically regulated.

PLANS AND CONTROLS

Plans and controls for hazardous waste management facilities include a waste analysis plan, closure and postclosure plans, emergency and contingency plans, a groundwater monitoring plan, and financial assurance. Plans and controls for chemicals in the workplace include medical surveillance, emergency and contingency plans, and a hazard communication program. Facilities that store oil or petroleum distillates aboveground in containers greater than 660 gallons or that store an aggregate amount of greater than or equal to 1320 gallons require a spill prevention, control, and countermeasure (SPCC) plan. Underground storage of more than 42,000 gallons also requires an SPCC plan.

RCRA Plans and Controls

A waste analysis plan is required for all permitted facilities and includes a list of all wastes generated at a facility, the constituents of each waste, and procedures used in determining a waste's hazard. When hazardous, the RCRA classification must be included. The plan is to be updated whenever a new waste is added.

A facility closure plan is required for all facilities that are permitted as hazardous waste TSD facilities. This plan includes procedures and cost estimates for closing a facility, including waste disposal, decontamination, and an evaluation of postclosure care requirements. If postclosure care is required, a separate plan addressing these activities and costs is needed.

A contingency plan and emergency procedures must also be developed by a TSD facility. These documents should detail methods of minimizing hazards to human health or the environment from fires, explosions, or any unplanned releases. Evacuation routes should be included in the plan, and the plan must be kept at the facility as well as provided to area police, fire department, hospitals, and state and local emergency response teams.

A groundwater monitoring plan for detecting any releases is required for hazardous waste surface impoundments, land treatment facilities, landfills, waste piles, and other regulated units, as required. The plan should detail the configuration of the monitoring system as well as parameters to be sampled and frequency of sampling and analysis.

Financial assurance is required of TSD facilities to provide proof that a facility can pay for or has insurance that covers any cleanup activity associated with a sudden or nonsudden release. RCRA also requires financial assurance for underground storage tanks that contain regulated substances. The financial assurance documentation must be submitted annually to EPA or the delegated state administrator.

OSHA Plans and Controls

Medical surveillance is required for employees who work in hazardous waste operations and emergency response, who are or may be exposed to hazardous substances or health hazards at or above the permissible exposure limits (disregarding the use of respirators) for 30 days or more per year, who wear a respirator for 30 days or more a year, or who are exposed to certain hazardous chemicals regulated by OSHA. Records of medical exams must include employee name and social security number, physicians' written opinions, employee medical complaints, and other information. Any medical results that indicate overexposure to a chemical must be reported to the employee. Additionally, all employees have the right to review their medical records, upon request.

An emergency response plan for hazardous materials incidents must be established for facilities that have employees participating in emergency response activities. This plan should include:

- Preemergency planning
- Personnel roles, lines of authority, and communications

- Emergency recognition and prevention
- Safe distances and places of refuge
- Site security and control
- Evacuation routes and procedures
- Decontamination procedures
- Emergency medical and first aid procedures
- Emergency alerting and response procedures
- Critique of response and follow-up
- Personal protective equipment and emergency equipment

An additional document detailing a facility's hazard communication program is required by facilities regulated under the "Hazard Communication Standard." This document must describe chemical labeling practices, use of MSDS information or equivalent, the location of this information for employee access, and training. Other items required include a list of hazardous materials known to be present in the workplace, methods for informing employees of hazards of nonroutine tasks, and methods of informing contractor employers of hazards to which their employees might be exposed.

NRC Requirements

A comprehensive quality control program must be documented and approved by the NRC as part of the licensing program for low-level radioactive waste disposal facilities (NRC 1991). The program should address aspects such as:

- Organization structure
- Quality assurance activities
- Facility design control
- Procurement control
- Control of processes
- Inspections
- Control of testing
- Audits, surveillance
- Corrective actions

Records of locations of the waste at the facility must be maintained, and other administrative requirements apply.

CHEMICAL/WASTE TRACKING, REPORTING, AND RECORD KEEPING

There are several key tracking, reporting, and record-keeping requirements associated with the management of hazardous materials. Most of these requirements are defined by EPA under RCRA, TSCA, SARA, CAA, and CWA. NRC also has requirements for radioactive waste tracking.

RCRA Requirements

Under RCRA regulations, all waste is tracked through a manifest. The manifest is a uniform, numbered form that includes information about a hazardous waste shipment, including quantities, waste codes, and generator, transporter, and disposal facility information.

The receiving facility must send the generator a copy of the manifest to show receipt of the waste. Any discrepancies in weight or content must be noted by the receiving facility. Errors or discrepancies must be reconciled between the generator and receiving facility and documented to file. If discrepancies cannot be reconciled, a formal report must be sent to EPA or the delegated state authority. In addition, if the receiving facility does not return a copy of the manifest to the generator within 45 days, the shipper must investigate the problem and file an exception notice with the agency. Under the regulations, the original manifest, returned copy, and any discrepancy reports must be kept at the facility for three years.

Hazardous waste regulations also require a large-quantity waste generating facility to submit a biennial (or, in some states, annual) report covering all hazardous waste activities. In addition, waste minimization reports are required for most large-quantity generators.

Medical wastes were tracked for a two-year period in states participating in the Medical Waste Demonstration Program, which was set up in 1989 by EPA. The manifest was utilized for tracking these wastes, including infectious and other medical wastes. Transporters of medical wastes in the participating states were required to submit to EPA an annual report of amounts of medical waste shipped. During this period, many states promulgated separate regulations that included medical waste tracking requirements.

TSCA Requirements

Under TSCA, the manufacturer of any new chemical, as defined under the act, is required to notify EPA of the intent to manufacture the chemical. Along with this premanufacture notice (PMN), scientific studies and other health and safety information pertaining to the hazards of the chemical must be submitted. Once submitted, the chemical receives a premanufacturing number and, when approved for manufacture, a Chemical Abstract System (CAS) number. These numbers serve as tracking mechanisms. In addition, once a chemical is manufactured and in use, any allegation from an employee or other person that the chemical is causing unexpected adverse reactions to health or the environment must be recorded by the facility, reported to EPA.

Asbestos abatement projects, including removal and enclosure or encapsulation of friable asbestos, are regulated under TSCA. Under the regulations, notification must be made to the regional asbestos coordinator for the EPA region in which the asbestos abatement project is located. Asbestos abatement projects also are regulated under the CAA and by OSHA.

PCB waste tracking and record-keeping requirements are specified under TSCA. Many of the requirements are similar to the waste tracking requirements under RCRA. The generator, storer, transporter, and disposer of PCB waste must have an EPA identification number. The waste must be manifested, and any manifest discrepancy must be resolved within 15 days or reported to EPA if resolution cannot be made. In addition, if the generator does not receive the return copy of the manifest from the storer or disposer within the specified period of 35 days, this must be reported to EPA.

In addition to manifesting requirements, TSCA regulations require the owner or operator of the disposal facility to prepare a certificate of disposal for each PCB shipment. This certificate is sent to the generator, and a copy is kept at the disposal facility. Additionally, storage and disposal facilities must maintain a written log on the disposition of all PCBs maintained at the facility.

SARA Requirements

Under the Emergency Planning and Community Right-to-Know Act, also known as SARA Title III, facilities have several reporting obligations, which were reviewed briefly in Chapter 1. First, under Section 312 of SARA Title III, the facility must notify local and state governments if it stores on site any hazardous chemical in a quantity that exceeds the threshold quantity listed in the regulations. In addition, the facility must submit a report to these same government entities. This report details chemical names, quantities, storage locations, associated hazards, and physical storage descriptions (i.e., type of container, pressure, temperature). This information is used for emergency planning purposes and citizen information. The submission must be updated annually.

Under Section 313 of SARA Title III, EPA has listed over 300 chemicals that the agency considers toxic. Any manufacturing facility that manufactures or processes 25,000 pounds or more or otherwise uses 10,000 pounds or more of a listed chemical must file the toxic release inventory (TRI) report specifying certain release information.

Included in the report are quantities of emissions to air, wastewater, stormwater, on-site disposal facilities, transfers to off-site facilities, and other information. This report is part of the Community-Right-to-Know portion of the law and provides citizens with information about chemical releases in their immediate area. In addition, the report also provides regulators with a summary of toxic releases, which can be used to determine future regulatory focus. TRI information is available in an on-line database accessible to the media and the public, as provided by EPA.

CAA Requirements

Notification to EPA or a delegated state authority is required 10 working days before an asbestos abatement project begins for jobs that contain friable asbestos and meet job-size criteria. Information pertaining to project location, area to be abated, date(s) of work, and other information is required. In addition, waste disposal sites accepting asbestos waste must maintain waste shipment records and meet other requirements. In addition, there are recordkeeping requirements for hazardous air pollutants. Specific sampling/reporting requirements vary from chemical to chemical.

CWA Requirements

Requirements for effluent discharge monitoring and reporting are defined in the NPDES permit. When the effluent is discharged to a publicly owned treatment works (POTW), pretreatment regulations and local requirements dictate monitoring and reporting frequency.

NRC Requirements

Transportation and tracking of radioactive material must conform to requirements in 10 CFR Part 71. Shipping papers, labeling, and packaging must meet NRC and DOT standards. Low-level radioactive waste must be accompanied by a manifest when sent to an off-site disposal facility. Transport of certain types of high-level and other dangerous radioactive waste and licensed materials outside the confines of the licensee's plant requires notification to the governors (or designees) of states through which the material will travel. The notification must be made in writing at least seven days before intended shipment. The licensee must retain a copy of the notification for three years. If the material being transported is a waste, tracking the waste becomes the responsibility of the licensee of the disposal site, once the waste is received at the waste disposal facility.

RELEASE AND OTHER NONCOMPLIANCE REPORTING

There are specific requirements for reporting accidental releases of chemicals or wastes under the various regulations. These include:

- *CERCLA*. Notification is required to the National Response Center upon the nonpermitted release of a listed chemical in an amount that exceeds reporting threshold.
- *RCRA*. Notification is required to the state or local agency when their help is needed. Notification is required to the government official des-

ignated as on-scene coordinator for the geographical region (as applicable) or the National Response Center when the facility has had a release, fire, or explosion that could threaten human health or the environment. Notification to the delegated state authority is required, as written into state laws or policies. For underground storage tanks, any release must be reported to the EPA or designated state agency.

- *SARA*. Notification is required to the Local Emergency Planning Committee (LEPC) or its designee and the State Emergency Response Commission (SERC) upon the occurrence of a nonpermitted release of a listed chemical that exceeds the reporting threshold quantity and that leaves the facility's boundaries.
- *DOT*. Notification is required to the Department of Transportation if an incident occurs during the course of transportation (including loading, unloading, and temporary storage) that results in a death or injury, carrier damage exceeding $50,000, evacuation of the general public, spillage of radioactive material, the shutdown of a transportation artery, or creates a danger. Notification is required to the Director, Centers for Disease Control, if the incident involves etiologic agents.
- *CWA*. Notification is required to the EPA Regional Administrator if a vessel or facility discharges in a single spill event reportable quantities of hazardous substances into or upon navigable waters of the United States or adjoining shorelines. Facilities that discharge wastewater must notify EPA or the delegated state authority of any activity that could result in a nonroutine or infrequent release of toxic pollutants not covered by an NPDES permit.
- *CAA*. Notification to the EPA Regional Administrator or delegated state authority is required for major upsets or releases that could threaten human health or the environment. Reporting requirements for noncompliance with respect to hazardous air pollutants also apply.
- *NRC*. Notification to the Nuclear Regulatory Commission of defects in packaging and other noncompliance issues is required by licensed radioactive waste disposal facilities.

Information required in the report includes such items as name and telephone number of reporter, name and address of facility, chemical or waste released, quantity released, possible hazards to human health or the environment, and the extent of injuries. For radioactive disposal facilities, the report will include information pertinent to the packaging defect or other noncompliance issue. In addition to these federal accidental release reporting requirements, individual permits may specify additional requirements for noncompliance reporting.

REFERENCES

Nuclear Regulatory Commission (1991). *Quality Assurance Guidance for a Low-Level Radioactive Waste Disposal Facility,* NRC, Washington, DC.

BIBLIOGRAPHY

Berger, D. A. and C. Harris (1990). *The SARA Title III Compliance Handbook,* Executive Enterprises Publications, New York.

Blattner, J. W. (1992). *The Clean Air Act Compliance Handbook,* 2nd ed., Executive Enterprises Publications, New York.

Bureau of National Affairs (1989). *Spill Reporting Procedures Guide,* BNA, Washington, DC.

Chemical Manufacturers Association (1987). *NPDES Discharge Permitting and Compliance Issues Manual,* CMA, Washington, DC.

——— (1987). *Understanding Title III: Emergency Planning and Community Right-to-Know,* CMA, Washington, DC, videotape.

——— (1989). *Overview of Resource Conservation and Recovery Act,* CMA, Washington, DC, videotape.

Chrismon, R. L., (1989). *The TSCA Compliance Handbook,* Executive Enterprises Publications, New York.

Environmental Protection Agency (1986). *EPA Guide for Infectious Waste Management,* EPA/530-SW-86-014, Office of Solid Waste, U.S. EPA, Washington, DC.

——— (1989). *Hospital Incinerator Operator Training Course,* EPA-450/3-89/003," Vols. 1–3, U.S. EPA Research Triangle Park, NC.

Frye, R. S. (1988). *The Clean Water Act Compliance Handbook,* Executive Enterprises Publications, New York.

Government Institutes (1986). *Clean Water Act Compliance/Enforcement Guidance Manual,* Rockville, MD.

——— (1989). *Hazardous Waste Manifests Videotape,* GI, Rockville, MD.

——— (1989). *RCRA Inspection Manual,* 2nd ed., GI, Rockville, MD.

——— (1991). *OSHA Field Operations Manual,* 4th ed., GI, Rockville, MD.

——— (1992). *Environmental Reporting and Recordkeeping Requirements,* 2nd ed., GI, Rockville, MD.

Jones, S. E., et al. (1989). *Occupational Hygiene Management Guide,* Lewis Publishers, Boca Raton, FL.

Kaufman, J. A., ed. (1990). *Waste Disposal in Academic Institutions,* Lewis Publishers, Boca Raton, FL.

Keith, L. H., and D. B. Walters (1992). *The National Toxicology Program's Chemical Data Compendium,* Vol. 8, "Shipping Classifications and Regulations," Lewis Publishers, Boca Raton, FL.

Lowry, G. T., and R. C. Lowry (1988). *Lowry's Handbook of Right-to-Know and Emergency Planning/SARA Title III,* Lewis Publishers, Boca Raton, FL.

Neizel, C. L. (1991). *The RCRA Compliance Handbook,* Executive Enterprises Publications, New York.

Phifer, M., Jr. (1988). *Handbook of Hazardous Waste Management for Small Quantity Generators,* Lewis Publishers, Boca Raton, FL.

Stensvaag, J. M. (1989). *Hazardous Waste Law and Practice,* Vol. 2, Wiley, New York.

——— (1991). *Clean Air Act 1990 Amendments: Law and Practice,* Wiley, New York.

Waldo, A. B., and R. deC. Hinds (1991). *Chemical Hazardous Communication Guidebook: OSHA, EPA, and DOT Requirements,* 2nd ed., Executive Enterprises Publications, New York.

Woodyard, J. P., and J. J. King (1992). *PCB Management Handbook,* 2nd ed., Executive Enterprises Publications, New York.

10

HAZARDOUS MATERIALS TRANSPORTATION

The Department of Transportation (DOT) has promulgated regulations to govern transportation of hazardous materials in intrastate, interstate, and foreign commerce, as authorized under the Hazardous Materials Transportation Act of 1974 and amended by the Hazardous Materials Transportation Uniform Safety Act of 1990. These regulations, promulgated under 49 CFR Parts 171–180 (DOT 1997), apply to classification of materials; packaging; hazard communication, including shipping paper requirements, package marking and labeling, placarding of vehicles and bulk packagings, and emergency response communication; transportation and handling; and incident reporting. Training is required for personnel who package and/or label hazardous materials, prepare the shipping documents, and load and unload vehicles. Additionally, any employee who offers a hazardous material for transportation, and each carrier by air, highway, rail, or water that transports a hazardous material, is responsible for complying with all DOT regulations governing the transport of hazardous materials.

GENERAL REQUIREMENTS OF THE STANDARD

Key parts of the hazardous materials regulations are presented in Table 10.1. The bulk of the most commonly used requirements are in Parts 172 and 173, and in late 1990 significant changes to these parts of the regulations were made as a result of the promulgation of HM 181 (so named after its docket number). The key features of the changes—almost all of which were phased in over a five-year period—included:

- Combining two hazardous materials tables into one
- Aligning the hazard classes for materials with the United Nations (UN) numerical system

TABLE 10.1 Key Parts of the Hazardous Materials Regulations

Part 171	Includes definitions, reporting requirements, reference materials, and procedural requirements
Part 172	Includes the Hazardous Materials Table, requirements for shipping papers,package marking and labeling, placarding of vehicles and bulk packagings, and emergency response communication
Part 173	Includes hazard class definitions, authorized DOT packaging for specific materials, and references the appropriate sections of part 178 when DOT specification packagings are required
Part 174	Includes requirements for transportation by railcar
Part 175	Includes requirements for transportation by aircraft
Part 176	Includes requirements for transportation by vessel such as ship, barge, or boat
Part 177	Includes requirements for transportation by motor vehicle
Part 178	Includes specifications for a wide variety of approved packagings
Part 179	Includes specifications for tank cars
Part 180	Includes requirements for continuing qualification and maintenance of packagings

- Assigning materials to packing groups that determine the level of regulatory requirements, based on their hazards and compatibility
- Specifying performance-oriented standards for packaging instead of composition/construction specification, based on the container's ability to contain and protect materials in realistic transportation conditions
- Setting standards for reuse of drum in shipment based on minimum wall thickness
- Setting new limits for filling of bulk packages
- New notification rules for manufacturers, who must now notify their customers of any failures of packages to meet the stated performance specifications

DEFINITION OF HAZARDOUS MATERIAL

As defined under 49 CFR §171.8, a hazardous material is a substance or material that has been determined by the Secretary of Transportation to be capable of posing an unreasonable risk to health, safety, and property when transported in commerce, and which has been so designated. These substances or materials can include hazardous substances, hazardous wastes, marine pollutants, and elevated temperature materials. The listing of materials governed under this regulation is found in Appendix A to 49 CFR §72.101, Hazardous Materials Table. Definitions of specific materials, as they relate to hazard class and division are discussed in the next section.

HAZARD CLASSES

Hazard classes are identified in 49 CFR Part 173 and are summarized in Table 10.2. As can be seen from the table, there are nine major hazard classes, as well as materials that are designated as "forbidden" and "other regulated material."

TABLE 10.2 DOT Hazard Classes

Class	Division	Name of Class or Division
N/A	N/A	Forbidden materials
N/A	N/A	Forbidden explosives
1		Explosive materials
	1.1	Explosives that have a mass explosion hazard
	1.2	Explosives that have a projection hazard but not a mass explosion hazard
	1.3	Explosives that have a fire hazard, and either a minor blast hazard or a minor projection hazard, but not a mass explosion hazard
	1.4	Explosive devices containing no more than 25 grams (9 oz) of detonating material presenting a minor explosion hazard
	1.5	Very insensitive explosives
	1.6	Extremely insensitive articles that do not have a mass explosive hazard
2		Gases
	2.1	Flammable gases
	2.2	Nonflammable gases, nonpoisonous compressed gases
	2.3	Poisonous gases (by inhalation)
3		Flammable liquids
4		Flammable solids
	4.1	Wetted explosives, self-reactive materials, and readily combustible solids
	4.2	Spontaneously combustible materials
	4.3	Dangerous when wet materials
5		Oxidizers
	5.1	Oxidizing materials
	5.2	Organic peroxides
6		Poisonous materials
	6.1	Poisonous liquids or solids
	6.2	Infectious substances (etiologic agents)
7		Radioactive materials
8		Corrosive materials
9		Miscellaneous hazardous materials
N/A	N/A	ORM-D: Other regulated material

Forbidden Materials and Forbidden Explosives

Although not considered a class, forbidden materials and forbidden explosives are designated as such in the Hazardous Materials Table. Examples of forbidden materials include unstable materials such as acetyl acetone peroxide or wet or hot aluminum dross; electrical devices that are likely to create sparks or generate a dangerous quantity of heat unless packaging precludes such an occurrence; incompatible materials in the same overpack or freight container that could cause a dangerous occurrence, such as evolution of heat or poisonous gases or vapors; and a package containing cigarette lighter, or similar device, equipped with an ignition element and containing fuel, unless examined by the Bureau of Explosives and approved by the Associate Administrator for Hazardous Materials Safety. Additional definitions pertaining to forbidden materials are delineated in 49 CFR §73.21.

Examples of forbidden explosives include liquid nitroglycerin that is not desensitized; a leaking or damaged package of explosives; propellants that are unstable, condemned, or deteriorated; a loaded firearm; fireworks containing yellow or white phosphorus; and other unstable explosive materials, devices, or conditions. Additional definitions pertaining to forbidden explosives is delineated in 49 CFR §173.54.

Explosives

Under hazardous material regulations defined in 49 CFR §173.50, explosives are any substance or article, including a device, that is designed to function by explosion (i.e., an extremely rapid release of gas and heat) or which, by chemical reaction within itself, is able to function in a similar manner even if not designed to function by explosion, unless otherwise classed under the regulations. The explosives class is divided into six divisions, as shown in Table 10.2, with the most dangerous explosives being listed first and the least dangerous being listed last (i.e., 1.1 is the most dangerous and 1.6 is the least dangerous). Many types of explosives—such as cartridges, detonators, flares, projectiles, rockets, and fireworks—may be classified in several divisions, depending on the explosive properties of the particular device. DOT definitions and examples of explosives within each division include:

- *Division 1.1:* consists of explosives that have a mass explosion hazard. A mass explosion is one that affects almost the entire load instantaneously. Examples include ammonium perchlorate, barium styphnate, trinitronaphthalene, electric detonators for blasting (for specific UN numbers), picryl chloride, and photo-flash bombs (for specific UN numbers).
- *Division 1.2:* consists of explosives that have a projection hazard but not a mass explosion hazard. Examples include smoke ammunition (for spe-

cific UN numbers), tear-producing ammunition (for specific UN numbers), and toxic ammunition (for specific UN numbers).

- *Division 1.3:* consists of explosives that have a fire hazard and either a minor blast hazard or a minor projection hazard or both, but not a mass explosion hazard. Examples include deflagrating metal salts of aromatic nitroderivatives, smoke ammunition (for specific UN numbers), tear-producing ammunition (for a specific UN number), toxic ammunition (for specific UN numbers), and practice ammunition (for a specific UN number).
- *Division 1.4:* consists of explosives that present a minor explosion hazard. The explosive effects are largely confined to the package and no projection of fragments of appreciable size or range is to be expected. An external fire must not cause virtually instantaneous explosion of almost the entire contents of the package. Examples include nonelectric detonator assemblies for blasting (for specific UN numbers), electric detonators for blasting (for a specific UN number), detonators for ammunition, tear-producing ammunition (for a specific UN number), practice ammunition (for a specific UN number), cartridges for weapons (for specific UN numbers), safety fuses, and fuse lighters.
- *Division 1.5:* consists of very insensitive explosives. This division is comprised of substances that have a mass explosion hazard but are so insensitive that there is little probability of initiation or of transition from burning to detonation under normal conditions of transport. (The probability of transition from burning to detonation is greater when large quantities are transported in a vessel.) Examples include blasting explosives (for specific UN numbers) and other very insensitive explosive substances.
- *Division 1.6:* consists of extremely insensitive articles that do not have a mass explosive hazard. This division is comprised of articles that contain only extremely insensitive detonating substances and demonstrate a negligible probability of accidental initiation or propagation. (The risk from articles of Division 1.6 is limited to the explosion of a single article.)

In addition to division, explosives are classed according to 13 compatibility groups, designated by the letters A through S. The compatibility group letters are used to specify the controls for the transportation, and storage related thereto, of explosives and to prevent an increase in hazard that might result if certain types of explosives were stored or transported together.

Gases

The hazard class of gases, as defined in 49 CFR §173.115, is divided into three divisions: flammable gas (Division 2.1); nonflammable gas, nonpoisonous compressed gas (Division 2.2); and poisonous gas (by inhalation) (Division 2.3).

Flammable Gas. A flammable gas is any material that is a gas at 20°C (68°F) or less and 101.3 kilopascal (kPa) (14.7 psi) of pressure [a material that has a boiling point of 20°C (68°F) or less and 101.3 kPa (14.7 psi) of pressure] and which (1) is ignitable at 101.3 kPa (14.7 when in a mixture of 13% or less by volume with air; or (2) has a flammable range at 101.3 kPa (14.7 psi) with air of at least 12% regardless of the lower limit. (Except for aerosols, materials are to be tested using ASTM E681-85.) Examples include propane, butane, compressed trifluoroethane, ethyl fluoride, anhydrous trimethyamine, and dissolved acetylene (liquefied acetylene is forbidden).

Nonflammable, Nonpoisonous Compressed Gas. The division for nonflammable, nonpoisonous compressed gases include compressed gas, liquefied gas, pressurized cryogenic gas, compressed gas in solution, asphyxiant gas, and oxidizing gas. For a material or mixture to be classified as a nonflammable, nonpoisonous compressed gas (1) it must exert in the packaging an absolute pressure of 280 kPa (41 psia) or greater at 20°C (68°F); and (2) it may not meet the definition of Division 2.1 or 2.3. Examples include compressed oxygen, carbon dioxide, compressed nitrogen, bromotrifluormethane, and nitrogen trifluoride.

Poisonous Gas (by Inhalation). A gas poisonous by inhalation is defined as a material that is a gas at 20°C (68°F) or less and a pressure of 101.3 kPa (14.7 psi) [a material that has a boiling point of 20°C (68°F) or less and 101.3 kPa (14.7 psi) of pressure] and that (1) is known to be so toxic to humans as to pose a hazard to health during transportation; or (2) in the absence of adequate data on human toxicity, is presumed to be toxic to humans because when tested on laboratory animals it has an LC_{50} value of not more than 5000 milliliters per cubic meter (mL/m^3), as defined under 49 CFR §173.116(a). Depending on the toxicity, the gas is also assigned a hazard zone of A, B, C, or D, with zone A being the most toxic and zone D being the least toxic. Examples of materials in this division include arsine (hazard zone A), compressed fluorine (hazard zone A), anhydrous hydrogen chloride (hazard zone B), methyl mercaptan (hazard zone C), liquefied sulfur dioxide (hazard zone C), and sulfuryl chloride (hazard zone D).

Flammable Liquids

Class 3 is comprised of flammable liquids as defined under 49 CFR §173.120. A flammable liquid is a liquid having a flash point of not more than 60.5°C (141°F), or any material in a liquid phase with a flash point at or above 37.8°C (100°F) that is intentially heated and offered for transportation or transported at or above its flash point in a bulk packaging. Exceptions include:

1. Any liquid meeting one of the definitions of a gas, as defined in 49 CFR §173.115.
2. Any mixture having one or more components with a flash point of 60.5°C (141°F) or higher, that make up at least 99% of the total volume of the mixture, if the mixture is not offered for transportation or transported at or above its flash point.
3. Any liquid with a flash point greater then 35°C (95°F) which does not sustain combustion, when tested according to the procedure specified in 40 CFR Part 173, Appendix H.
4. Any liquid with a flash point greater than 35°C (95°F) and with a fire point greater than 100°C (212°F) according to ISO 2592.
5. Any liquid with a flash point greater then 35°C (95°F) which is in a water-miscible solution with a water content of more than 90% by mass.

Examples of flammable liquids include gasoline, gas oil, acetone, turpentine, and liquid flavoring extracts.

Combustible liquid is defined as a material that has a flash point of above 60.5°C (141°F) and below 93°C (200°F) and that does not meet the definition of any other class, except class 9. This classification is used in the United States only and is not an international definition. [*Note:* A flammable liquid with a flash point at or above 38°C (100°F), which does not meet any other class except class 9, may be reclassed as a combustible liquid for ground transport.] Examples of combustible liquids include heavy oils and mixtures that meet this definition.

Flammable Solids

Flammable solids are defined in 49 CFR §173.124. There are three divisions of flammable solids including flammable solid (Division 4.1); spontaneously combustible material (Division 4.2); and dangerous when wet material (Division 4.3).

Flammable Solids. Division 4.1 can be any of three types of materials, defined below.

- *Wetted explosives:* materials that when dry are explosives of class 1, other than those of compatibility group A, which are wetted with sufficient water, alcohol, or plasticizer to suppress explosive properties; and that are specifically authorized by name either in the §172.101 table or have been assigned a shipping name and hazard class by the associate Administrator for Hazardous Materials Safety; or are otherwise approved under the regulations. Examples include gelatine-coated, nitrocellulose-based films; and trinotrotoluene, wetted with not less than 30% water by mass.

- *Self-reactive materials:* materials that are thermally unstable and that can undergo a strongly exothermal decomposition even without participation of oxygen (air). (Certain exclusions apply.) Additionally, self-reactive materials are assigned to a generic system consisting of seven types, designated by the letters A through G, based on the material's physical state, determination as to its control temperature and emergency temperature, testing, and other factors. These materials are specified in 49 CFR §173.224. Examples of self-reactive materials include azodicarbonamide formulation (types B, C, and D, solid, temperature controlled) and benzene sulphohydrazide (type D, solid).
- *Readily combustible solids:* materials that are solids which may cause a fire through friction, such as matches; or that show a burning rate faster than 2.2 mm (0.087 inch) per second, when tested using a test method defined in Part 273, Appendix E; or any metal powders that can be ignited and react over the entire length of a sample in 10 minutes or less, when tested using a test method defined in Part 273, Appendix E; or any metal powders that can be ignited and react over the entire length of a sample in 10 minutes or less, when tested using a test method defined in Part 273, Appendix E. Examples include strike-anywhere matches, aluminum resinate, wetted ammonium picrate, and paraformaldehyde.

Spontaneously Combustible Material. Division 4.2 is comprised of spontaneously combustible material, including pyrophoric materials or self-heating materials. A pyrophoric material is a liquid or solid that, even in small quantities and without an external ignition source, can ignite within 5 minutes after coming in contact with air when tested using the test method specified in 40 CFR Part 273, Appendix E. A self-heating material is one that, when in contact with air and without an energy supply, is likely to self-heat when tested using a test method specified in 40 CFR Part 273, Appendix E. Examples of materials in this division include pyrophoric titanium trichloride, sodium hydrosulfite, seed cake with more than 1.5% oil and not more than 11% moisture, and animal or vegetable fibers or fabrics.

Dangerous When Wet. The third division of flammable solids, Division 4.3, is made up of materials that are dangerous when wet. These are materials that, by contact with water, are likely to become spontaneously flammable or give off a flammable or toxic gas at a rate greater than 1 liter per kilogram of material per hour when tested using the test method defined in 49 CFR Part 173, Appendix E. Examples include lithium hydride, magnesium powder, and alkaline earth metal alloys and amalgams.

Oxidizers

There are two divisions of oxidizers: Division 5.1, which is made up of oxidizing materials, and Division 5.2, which is made up of organic peroxides.

Definitions for these materials are found in 49 CFR §173.127 and §173.128, respectively.

Oxidizer. An oxidizing material is a material that may, generally by yielding oxygen, cause or enhance the combustion of other materials, when tested using a method specified in 49 CFR Part 173, Appendix F. Examples of oxidizing materials include calcium permanganate, sodium chlorite, and lead perchlorate.

Organic Peroxide. An organic peroxide is any organic compound containing oxygen (O) in the bivalent —O—O— structure which may be considered a derivative of hydrogen peroxide, where one or more of the hydrogen atoms have been replaced by organic radicals. Some exceptions apply. Additionally, materials in this division are assigned to seven generic types, designated by the letters A through G, as defined in 49 CFR §173.225. Examples include methyl ethyl ketone peroxides (types B, C, and D, liquid) and diacetyl peroxide (type D, liquid, temperature controlled).

Poisonous Materials

The poisonous materials hazard class is defined in 49 CFR §173.132 and §173.134. It is comprised of poisonous liquids or solids (Division 6.1) and infectious substances (etiologic agent) (Division 6.2).

Poisonous Liquid or Solid. A material classed within Division 6.1 is defined as a material, other than a gas, that is known to be so toxic to humans as to afford a hazard to health during transportation; or which, in the absence of adequate data on human toxicity:

1. Is presumed to be toxic to humans because it falls into any one of three categories (oral toxicity, dermal toxicity, or inhalation toxicity) when tested on laboratory animals.
   a. *Oral toxicity* is defined as a liquid with an LD_{50} for acute oral toxicity of not more than 500 milligrams per kilogram (mg/kg) or a solid with an LD_{50} for acute oral toxicity of not more than 200 mg/kg.
   b. *Dermal toxicity* is defined as a material with an LD_{50} for acute dermal toxicity of not more than 1000 mg/kg.
   c. *Inhalation toxicity* is a dust or mist with an LC_{50} for acute toxicity on inhalation of not more than 10 mg/L; or a material with a saturated vapor concentration in air at 20°C (68°F) of more than one-fifth of the LC_{50} for acute toxicity on inhalation of vapors and with an LC_{50} for acute toxicity on inhalation of vapors of not more than 5000 mL/m^3.
2. Is an irritating material, with properties similar to tear gas, which causes extreme irritation, especially in confined spaces. Examples of materials

in this division include liquid triazine pesticides, thioglycol, thallium nitrate, liquid or solid isocyanates, and carbon tetrachloride.

Infectious Substance (Etiologic Agent). A material classed within Division 6.2 is defined as a viable microorganism, or its toxin, which causes or may cause disease in humans or animals, and includes agents listed by the regulations of the Department of Health and Human Services (42 CFR §72.3) or any other agent that causes or may cause severe, disabling, or fatal disease. The terms *infectious substance* and *etiologic agent* are synonymous. Examples include biological cultures and medical waste.

Radioactive Material

Radioactive material (class 7) is defined in 49 CFR §173.403 as having a specific activity greater than 0.002 microcurie per gram (μCi/g). Examples include cobalt, gallium, plutonium, and other radionuclides.

Corrosive Material

Corrosive material (class 8) is defined in 49 CFR §173.136 as a liquid or solid that causes full-thickness destruction of human skin tissue at the site of contact, within a specified period of time. Additionally, a liquid that has a severe corrosion rate on steel or aluminum, as defined under 49 CFR §173.36, is also a corrosive. Examples include zinc chloride, sodium hydrogen fluoride solution, and calcium oxide.

Miscellaneous Hazardous Material

Miscellaneous hazardous materials are grouped under hazard class 9. They are defined in 49 CFR §173.140 as presenting a hazard during transportation but not meeting the definition of any other hazard class. This class includes any material that has an anesthetic, noxious, or other similar property that could cause extreme annoyance or discomfort to a flight crew so as to prevent the correct performance of assigned duties; or which meets the DOT definition of a hazardous substance, a hazardous waste, a marine pollutant, or elevated temperature material (49 CFR §171.8). Examples include plastic molding material in dough, sheet, or rope form; expandable polystyrene bead, evolving flammable vapor; and electric wheelchair with spillable or nonspillable batteries.

Other Regulated Material (ORM-D)

An ORM-D material is defined in 49 CFR §173.144 as a material such as a consumer commodity, which although otherwise subject to hazardous materials regulations, presents a limited hazard during transportation due to its form, quantity, and packing. It must be a material for which exceptions are provided in the Hazardous Materials Table.

OTHER IMPORTANT DEFINITIONS

There are several other definitions found in 49 CFR §171.8 that pertain to the transportation of hazardous materials that are worth review.

- *Hazardous substance:* a material, or its mixture or solutions, found in the Hazardous Materials Table in a quantity, in one package, that equals or exceeds the reportable quantity (RQ) in Appendix A of §172.101 (which is a reprint of EPA's CERCLA reportable quantity list). The list contains hundreds of materials and wastes, with reportable quantities ranging from 1 pound (examples include benzidine and silver nitrate) to 5000 pounds (examples include benzoic acid and sodium hydrosulfide).
- *Hazardous waste:* any material that is subject to EPA's hazardous waste manifest requirements under 40 CFR Part 262. This includes listed and characteristic wastes.
- *Elevated temperature material:* a material which, when offered for transportation or transported in a bulk packaging is in a liquid phase and at a temperature at or above 100°C (212°F); is in a liquid phase with a flash point at or above 37.8°C (100°F) that is intentionally heated and offered for transportation or transported at or above its flash point; or is in a solid phase and at a temperature at or above 240°C (464°F).
- *Marine pollutant:* a hazardous material that is listed in Appendix B of §172.101 and when in a solution or mixture of one or more marine pollutants is packaged in a concentration that equals or exceeds 10% by weight of the solution or mixture for the materials listed in the appendix; or 1% by weight of the solution or mixture for materials that are identified as severe marine pollutants in the appendix. Examples include metal compounds such as cadmium and mercury compounds, and pesticides such as heptochlor and parathion.

PACKAGING

Hazardous materials and hazardous waste must be transported in bulk and in nonbulk packages that meet the requirements of DOT. Package specifications are based on international performance-oriented packaging standards.

Requirements for packaging of bulk and nonbulk hazardous materials are defined under 49 CFR §173.24. There are three packing groups (PG I, PG II, and PG III) that indicate the degree of danger of the material being shipped, with PG I being great danger; PG II being medium danger; and PG III being minor danger. Packing groups and packaging requirements for hazardous materials can be located in the Hazardous Materials Table.

Types of packages used to transport nonbulk materials include those made of combination packagings (inner and outer packagings), composite packagings, and single packagings. Examples of acceptable inner packagings (depending on the hazardous material shipped) include glass or earthenware receptacles, plastic receptacles, metal receptacles, and glass ampoules. Examples of outer packagings that might be acceptable (depending on the hazardous material shipped) include steel drum, aluminum drum, plywood drum, fiber drum, aluminum box, solid plastic box, fiberboard box, and steel jerrican. An example composite packaging is a plastic-lined drum or a plastic-lined fiberboard box. Single packagings are typically made of sturdy materials. With the exception of transport by passenger aircraft, single packagings may be acceptable if they meet performance requirements.

Under DOT regulations, containers in good condition can be reused, except for those made of paper, plastic film, or textiles. For the container to be reused, there can be no evidence of a reduction in container integrity unless the container is then reconditioned. Further, packages that are subject to the leakproof test must be retested and marked with the letter "L". Metal and plastic containers may be reused only if they are marked with a minimum thickness for reuse and meet that thickness specification. Nonbulk containers can also be reused for hazardous waste if certain requirements are met.

Empty containers are also regulated by DOT as well as by EPA. A specialist at the facility should review all requirements pertaining to empty containers before shipping. EPA allows for a container to be "empty" under certain conditions, such as when there is no more than 1 inch of residue remaining on the bottom of the container, but the container will still be regulated under DOT and subject to numerous requirements.

LABELING AND MARKING

Any person or employee who offers a package for transportation that contains a hazardous material or hazardous waste must affix the appropriate DOT warning label. This label is designated in the Hazardous Materials Table. Examples of these labels are presented in Figures 10.1 and 10.2 In some cases, a primary and a subsidiary label may be needed. The subsidiary label may not display the hazard class number or division number. For preprinted labels, the number can be marked out or covered with tape of the same color as the area of the label containing the number.

In addition to affixing labels, other markings are required, including proper shipping name, UN identification number, and UN packaging specification number. Additional information may be required for transport of some materials, including "This End Up" verbiage and arrows, technical names of the two most hazardous constituents in a solution or mixture, indication of reportable quantity [by means of the letters "RQ" and associated technical name(s)], indication of material poisonous by inhalation, indication of ele-

Note: See 49 CFR 172.411 for specific requirements pertaining to explosive labeles.

Figure 10.1 DOT labels for explosives.

vated-temperature materials, indication of marine pollutant, EPA Hazardous Waste Code, and other information. Mode of transport also affects marks and labeling on a package. Additionally, some materials are forbidden to be transported by air altogether, such as poison aerosols, and many are not allowed on passenger aircraft, only on air cargo flights. These and all other require-

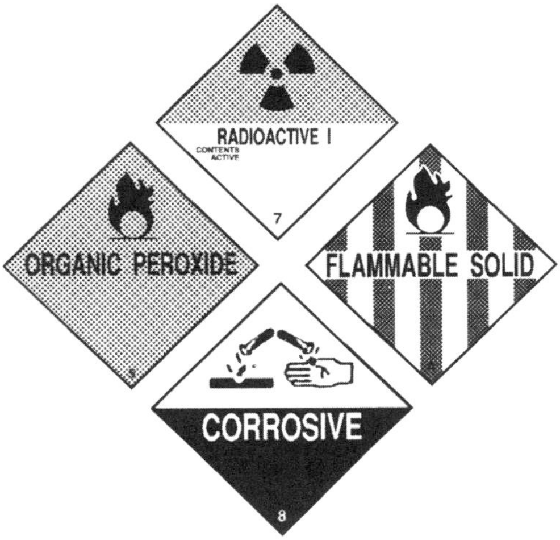

Figure 10.2 DOT labels.

ments related to transport of hazardous materials should be investigated carefully by a trained specialist prior to offering any materials for transport.

PLACARDING

Another requirement under DOT regulations is the use of placards for each bulk packaging, freight container, unit load device, transport vehicle, or rail car containing hazardous material. Any person offering a hazardous material for transport must offer placards to the carrier unless exemptions apply. A placard must be placed on each side and at each end. Placarding is regulated under 49 CFR §172.520–560. Placard names for each of the hazard classes and divisions are presented in Table 10.3. As with labeling, primary and subsidiary placards will be required, depending on material class and quantities.

Placards are not required for the following:

- Infectious substances
- Hazardous materials classes as ORM-D
- Hazardous materials authorized to be offered as "limited quantities"
- Hazardous materials that are packaged as small quantities
- Combustible liquids in nonbulk packagings

SHIPPING PAPERS

It is the shipper's responsibility to complete the appropriate shipping documents properly. Contents of the shipping papers, at a minimum, include:

- Name and address of the consignor and consignee
- Proper shipping name of the hazardous material
- Hazard class
- UN/NA identification number
- Packing group
- Total quantity of the shipment by weight or volume
- Emergency response telephone number
- Shipper's certification

In addition, some of the information must be written in a specified sequence: proper shipping name, hazard class, identification number, and packing group. Other requirements apply, including the identification of hazardous substance reportable quantities, identification of poison by inhalation, identification of dangerous when wet materials, and other marks. Manifests are shipping documents for hazardous waste and are regulated under EPA. This topic is discussed in Chapter 9.

TABLE 10.3 Placards Required Under DOT

Hazard Class or Division Number	Placard Name[a]	Comment
1.1	EXPLOSIVES 1.1	All quantities must be placarded.
1.2	EXPLOSIVES 1.2	All quantities must be placarded.
1.3	EXPLOSIVES 1.3	All quantities must be placarded.
1.4	EXPLOSIVES 1.4	1000 pounds or more must be placarded.
1.5	EXPLOSIVES 1.5	1000 pounds or more must be placarded.
1.6	EXPLOSIVES 1.6	1000 pounds or more must be placarded.
2.1	FLAMMABLE GAS	1000 pounds or more must be placarded.
2.2	NONFLAMMABLE GAS	1000 pounds or more must be placarded.
2.3	POISON GAS	All quantities must be placarded.
3	FLAMMABLE	1000 pounds or more must be placarded.
Combustible liquid	COMBUSTIBLE	1000 pounds or more must be placarded.
4.1	FLAMMABLE SOLID	1000 pounds or more must be placarded.
4.2	SPONTANEOUSLY COMBUSTIBLE	1000 pounds or more must be placarded.
4.3	DANGEROUS WHEN WET	All quantities must be placarded.
5.1	OXIDIZER	1000 pounds or more must be placarded.
5.2	ORGANIC PEROXIDE	1000 pounds or more must be placarded.
6.1 (PG I, inhalation hazard only)	POISON	All quantities must be placarded.
6.1 (PG I or II, other than PG I inhalation hazard)	POISON	1000 pounds or more must be placarded.
6.1 (PG III)	KEEP AWAY FROM FOOD	1000 pounds or more must be placarded.
6.2		No placard required.
7	RADIOACTIVE	Radioactive Yellow III label only; any quantity must be placarded.
8	CORROSIVE	1000 pounds or more must be placarded.
9	CLASS 9	1000 pounds or more must be placarded.
ORM-D		No placard required.

[a]The DANGEROUS placard is allowed when transporting two or more materials (excluding those that must be placarded in any quantity) instead of using each individual placard. However, when 5000 pounds or more of one hazard category is loaded at one facility, the placard for that category must be applied.

TRAINING

Employees who offer hazardous materials for transportation must be trained at least every two years, and certification of such training must be obtained.

REFERENCES

Department of Transportation (1997). 49 CFR Parts 171–180, U.S. DOT, Washington, DC.

BIBLIOGRAPHY

Environmental Resource Center (1994). *HM-181 and HM-126F: A Compliance Guide for DOT's New Hazmat Transportation Regulations,* Van Nostrand Reinhold, New York.

Government Institutes (1992). *Transportation of Hazardous Materials: A Compliance and Practice Guide for Safe Transportation of Hazardous Materials,* 2nd ed., GI, Rockville, MD.

GSI Training Services (1996). *10 Steps to Safety and DOT Compliance,* GSI, Branson, MO.

PART III

ENVIRONMENTAL ASSESSMENT AND MANAGEMENT OF HAZARDOUS MATERIALS AND HAZARDOUS WASTE

11

TANK SYSTEMS

Any permitted hazardous waste tank system must meet design requirements defined under RCRA in 40 CFR Part 264, Subpart J. Design criteria were revised in 1986 extensively and include the requirement for secondary containment and other technology requirements. Regulation of underground storage tanks containing petroleum products and hazardous substances listed under CERCLA was initiated by EPA as a result of the Hazardous and Solid Waste Amendments (HSWA). Regulations pertaining to these units are defined under 40 CFR Part 280. Aboveground process material and product tank systems and design criteria are not regulated federally but may be regulated at the state or local level.

Required or recommended design standards for aboveground and underground hazardous waste and hazardous materials tank systems include design standards not only for the tank but also for piping and ancillary equipment such as spill and overfill prevention equipment. In this chapter we discuss design elements for aboveground and underground tank systems. Additionally, testing methodologies for corrosion resistance and tank integrity testing are provided, and information about tank materials and chemical comp compatibilities is presented.

BASIC DESIGN CONSIDERATIONS

Basic considerations in designing a tank system include material selection, structural support, operational pressure, corrosion protection, and overfill and spill prevention. Material selection is based on compatibility with wastes or process or product materials. A general overview of materials and chemical compatibilities is provided later in the chapter.

Structural support considerations include soil load-bearing data, piling requirements, foundation properties, tank wall thickness requirements, integrity of seams (particularly on raised systems), and support of ancillary piping and connections. Other considerations affecting structural design include climatic conditions such as earthquake, hurricane, or flood risks and extremes of temperature. For underground tank systems, considerations include the weight of vehicular traffic, weight of structures above tanks, subsurface saturation characteristics, depth to groundwater, and depth of freeze zone.

The tank system should include pressure controls supplemented with devices such as rupture disks, pressure relief valves, and automatic shutoff devices at preset pressure changes to protect the tank against overpressurizing. Liquid-level controls for loading and unloading operations should include a preset high-level shutoff to prevent overfill and vacuum relief valves to protect the tank in negative pressure situations. Alarms for these and other abnormal operating conditions are also appropriate.

Factors affecting the potential for corrosion of an underground tank system are varied. These include soil properties such as pH, moisture content, sulfide level, and the presence of metal salts. Additionally, corrosion rates can be affected by other metal structures and interference (stray) electric current. Corrosion can occur inside the tank if the tank metallurgy is not adequate for the material stored.

Types of Aboveground Storage Tanks

There are several broad categories of aboveground storage tanks, including atmospheric tanks, low-pressure tanks, pressure tanks, and cryogenic tanks. Atmospheric tanks operate at atmospheric pressure, although some field applications operating at pressures below 0.5 pound per square inch gauge (psig) are referred to as atmospheric applications. Low-pressure tanks can operate at pressures ranging from 0.5 to 15 psig. Pressure tanks operate at pressures greater than 15 psig.

Tank designs for atmospheric tanks include the open top tank, fixed roof tank, and the floating roof tank. The open top tank is often used in wastewater treatment applications for equalization or neutralization. The fixed roof tank is used for storing materials with low vapor pressure, such as methanol, ethanol, and kerosene. Floating roof tanks are used mainly for petroleum products such as crude oil and gasoline. The floating roof design reduces evaporative losses while providing increased fire protection.

Low-pressure tank designs include hemispheroid and noded hemispheroid tanks and spheroid tanks. Low-pressure tanks are used for storage of volatile materials, which are gases at ambient temperature and pressure, such as pentane. In addition, combustible chemicals such as benzene and large-volume volatile liquids, such as butane, can be stored under low pressure.

Pressure tanks are typically cylindrical and are supported vertically or horizontally. These tanks are used to store high-volatility materials such as liquefied petroleum gas.

Cryogenic tanks are made of specialty steels with rigid low-temperature specifications and are heavily insulated. Examples of materials stored in this type of tank include ammonia, liquid nitrogen, liquid propane, and other compounds that are gaseous at atmospheric temperature and pressure.

Basic Design Considerations for Underground Storage Tanks

Underground storage tanks typically are used for storing high-volume flammable materials such as gasoline, diesel fuel, and heating oils. These tanks normally range from 5000 to 20,000 gallons in storage capacity and are horizontal cylindrical tanks. Underground storage tanks are especially suitable for flammable materials since diurnal temperature changes and tank breathing losses are minimized in underground applications. Other applications for underground tanks include storage of flammable product or process chemicals.

An integral part of underground storage tank design is the use of cathodic protection for prevention of tank surface corrosion. Guidelines for cathodic protection for underground tanks systems are published by several standards organizations and include:

- American Petroleum Institute (API) Publication 1632 (1987)—Cathodic Protection of Underground Petroleum Storage Tanks and Piping Systems, 2nd ed.
- National Association of Corrosion Engineers (NACE) Standard RP0285 (1985)—Control of External Corrosion on Metallic Buried, Partially Buried, or Submerged Liquid Storage Systems
- NACE Publication 2M363 (1963)—Recommended Practice for Cathodic Protection
- Underwriters' Laboratories (UL) Standard 1746 (1989)—Corrosion Protection Systems for Underground Storage Tanks

These guidelines address design considerations, coatings, cathodic protection voltage criteria, and installation and testing of cathodic protection systems.

Corrosion is an electrochemical phenomenon. In an electrochemical cell, the positive terminal—the anode—undergoes an oxidation reaction, while the negative terminal—the cathode—undergoes a reduction reaction. Corrosion is an oxidation reaction. Cathodic protection is a standard engineering method used to prevent external corrosion on the surface of a buried, partially buried, or submerged metal tank by allowing the metal surface to become the cathode of the electrochemical cell. A protective array of anodes is set in the soil at calculated distances from the tank to protect the tank against corrosion.

Two recognized types of cathodic protection systems are galvanic anode systems and impressed current anode systems. Galvanic anodes usually are made of magnesium or zinc and, depending on soil properties, may require use of special backfill material such as mixtures of gypsum, bentonite, and sodium sulfate (NACE, 1985a). The anodes are made of a less noble metal than the tank so that they will be more susceptible to undergoing an oxidation reaction. Once installed, the system provides sacrificial protection to the tank by inducing a continuous current source to inhibit electrical corrosion of the tank. Galvanic anode cathodic protection systems are limited in electrical current output. Thus applications for these systems are limited to tanks well insulated by a nonconductive coating that minimizes the exposed surface area of the tank.

Impressed current anode systems are made of higher-grade materials such as graphite, platinum, and steel. As with galvanic anode systems, soil properties may require the use of special backfill material such as coke breeze and calcined petroleum coke (NACE 1985a). A direct current source is used in this system. The anodes are connected to a positive terminal of the power source, while the structure is connected to the negative terminal.

A survey of a cathodic protection system to verify proper operation includes measurements such as structure to soil potential, anode current, structure to structure potential, piping to tank isolation (if protected separately), or effect on adjacent structures. A regulated system must be tested within six months after it is installed and at least every three years thereafter. Some state agencies and standards organizations require or recommend more frequent testing.

Underground tanks for petroleum products are often constructed of carbon steel that has been coated at the factory. Desirable characteristics of a coating include resistance to deterioration when exposed to products stored in the tank, high dielectric resistance, resistance to moisture transfer and penetration, and others. Testing using ASTM standard methods for determining cathodic disbonding of coatings can be performed on coated tanks and piping. These methods are presented in Appendix C. In addition to tests for disbonding of the coating, the coating should be tested for pinholes or other defects. Further, precautions must be taken by the construction personnel to ensure that the protective coat is not damaged during tank installation.

REGULATORY STANDARDS AND OTHER DESIGN CONCEPTS

Regulatory standards apply to hazardous waste tank systems and underground tank systems that store petroleum products and CERCLA hazardous substances. These standards and other design concepts that maximize environmental protection are described in this section.

Hazardous Waste Tank Systems

New tank systems for hazardous waste storage, including aboveground and underground tank systems, must be designed and installed to meet current regulatory requirements. These requirements include structural integrity assessment, inspection of installation, backfill material specifications, tightness testing, proper support of ancillary piping, corrosion protection, certification statements by tank system experts, and containment and detection systems that meet design criteria listed in the regulations.

Some of the requirements for containment and detection systems are:

- The system must be designed, installed, and operated to prevent any migration of wastes or accumulated liquid out of the system and into the environment.
- The system must be capable of detecting and collecting releases and accumulated liquids until the collected material is removed.
- The secondary containment for tanks must include either a liner external to the tank, a vault, a double-walled tank, or an equivalent device.
- A liner or vault must have enough capacity to contain the contents of the largest tank when full plus precipitation from a 25-year, 24-hour rainfall event.
- Liners, vault systems, and double-walled tanks must meet specified design criteria.
- Ancillary equipment must be provided with full secondary containment such as a trench, jacketing, double-walled piping, or other containment method that prevents migration of wastes into the environment and provides leak detection. Some exceptions apply to this requirement (i.e., if the aboveground piping is inspected daily and piping systems are equipped with automatic shutoff devices).

Additional requirements apply to underground hazardous waste tank systems. Any existing tank system that fails must be retrofitted to the established standards before the tank can be put back into service. Other existing tanks must be retrofitted to the standards by the time the tank reaches 15 years of age.

Design Concepts for Aboveground Tank Systems

In addition to RCRA requirements, there are several other design concepts that can be incorporated into any new aboveground double-containment project to maximize environmental protection (Langlois et al. 1990). These include:

- Maximum inspectability/testability
- Greater than 100% capacity of secondary containment for indoor applications as well as outdoor applications
- Specialized leak detection
- Elimination of penetrations through containment

These concepts are not detailed specifically in current regulations but are good management practices where economically achievable.

Inspectability/Testability. Installing tanks and ancillary piping aboveground is the first step toward making a tank system inspectable and testable. Raising the primary tank on beams or saddles inside the secondary containment allows for daily inspection under the tank. If the secondary containment is also raised, the system becomes 100% inspectable.

Similarly, if ancillary piping is run aboveground whenever possible, leaking joints or material failures can easily be detected. For cases where pipe-in-a-pipe is installed, ports with leak detectors and drain valves can be installed every 50 to 200 feet for assessment of primary pipe integrity.

Tanks and piping can be tested for leaks using the methods described in the next section of this chapter. A good engineering practice is to set up a schedule for periodic testing of all tanks at the facility. The length of time between tests will vary depending on application and materials of construction.

Greater Than 100% Capacity of the Secondary Containment. There are required RCRA standards for greater than 100% capacity of the secondary containment for outdoor liners and vaults used as secondary containment. In addition to these requirements, environmental protection can be maximized if indoor secondary containment is sized for catastrophic failure of the primary unit, plus spillage from ancillary piping. A capacity of 125% containment typically will be adequate. Use of this design concept will negate the possibility of spilled materials overflowing secondary containment and penetrating through the building floor.

Specialized Leak Detection. Standard leak detection systems, such as electronic moisture sensors or sensor floats, are used commonly in double containment projects. Although usually adequate for indoor applications, outdoor leak detection can be maximized with the use of specialized systems that differentiate chemical leakage from rainwater. Examples of these devices are detectors that alarm based on pH, oxidation/reduction potential, or conductivity.

Elimination of Penetrations Through Secondary Containment. Joints and penetrations are usually the weakest points in secondary containment and are the most likely to leak should a catastrophic failure occur. Locating pumps inside the secondary containment is one way to minimize penetrations. In addition, overhead trestles can be used to route piping over, instead of through, secondary tank walls.

Optimum Containment for Tanks, Pipes, and Pumps. As discussed earlier, optimum systems include inspectable/testable primary and secondary containment. An example of this type of tank system is a tank in a tank, with both tanks raised on beams. A tank in a pan, with tank and pan raised on beams, is another option. Maximum environmental protection is provided by double-contained piping options including pipe-in-a-pipe on an overhead trestle. Another piping option is pipe on a rack inside a pan, building, or trench.

As mentioned previously, pumps placed inside containment is the optimum design. When this design is not practical, pump drip pans can be used and are particularly useful for smaller applications. Drip pans are not adequate, however, for failures of pressurized pipes or other major failures such as diaphragm tears or broken seals. They are also less useful for outdoor applications, since rainwater must be drained or absorbed after each rain event.

Double Containment for Portable Chemical and Waste Containers. Containment of portable chemical and waste containers during storage and when in use is becoming a widespread practice in industry. Containment options vary widely, but 100% inspectable/testable secondary containment typically is not achievable. For most drum storage areas, such as container storage rooms and chemical distribution centers, curbed and coated concrete can be used. For these applications, a coating should be selected not only for chemical compatibility, but also for its ability to withstand fork truck and other equipment loadings. Physical inspections of the coating should be made regularly to ensure that cracking or delaminating has not occurred.

Smaller staging areas for chemical and waste containers can utilize bermed and coated concrete or a containment pan. For these applications, fork truck loadings are not usually a factor. The containment area should be large enough to contain, at a minimum, spillage from the largest container stored.

Routine barrel pumping operations should be contained using, as a minimum, a drip pan under the barrel and, preferably, a pan sized to contain at least the contents of one barrel. Relatively inexpensive, open-grated portable containment pallets are marketed that hold up to four drums and have enough spill capacity to contain a drum failure easily. These pallets can be moved with a fork truck. Another portable containment unit that holds one or two drums and is moved easily from one job site to another is also available.

Finally, specially designed, very durable, portable chemical containers that hold approximately 220 to 440 gallons (four to eight drums) are available. Although spill containment per se is not part of the container design, some containers are enclosed with a metal casing to ensure that dropping or excessive bumping does not crack the container. This design can reduce the risk of spills during transport.

Underground Storage Tanks

RCRA Requirements for Underground Storage Tanks Containing Regulated Materials. The problem of leakage from underground storage tanks (USTs) was recognized in the 1980s, and regulations were promulgated to identify leaking tanks, initiate cleanup activities, and set forth design standards for new USTs. RCRA performance standards for new USTs include the following:

- Tanks must be properly designed and constructed, and any portion underground that routinely contains product must be protected from corrosion.
- Piping that routinely contains regulated substances and is in contact with the ground must be properly designed, constructed, and protected from corrosion.
- Spill and overfill prevention equipment associated with product transfer to the UST must be designed to prevent a release of the product to the environment during transfers (including use of catch basins for hoses, automatic shutoffs, flow restriction, alarms, and other equipment).
- All tanks and piping must be properly installed in accordance with a nationally recognized code of practice.
- The installation must be performed by a certified installer and completed properly.

Standards for proper design and installation of underground tanks and piping are documented by several standards organizations, including:

- API RP 1615 (1987)—Installation of Underground Petroleum Storage Systems, 4th ed.
- American Society of Testing and Materials (ASTM) D4021-86—Standard Specification for Glass-Fiber-Reinforced Polyester Underground Petroleum Storage Tanks
- UL Standard 1316—Standard for Glass-Fiber-Reinforced Plastic Underground Storage Tanks for Petroleum Products

In addition to new tank performance standards, there are requirements for existing tanks. These include monthly release monitoring of tanks using an approved release detection method and monitoring of piping on a specified basis, depending on the pipe type and existing leak detection equipment in the pipe. Any leaks found must be reported immediately and corrective action must be taken to prevent further leakage and to remediate the site. Existing, nonleaking USTs that do not meet the performance standards must be upgraded to meet these or alternative standards no later than December 22, 1998.

Additional requirements are defined for UST systems that contain CERCLA hazardous substances. These include secondary containment performance that will prevent the release of these substances to the environment at any time during operational life of the UST system.

Maximizing Environmental Protection. EPA's guidance manual entitled *Detecting Leaks: Successful Methods Step by Step* (EPA 1989) provides information on design concepts that maximize environmental protection for USTs. Information is provided on acceptable types of secondary containment for underground tanks, interstitial monitoring, and underground piping. In addition, release detection methods are addressed.

Secondary Containment. Secondary containment systems should be designed to provide an outer barrier between the tank and the soil and backfill material that is capable of holding the material long enough so that a release can be detected. For tank systems containing hazardous substances, preventing migration of substances is also a requirement. A double-walled tank, a jacketed tank, and fully enclosed external liners can provide this type of protection. In addition, the secondary containment must have enough capacity to hold the contents of the tank (or the largest tank if more than one tank is included in the system) and should prevent precipitation or groundwater intrusion.

To maximize environmental protection for double-walled tanks, the second tank should be made of the same material as the primary tank. For jacketed tanks or tanks with liners, the material used should be compatible with the product stored. The material should also be sufficiently thick and impermeable to direct a release to the monitoring point and permit its detection. The permeation rate of the stored substance through the jacket or liner should be 10 to 6 centimeters per second (cm/s) or less (EPA 1989).

Interstitial Monitoring. An interstitial monitoring system should be designed to detect any leak from the underground tank under normal operating conditions. In most cases, the leak detection system does not measure the leak rate, only the presence of a leak. Interstitial monitoring systems operate to detect leaks based on one of several mechanisms, such as electrical conductivity, pressure sensing, fluid sensing, hydrostatic monitoring, manual inspection, and vapor monitoring. Conductivity, fluid sensing, manual inspection, and vapor monitoring are suitable for all applications. Pressure sensing and hydrostatic monitoring are applicable only to double-walled installations.

Because of the inability to inspect physically tanks that are underground, the interstitial monitoring system becomes the primary means of verifying tank integrity. For this reason, potential problems with the monitoring system should be identified before installation and reflected in the system's design. These potential problems might include groundwater or aboveground runoff penetration into the secondary containment from pinhole leaks in the secondary containment system or from inadequate lining. Further, faulty electrical installation can render some monitoring devices inoperable.

Other Release Detection Methods. For existing UST systems that do not have interstitial monitoring, other release detection methods are required. These methods include inventory control, vapor monitoring, groundwater monitoring. manual or automatic tank gauging, and tank tightness testing using volumetric tests.

Product inventory control is a technique effective in finding leaks over 1 gallon per hour. Because of its low sensitivity, it must be combined with tank tightness testing, with the tank tightness testing occurring at least every five years until the system is retrofitted to performance standards. In general, this method requires careful tracking of inputs, withdrawals, and amounts still remaining in the tank. API RP 1621, "Recommended Practice for Bulk Liquid Stock Control at Retail Outlets" (API 1987), provides guidance for meeting inventory control requirements.

Vapor monitoring is used predominantly with USTs that store petroleum products. Success of this type of release detection system is dependent on site geology and soil types. Vapor monitoring systems consist of a vapor monitoring well and a vapor sensor. If a product leaks from an UST, the liquid and vapors spread throughout the surrounding soil. If the vapor sample reaches a sensor in a concentration above a predetermined set point, the system responds with an alarm.

Groundwater monitoring is a release detection method effective for petroleum and other products that have a specific gravity of less than 1.0. The method can be used in areas where the groundwater table is very near to the excavation zone of the tank. Permanent observation wells are placed close to the tank and the wells are checked periodically for evidence of free phase product on top of the water in the well.

Manual and automatic tank gauging and tank tightness testing using volumetric tests are also acceptable methods of leak detection. These methods are described in the next section of this chapter.

Design of Underground Piping. Double-contained piping with interstitial monitoring maximizes environmental protection in underground systems and is required for uses with hazardous substances. For pressurized lines, the use of automatic flow restrictors to restrict flow in the event of a leak and automatic flow shutoff devices to stop flow when a pressure drop occurs also provide environmental protection.

Underground double-contained piping options include the use of trench liners as well as double-walled pipe. For the trench liner application, the pipe trench is lined with a flexible membrane that is impervious to the stored product. Often, the liners are thermoplastic or polymeric sheets and are at least 50 mils thick. The trench is designed sloping away from the tank so that pipe leaks can be differentiated from tank leaks.

For applications of double-walled piping, the inner and outer pipes can be made of the same material or the outer pipe can be made of a cheaper material, such as fiberglass-reinforced plastic. Fiberglass-reinforced plastic is not used for highly pressurized applications, but if compatible with the product is functional in a nonpressurized or less pressurized state. Double-walled piping is usually sloped to a containment structure, sump, or observation well that can be monitored for the presence of liquids or vapors.

TANK TESTING METHODS

Testing of tanks for corrosion resistance on a periodic basis can indicate the need for repairs and can aid in averting catastrophic failure of the system through early warning of material thinning or fatigue. Similarly, tank integrity testing, if performed according to a preset periodic schedule, can allow for early detection of leaks or pending failure. In this section we discuss test methodologies used for corrosion resistance and tank integrity testing.

Testing for Corrosion Resistance

There are numerous testing procedures for determining corrosion resistance of common materials used in tank systems. Some of these tests may be suitable for field use to determine the extent of corrosion in existing equipment. Others are useful for material screening during the system design phase. Tank manufacturers usually perform a variety of corrosion resistance and other tests on a tank material and make the test data available to potential customers.

Testing for Pitting and Crevice Corrosion Resistance of Stainless Steels and Related Alloys. This test is defined in ASTM G 48-76 (1980). The ASTM standard describes two test methods: method A, which is a total immersion ferric chloride test, and method B, which is a ferric chloride crevice test. Method A is used to determine the relative pitting resistance of stainless steels and other alloys, particularly those that have been heat treated or have had surface finishes applied. Method B is useful for determining both pitting and crevice corrosion resistance of those metals.

Standard specimens are immersed in a 10% ferric chloride solution in both tests for approximately 72 hours. For crevice testing, the specimens are attached to TFE–fluorocarbon blocks with O-rings or rubber bands to form crevices at the points of contact. The test specimens are evaluated by visual inspection, measurement of pit and crevice depth, and weight loss.

Guide for Conducting Corrosion Coupon Tests in Plant Equipment. This guide, ASTM G4-84, defines a method for evaluating corrosion of engineering materials under the varying conditions present in actual service. The method typically is used to evaluate materials of construction for use in similar service or as replacement or modification materials.

Size and shape of test specimens is dependent on the specific test application. The duration of exposure is based on known rates of corrosion of the materials in use, or by the convenience by which plant operations can be interrupted to introduce and remove test specimens. Evaluation of the specimen includes microscopic inspection for etching, pitting, tarnishing, scaling, and other defects. The depth of the pits can be measured and number, size, and distribution noted. A metallographic examination for intergranular corrosion or stress-corrosion cracking may also be performed.

Standard Test Method for Chemical Resistance of Protective Linings. This method is defined in ASTM C868-85(1990). The standard defines a procedure for testing and evaluating the chemical resistance of a protective lining applied to steel or other metals. The liner is applied and cured, and then immersed in the service solution for six months. As necessary to simulate actual conditions, heat may be applied. Color, surface gloss, surface texture, and blisters are all visually evaluated before, at interim periods, and after the completion of the test. In addition, lining thickness is measured before and after the test to quantify the chemical attack on or the dissolution of the lining material.

Standard Methods of Testing Vulcanizable Rubber Tank and Pipe Lining. This test method is documented in ASTM D3491-85, which describes testing and evaluation of chemical resistance of vulcanizable rubber tank lining. A test specimen of a rubber component applied to a steel plate is immersed in the service solution and heated, if required. The duration of the test should be a minimum of six months, with inspections performed every month. At the end of the test period, a visual inspection is made for changes in surface texture, evidence of cracking, blistering, swelling, delaminating, and permeation. In addition, substrate attack or corrosion such as rusting or metal darkening is noted. Because of the length of time required for this test, prescreening of lining materials prior to this test can be performed using ASTM D471-79(1991), which is a more convenient method.

X-Ray Fluorescence. The x-ray fluorescence (XRF) test method for determining coating thickness is defined in ASTM A754-79(1990). A radiation detector that can discriminate between the energy levels of all radiations is used to measure the thickness of a coating and substrate exposed to an intense beam of radiation generated by a radioisotope source or an x-ray tube. The combined interaction of the coating and substrate with the beam of radiation generates x-rays of well-defined energy, which are singularly characteristic of that element.

If XRF thickness testing is performed in the field, environmental factors such as temperature, humidity, and surface cleanliness must be taken into account. Other factors, including specimen size, specimen uniformity, radiation source, and radiation detector, can also affect the test results.

Other ASTM Standards. Other ASTM standards, guides, and practices that address corrosion testing are documented in Volume 3.02, "Wear and Erosion; Metal Corrosion," *Annual Book of ASTM Standards* (ASTM 1992b). Examples of these standards include:

- G1-90—Recommended Practice for Preparing, Cleaning, and Evaluating Corrosion Test Specimens

- G15-90—Terminology Relating to Corrosion and Corrosion Testing
- G46-76(1986)—Recommended Practice for Examination and Evaluation of Pitting Corrosion
- G50-76(1984)—Recommended Practice for Conducting Atmospheric Corrosion Tests on Metals
- G78-89—Guide for Crevice Corrosion Testing of Iron Base and Nickel Base Stainless Alloys in Seawater and Other Chloride-Containing Aqueous Environments
- G82-83(1989)—Guide for Development and Use of a Galvanic Series for Predicting Galvanic Corrosion Performance
- G96-90—Practice for On-line Monitoring of Corrosion in Plant Equipment (Electrical and Electrochemical Methods)
- G104-89—Test Method for Assessing Galvanic Corrosion Caused by the Atmosphere

Additional ASTM standards pertaining to corrosion resistance and other tank testing are presented in Appendix C.

Tank Integrity Testing

Once a tank system is installed, periodic tank integrity testing is a necessary part of the system's operation and maintenance program. A guide for selection of a leak testing method for tank and material testing and other applications is presented in ASTME 432-91. In addition, commonly used nondestructive test methods pertaining to tank testing are addressed in the following:

- *Annual Book of ASTM Standards,* Vol. 3.03, "Nondestructive Testing" (ASTM 1992a)
- *ASME Boiler and Pressure Vessel Code,* Section V, "Nondestructive Testing" (ASME 1986)
- "Design and Construction of Large, Welded, Low-Pressure Storage Tanks" (API 1990)
- *Detecting Leaks: Successful Methods Step by Step,* (EPA 1989)

Most methods for tank system testing are applicable to aboveground tanks. Tests applicable to underground tanks are limited because of inaccessibility of the tank.

Holiday Test. The holiday test, defined in ASTM G62-87, is used to detect pinholes, voids, or small faults that allow current drainage through protective coatings on steel pipe or polymeric precoated corrugated steel pipe. Although this test method defines the holiday test for pipeline coatings, it is also applicable to the walls and bottom of a coated tank. A highly sensitive electrical device is used in conjunction with water or another electrically conductive wetting agent to locate pinholes and thin spots (defined as holidays) in coatings of steel pipes and tanks. If electrical contact is made on the metal surface through a holiday, an alarm is activated to alert the operator of the coating flaw.

There are two methods defined for performing the test, and the thickness of the coating determines which method should apply. For thin-film coatings of 1 to 20 mils, a low-voltage holiday detector is used, which has an electrical energy source of less than 100 V dc. This method detects pinholes and other voids but will not detect thin spots in the coating. The high-voltage detector has an electrical energy source of 900 to 20 000 V dc, and can be used to detect holidays and thin spots in the coating. The high-voltage detector normally is used on materials that have a coating thickness of greater than 20 mils.

Magnetic Particle Testing. Magnetic particle testing is used to detect cracks and other discontinuities near the surface of ferromagnetic materials. Applications include tank walls, tank bottom, and welds. The area to be tested is cleaned and then magnetized. Magnetic particles are applied to the surface. The particles form patterns where there are disturbances in the normal magnetic field and indicate cracks or flaws. The method is sensitive to very small discontinuities. ASTM standards that relate to this method include:

- A275/A275M-90—Method for Magnetic Particle Examination of Steel Forgings
- E125-63(1985)—Reference Photographs for Magnetic Particle Indications on Ferrous Castings
- E709-80(1985)—Practice for Magnetic Particle Examination

Ultrasonics Testing. Ultrasonics testing is defined in ASTM E1002-86 and in the ASME Code, Section V, Article 5. This test can be used to locate pressurized gas leaks and estimate leak rates. In general, this test is considered a screening tool to be used prior to other, more sensitive and time-consuming tests.

This test method uses an acoustic leak detection system to detect impulsive signals that are much larger than background noise level. The ultrasonic test system provides for detection of acoustic energy in the ultrasonic range and translates energy into an audible signal that can be heard by use of speakers or earphones. The detected energy is indicated on a meter readout. Leak rates can be approximated from a formula that uses the maximum detection distance at calibrated sensitivity.

Acoustic Emission Testing. This test method can be used to monitor vessels and piping during operation to detect defects such as flaws and cracks. The tank or pipe is put under pressure, which results in a stress concentration, causing the defect to enlarge. This enlargement generates sound vibrations that can be detected using sensors located along the surface of the structure. Arrival times at the sensors are used to pinpoint the defect. To evaluate the acoustical vibrations properly, the tank should be in a quiescent state.

ASTM standards that pertain to this test include:

- E750-88—Practice for Measuring Operating Characteristics of Acoustic Emission Instrumentation
- E976-84(1988)—Guide for Determining the Reproducibility of Acoustic Emission Sensor Response
- E1067-89—Practice for Acoustic Emission Testing of Fiberglass Reinforces Plastic Resin (FRP) Tanks/Vessels
- E1139-87—Practice for Continuous Monitoring of Acoustic Emission from Metal Pressure Boundaries
- E1211-87—Practice for Leak Detection and Location Using Surface-Mounted Acoustic Emission Sensors

Hydrostatic Leak Testing. Hydrostatic testing, documented in ASTM E1003-84(1990) and API Standard 620, is a method for testing tanks with pressurized liquid. The method requires that a component be filled completely with a liquid, preferably water. Pressure is applied slowly to the liquid until the required pressure, usually between 75 and 150% of the designed operating pressure, is reached. The pressure is held for a designated period of time. Leakage can be determined by visual inspection or pressure drop indication.

Since liquid may clog small leaks, this method of testing is performed after pneumatic testing. In addition, the test liquid temperature must be equal to or above ambient temperature or condensation can form on the outside of the tank, making visual leak detection difficult.

Pneumatic Pressure Testing. Pneumatic pressure testing is described fully in the ASME Code. The method outlines a procedure for testing tanks and metal piping with air pressure. The test is not appropriate for fiberglass or other low-pressure materials. The empty tank or pipe is pressurized to the operating design pressure and held for a specified period of time, during which any drop in pressure is noted. Since air at high pressure can be an explosive hazard if a contaminant such as oil vapor is present, inspections should be made at a reasonable distance from the tank, using field glasses, as required.

Liquid Penetrant Testing. Liquid penetrant test methods and practices are described in several ASTM standards and in the ASME Code, Section V, Article 6. ASTM standards include:

- E165-91—Practice for Liquid Penetrant Inspection Method
- E433-71(1985)—Reference Photographs for Liquid Penetrant Inspection

- E1208-91—Test Method for Fluorescent Liquid Penetrant Examination Using the Lipophilic Post-Emulsification Process
- E1209-91—Test Method for Fluorescent Penetrant Examination Using the Water-Washable Process
- E1210-91—Test Method for Fluorescent Penetrant Examination Using the Hydrophilic Post-emulsification Process
- E1219-91—Test Method for Fluorescent Penetrant Examination Using the Solvent-Removable Process
- E1220-91—Test Method for Visible Penetrant Examination Using the Solvent-Removable Process

The test allows for visible penetrant examination for detecting discontinuities such as cracks, openings in seams, and isolated porosity. Testing involves spreading a liquid penetrant, typically a light oil with visible or fluorescent dyes in it, over the surface to be tested. The penetrant is given time to enter open discontinuities. Excess penetrant is removed, and the liquid in any discontinuity is drawn out. The discontinuity and near surfaces are stained by the penetrant in the process. The surface can then be visually inspected for indications of surface discontinuities.

Partial-Vacuum Tests. Partial vacuum tests on closed tanks are defined in API Standard 620. These tests are performed to ensure that the tank walls and roof meet design specifications. During the test, water is withdrawn from the tank with all vents closed until the design partial vacuum is developed at the top of the tank. Observations are made to as to when the vacuum relief valves start to open. These valves should open before the design pressure is reached. These tests are performed with the tank full, half full, and empty.

Partial vacuum testing using a vacuum box can be performed on welds of a tank bottom that rests directly on the ground. This test is accomplished by applying a solution film at the joints and pulling a partial vacuum of at least 3 psig. For this application, the vacuum box must have a transparent top.

Tank Gauging. Manual tank gauging, commonly called *static testing,* is defined in EPA's guidance manual for release detection (EPA 1989). The test is effective for small-volume tanks of less than 550 gallons. The liquid level is measured in a quiescent tank at the beginning and end of a 36-hour period or other specified length of time. Any change in liquid level can be used to calculate the change in volume. Unless dramatic temperature changes occur, the liquid level change can be compared against established guidelines to determine whether any differences in the measurements are significant enough to indicate a leak.

Automatic gauging systems can be permanently installed in tanks to provide both tank integrity testing and inventory information. The system can measure the change in product level within the tank over time and can detect drops in level not associated with tank withdrawals.

Volumetric Tank Testing. This test, defined in EPA's guidance manual (EPA 1989) is applicable mainly to underground storage tanks. The tank is tested for tightness by placing a known volume of liquid, usually water, into the tank for a period of time. The tank level is monitored for any changes during the test that might indicate leakage. For maximum sensing of level change, the tank should be overfilled so that the liquid reaches the fill tube or standpipe located above grade. Since the level changes occur in a small area, small changes in volume can be detected readily with gauging equipment.

Temperature variations must be taken into account during testing since they can affect the volume. In addition, structural deformation resulting from filling the tank can occur, so a waiting period must be observed to ensure that this effect has stabilized.

MATERIAL SELECTION FOR TANK SYSTEMS

When selecting materials for a tank system, chemical compatibility and material cost are the two key considerations. Material selection must be made carefully for each application and must include expected variations in chemical or waste solutions and storage temperature and pressure ranges. Other factors to consider include normal atmospheric conditions, as well as hazardous climatic conditions such as the possibility of earthquakes and hurricanes.

Materials selected for the primary tank should be as optimum as the budget will allow. To keep costs down, the secondary containment, which is used only in an emergency, often can be made of less expensive materials.

Material selections for tank systems must be made based on the parameters of the individual application. Test data for compatibility of specific chemicals with various metal and nonmetallic materials have been compiled by NACE (1975, 1985b) into two volumes entitled *Corrosion Data Survey*. Included in metal tests are:

- Iron-based metals such as carbon steel and stainless steel
- Copper-based metals such as copper and copper alloys
- Nickel alloys such as nickel–chrome–iron and nickel–chrome–molybdenum
- Other metals and alloys, such as aluminum, silver, and titanium

Included in the nonmetal tests are materials such as carbon, glass, synthetic and natural rubber, epoxy fiberglass and other fiberglass materials, and some plastics, such as polyethylene, polypropylene, and polyvinyl chloride. The temperature range for use of these materials is generally 70 to 140°F.

Presented next is a general overview of types of chemical families that typically are used in industry and a general discussion of compatibility requirements for these chemical types, based on NACE data. Both metals and nonmetallic materials are addressed. All material selections should be investigated thoroughly and/or tested before being put into service.

Acids

The iron-based metals—cast iron, carbon steel, and stainless steel—are not recommended for use with weak acids. Austenitic stainless steels, such as AISI 316 and 317, and several nickel-based alloys with molybdenum are suitable for some acids at varying concentrations and at low temperatures (<200°F). Other alloys and metals, such as gold and platinum alloys, silver, tantalum, titanium, and Hastelloy, are also resistant to specific acids at somewhat higher temperatures. These alloys and metals typically are not used in large applications because of expense. Gaskets, valves, and other small but critical parts, however, often are made of these stronger materials since fittings and joints are particularly subject to chemical attack.

Acid-compatible nonmetals include polyester–fiberglass, glass-lined steel, and synthetic (butyl and fluorine) and natural rubber, depending on the application. Epoxy fiberglass and carbon are also compatible for some applications. Plastics such as polyethylene and polypropylene are adequate for use with some weak acids. These materials are not recommended for use with high concentrations of strong acids such as sulfuric acid, nitric acid, and hydrofluoric acid.

Alcohols

Carbon and stainless steel, cast iron, aluminum, and other metal alloys are compatible with most alcohols at low temperatures. Synthetic or natural rubber sometimes can be applied with success, but the type of rubber varies according to specific alcohol type. If the rubber is incompatible, the alcohol will dissolve it. Isopropyl alcohol is most compatible with fluorine rubber, but butyl and natural rubber also can be used at low temperatures. Other acceptable nonmetals data varied significantly according to type of alcohol.

Aldehydes

Generally, stainless steels of AISI 304 and higher are compatible with aldehydes, depending on the specific chemical. Alloys such as nickel-based alloys and copper-based alloys can also be used successfully in most applications.

Nonmetallic materials are not as suitable. Furfuryl alcohol–glass, glass-lined steel, and epoxy–asbestos–glass are compatible for various applications, but material costs are generally prohibitive. Compatibility with other nonmetals varies. Plastics such as polypropylene soften at higher concentrations and temperatures. Applications for synthetic and natural rubber are very limited.

Ammonium Solutions

Ammonium solutions are not compatible with copper-based metals. Stainless steel and nickel-based alloys are compatible with most ammonium solutions. Many solutions are compatible with aluminum.

Synthetic and natural rubber are resistant to almost all ammonium solutions at low temperatures. Glass-lined steel is resistant to the more aggressive solutions, such as ammonium fluoride. Polychloroprene, polyethylene, and polypropylene are compatible for many less aggressive solutions.

Ammonia is compatible with most metals and is commonly handled in carbon steel. Urea is compatible with AISI 304 stainless steel and above.

Caustics

Carbon steel and AISI 304 and 316 stainless are acceptable for most caustic solutions, particularly at low temperatures. Nickel and nickel-based alloys can also be used.

Natural and butyl rubbers are generally compatible with caustics. Other materials, such as polychloroprene, polypropylene, epoxy fiberglass, and polyvinyl chloride, also are generally acceptable.

Petroleum Distillates and Offshore Applications

Most petroleum distillates are compatible with a wide variety of metals, with carbon steel being the preferred material for tank systems throughout a petroleum facility. Synthetic and natural rubbers generally are inadequate. Polyethylene and polypropylene also are inadequate because the aromatic hydrocarbons tend to cause varying degrees of material swelling, softening, and stress cracking. Some glass, epoxy, or fiberglass materials have proven compatible, but petroleum distillate tank applications normally are too large for these materials to be selected from a structural standpoint.

Offshore piping and other applications must be resistant to seawater. Cupronickel alloys are compatible for these uses.

REFERENCES

American Petroleum Institute (1987). *Recommended Practice for Bulk Liquid Stock Control at Retail Outlets,* 4th ed., API, RP 1621, API, Washington, DC.

——— *Design and Construction of Large, Welded, Low-Pressure Storage Tanks* (1990). 8th ed., API Standard 620, API, Washington, DC.

American Society for Testing and Materials (1992a). *Annual Book of ASTM Standards,* Vol. 3.03, "Nondestructive Testing," ASTM, Philadelphia.

——— (1992b). Annual Book of ASTM Standards, Vol. 3.02, "Wear and Erosion: Metal Corrosion," ASTM, Philadelphia.

American Society of Mechanical Engineers (1986). *ASME Boiler and Pressure Vessel Code,* Section V, "Nondestructive Testing," ASME, New York.

Environmental Protection Agency (1989). *Detecting Leaks: Successful Methods Step by Step,* EPA/530/UST-89/012, Office of Underground Storage Tanks, Office of Solid Waste and mergency Response, U.S. EPA, Washington, DC.

Langlois, K. E., C. Bauer, and G. Woodside (1990). "Double Contained Wastewater Treatment Tanks," *Proceedings of the Water Pollution Control Federation Annual Conference,* Washington, DC, October.

National Association of Corrosion Engineers (1975). *Corrosion Data Survey, Metals Section,* 5th ed., NACE, Houston, TX.

——— (1985a). *Control of External Corrosion on Metallic Buried, Partially Buried, or Submerged Liquid Storage Systems,* NACE Standard RP0285-85, NACE, Houston, TX.

——— (1985b). *Corrosion Data Survey: Metals Section,* 6th ed., NACE, Houston, TX.

BIBLIOGRAPHY

American Society of Mechanical Engineers (1986). *ASME Boiler and Pressure Vessel Code,* Section VIII, "Rules for Construction of Pressure Vessels," ASME, New York.

American Society for Testing and Materials (1988). *Galvanic Corrosion,* Special Technical Publication 978, H. P. Hack, ed., ASTM, Philadelphia.

——— (1989). *Effects of Soil Characteristics on Corrosion,* Special Technical Publication 1013, Chaker and Palmer, eds., ASTM, Philadelphia.

——— (1990). *Corrosion Testing and Evaluation: Silver Anniversary Volume,* Special Technical Publication 1000, Baboian and Dean, eds., ASTM, Philadelphia.

——— (1991). *Acoustic Emission: Current Practice and Future Directions,* Special Technical Publication 1077, Sachse, Roget, and Yamaguchi, eds., ASTM, Philadelphia.

American Society of Mechanical Engineers and American National Standards Institute, (1987). *Chemical Plant and Petroleum Refinery Piping,* ASME/ANSI Standard B31.1, ASME/ANSI, New York.

Cole, M. G. (1992). *Underground Storage Tank Installation and Management,* Lewis Publishers, Boca Raton, FL.

De Renzo, D. J. (1985). *Corrosion Resistant Materials Handbook,* 4th ed., Noyes Data Corporation, Park Ridge, NJ.

Ecology and Environment, Inc., and Whitman, Requardt, and Associates (1985). *Toxic Substance Storage Tank Containment,* Noyes Data Corporation, Park Ridge, NJ.

Environmental Protection Agency (1986). *Underground Tank Leak Detection Methods: A State-of-the-Art Review,* EPA/600/2-88/001, prepared by IT Corporation for Hazardous Waste Engineering Research Laboratory, Office of Research and Development, U.S. EPA, Washington, DC.

——— (1987). *Soil-Gas Measurement for Detection of Subsurface Organic Contamination,* Environmental Monitoring Systems Laboratory, U.S. EPA, Washington, DC.

——— (1988). *Analysis of Manual Inventory Reconciliation,* prepared by Midwest Research Institute for the Office of Underground Storage Tanks, Office of Solid Waste and Emergency Response, U.S. EPA, Washington, DC.

——— (1988). *Common Human Errors in Release Detection Usage,* prepared by Camp Dresser & McKee, Inc., for U.S. EPA, Washington, DC.

——— (1988). *Evaluation of Volumetric Leak Detection Methods for Underground Fuel Storage Tanks,* Vol. 1, EPA/600/2-88/068a, prepared by Vista Research, Inc., for U.S. EPA, Washington, DC.

——— (1988). *Review of Effectiveness of Static Tank Testing,* prepared by Midwest Research Institute for the Office of Underground Storage Tanks, Office of Solid Waste and Emergency Response, U.S. EPA, Washington, DC.

——— (1988). *Standard Practice for Evaluating Performance of Underground Storage Tank External Leak/Release Detection Components and Systems,* prepared by Radian Corporation for Environmental Monitoring Systems Laboratory, U.S. EPA, Washington, DC.

——— (1989). *Soil Vapor Monitoring for Fuel Tank Leak Detection: Data Compiled for Thirteen Case Studies,* prepared by On-Site Technologies for Environmental Monitoring Systems Laboratory, U.S. EPA, Washington, DC.

——— (1989). *Volumetric Tank Testing: An Overview,* EPA/625/9-89/009, U.S. EPA, Washington, D.C.

Gangadharan, et al. (1988). *Leak Prevention and Corrective Action for Underground Storage Tanks,* Noyes Data Corporation, Park Ridge, NJ.

Jawad, M. H., and J. R. Farr (1989). *Structural Analysis and Design of Process,* Wiley, New York.

LeVine, R., and Arthur D. Little, Inc. (1988). *Guidelines for Safe Storage and Handling of High Toxic Hazard Materials,* American Institute of Chemical Engineers–Center for Chemical Process Safety, New York.

National Association of Corrosion Engineers (1975). *Control of Internal Corrosion in Steel Pipelines and Piping Systems,* NACE Standard RP0175, NACE, Houston, TX.

Rizzo, J. A. (1991). *Underground Storage Tank Management: A Practical Guide,* Government Institutes, Rockville, MD.

Rizzo, J. A., and A. D. Young (1990). *Aboveground Storage Tanks: A Practical Guide,* Government Institutes, Rockville, MD.

Schwendeman, T. G., and H. K. Wilcox (1987). *Underground Storage Systems: Leak Detection and Monitoring,* Lewis Publishers, Boca Raton, FL.

Young, A. D. (1990). *Corrective Response Guide for Leaking Underground Storage Tanks,* Government Institutes, Rockville, MD.

Whitlow, R. (1990). *Basic Soil Mechanics,* 2nd ed., Wiley, New York.

Woodside, G., and J. J. Prusak (1992). "Above Ground Storage: State-of-the-Art Systems," *Proceedings of the Annual Air and Waste Management Association Annual Conference,* Kansas City, MO, June.

12

WASTE TREATMENT AND DISPOSAL TECHNOLOGIES

Waste treatment and disposal systems can incorporate many different unit processes. For example, a wastewater treatment system may include oil–water separation, neutralization, biological treatment, sedimentation, and sludge dewatering. There are many well-known and widely used technologies, such as activated sludge, incineration, carbon adsorption, air stripping, metal precipitation, ion exchange, reverse osmosis, distillation, landfilling, land treatment, and deep well injection. However, more stringent environmental regulation and an increasing knowledge of chemical hazards continually challenge the capabilities of waste treatment and monetary resources. These challenges have given rise to a keen interest in the development of new, innovative technologies. Many of these new technologies use familiar unit processes, but in innovative ways. Recycling technology, an important part of pollution prevention, incorporates both old and new unit processes.

In this chapter, the discussion of waste treatment and disposal technologies is divided into five major groups reflecting either the type of technology or major environmental area: physical/chemical processes, biological processes, thermal processes, soil and groundwater treatment, and land-based systems. To make it easy for the reader to locate a particular technology, technologies are listed alphabetically within each group.

PHYSICAL AND CHEMICAL PROCESSES

Carbon Adsorption

Carbon adsorption is a well-established and widely used technology for the removal of organics from wastewaters and gaseous streams. Carbon adsorption is identified as a "best available technology" for meeting maximum contaminant levels for organic contaminants in drinking water at 40 CFR

§141.61(b). Thus, as a proven technology for drinking water, carbon adsorption can reduce concentrations to very low or nondetectable levels.

In carbon adsorption, contaminants are physically attracted or adsorbed on the surface of the carbon. Adsorption capacities are high for carbon because its porous nature provides a large surface area relative to its volume. *Activated carbon* is prepared from lignite, bituminous coal, coke, wood, or other organic materials, such as coconut shells.

Carbon adsorption is most effective at removing organic compounds that have low polarity, high molecular weight, low water solubility, and high boiling point. Some general guidelines for assessing carbon adsorption efficiency are (WEF 1990):

- Branched-chain compounds are more sorbable than straight-chain compounds.
- Compounds low in polarity and water solubility [less than 0.1 milligram per liter (mg/L)] are more sorbable.
- For compounds that have similar chemical characteristics, those of larger molecular size are more sorbable.
- There is a wide range in adsorption efficiency for inorganics.
- A low pH value promotes adsorption of organic acids.
- A high pH values promotes adsorption of organic bases.

Common examples of compounds that are amenable to carbon adsorption are aromatics such as benzene and toluene; and chlorinated organics such as trichloroethene, trichloroethane, tetrachloroethene, polychlorinated biphenyls (PCBs), DDT, and pentachlorophenol. Compounds that are not adsorbed effectively by carbon include ethanol, diethylene glycol, and numerous amines (butylamine, triethanolamine, cyclohexylamine, hexamethylenediamine) (WEF 1990). In general, to be suitable for carbon adsorption, the organic concentration of a wastewater should be less than 5000 mg/L.

Most carbon adsorption units use granular activated carbon (GAC), which has an average particle diameter of about 1500 μm. The powdered form of activated carbon (PAC) typically is less than 100 μm in diameter. PAC is used in the treatment of drinking water and wastewater treatment (see the section on "Activated Sludge" under the heading "Biological Processes"). It may also be used to reduce dioxins in incinerator emissions (Roeck and Sigg 1996).

GAC may be used in fixed or "moving" beds and in downflow and/or upflow mode. Fixed beds are operated in downflow or upflow (but not both) mode and as such, provide some amount of solids filtration, as shown in Figure 12.1. In fixed beds, influent solids concentration must be kept low (less than 50 mg/L suspended solids) to prevent rapid plugging of the bed. Filtered solids are removed periodically by backwashing. Upflow moving beds are more tolerant of solids because they are fluidized and expanded by

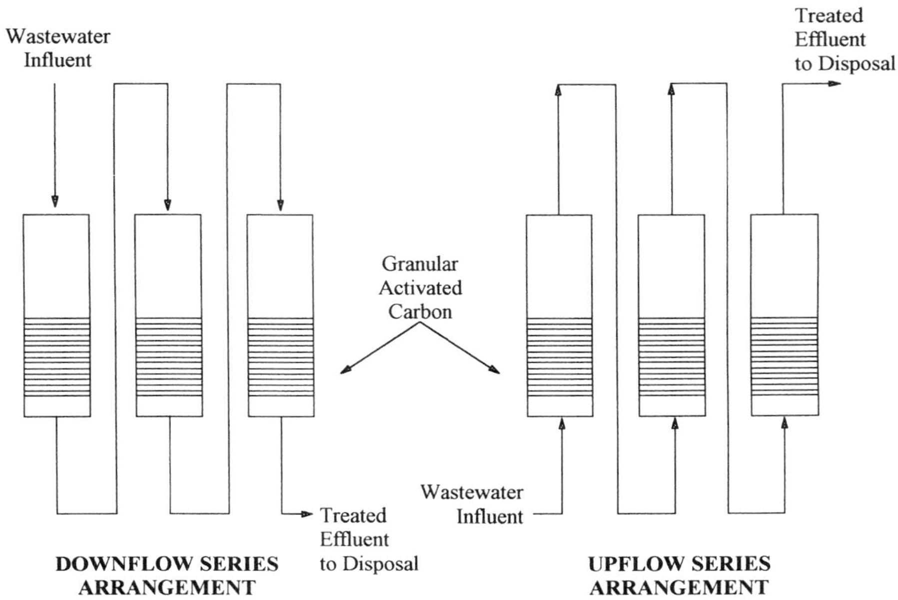

Figure 12.1 Carbon adsorption systems.

the wastewater entering at the bottom. In moving beds the flow is countercurrent and makeup, fresh carbon is added continuously at the top of the unit while an equal amount of spent carbon is removed from the bottom.

When the adsorption capacity of a carbon unit is exceeded, there is breakthrough of the contaminant in the treated stream. Fixed beds may be operated in series to allow continuous treatment while spent or exhausted units are recharged with fresh carbon. In series operation, there are two or more units. The majority of the contaminant is removed by the first unit in the series with the downstream units acting as polishing units. When breakthrough occurs in the first or primary unit, it is recharged with fresh carbon and becomes a polishing unit while the next unit in the series takes over and becomes the primary treatment unit. This concept is illustrated in Figure 12.2.

Spent carbon is usually regenerated, but if the quantity is small, it may be disposed instead by incineration or landfilling. Regeneration can be done by several different methods and may actually be part of the treatment process. Regeneration technologies include steam or thermal desorption, incineration, acid and base washing, and solvent extraction. Organic-laden streams from the regeneration process are treated, recovered, or incinerated. Regeneration by incineration results in some loss of carbon and deterioration and loss of active surface area.

Design criteria for carbon adsorption include type and concentration of contaminant, hydraulic loading, bed depth, and contact time. Typical ranges are 2 to 10 gallons per minute per square foot (gpm/ft^2) for hydraulic loading,

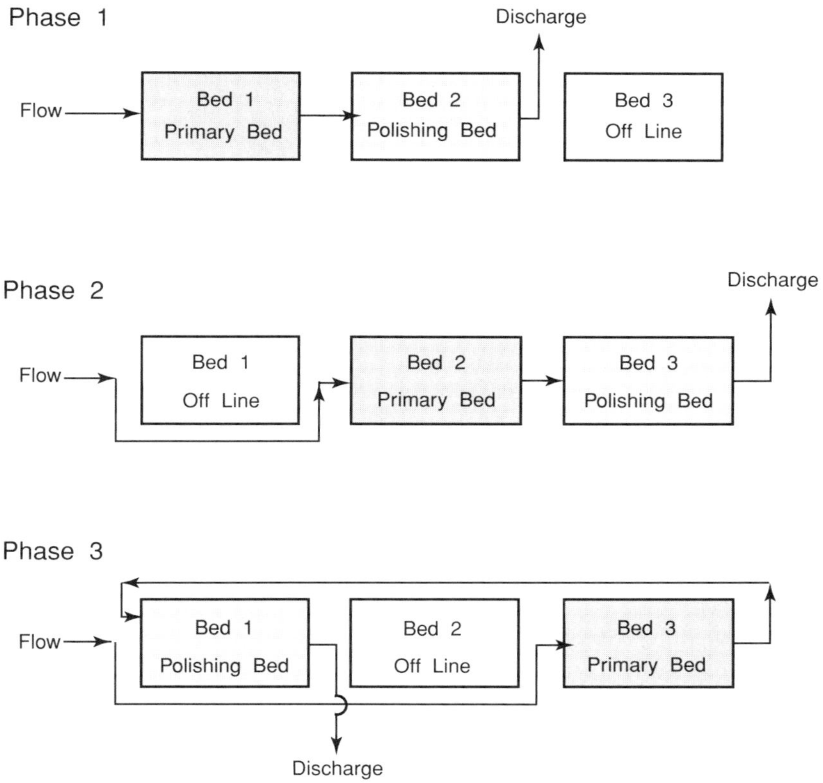

Figure 12.2 Round-robin operation of a three-bed carbon adsorption unit.

5 to 30 feet for bed depth, and 10 to 50 minutes for contact time (WEF 1990). The adsorption capacity for a particular compound or mixed waste stream can be determined as an adsorption isotherm and tested in a pilot unit. The adsorption isotherm relates the observed effluent concentration to the amount of material adsorbed per mass of carbon.

New areas in adsorption technology include carbonaceous and polymeric resins (Musterman and Boero 1995). Based on synthetic organic polymer materials, these resins may find special uses where compound selectivity is important, low effluent concentrations are required, carbon regeneration is impractical, or the waste to be treated contains high levels of inorganic dissolved solids.

Decantation

Decantation is a process that can be used to separate two immiscible liquids of different densities. A gravity decanter that can be used to separate two phases in a continuous process is depicted in Figure 12.3. The denser liquid

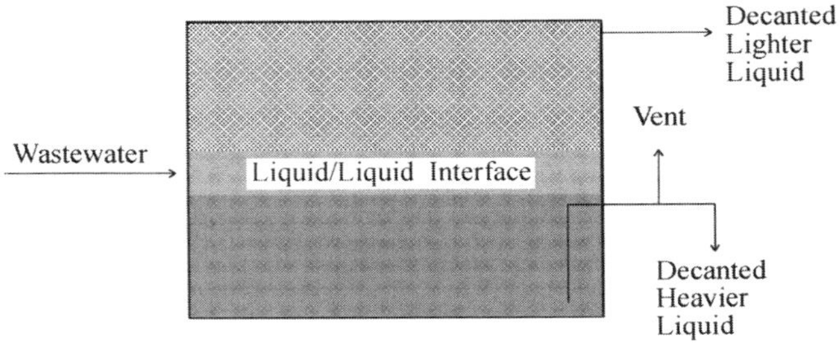

Figure 12.3 Decantation system.

phase settles to the bottom of the tank and is drawn off through a bottom drain while the lighter phase floats on top and is drawn off via an overflow drain. For optimum performance, the liquid–liquid interface is equidistant from the bottom and the overflow drains and the influent line is located at the midpoint of the system near the interface.

Dissolved Air Flotation

Dissolved air flotation (DAF) is used to separate suspended solids and oil and grease from aqueous streams and to concentrate or thicken sludges. It is especially effective for emulsions after they have been chemically broken. DAF is capable of producing an effluent with low oil concentration. DAF is used in many wastewater treatment systems, but it is perhaps best known with respect to hazardous waste in its association with the listed waste, K048, DAF float solids from petroleum refining wastewaters. Of course, the process itself is not what is hazardous but the materials it helps to remove from refining wastewaters.

With DAF, air bubbles carry or float the solids and oils to the surface, where they can be removed. The air bubbles are formed by pressurizing either the influent wastewater or a portion of the effluent in the presence of air. When the pressurized stream enters the flotation tank, which is at atmospheric pressure, the dissolved air comes out of solution as tiny, microscopic bubbles. Coagulant and flocculant chemicals are often needed to improve process removal efficiencies as well as the dewatering characteristics of the DAF float.

Distillation

Distillation separates volatile components from a waste stream by taking advantage of differences in vapor pressures or boiling points among volatile fractions and water. There are two general types of distillation, batch, or differential distillation and continuous fractional or multistage distillation.

Batch Distillation. This process typically is used for small amounts of solvent wastes that are concentrated and consist of very volatile components that are easily separated from the nonvolatile fraction. Batch distillation is amenable to small quantities of spent solvents which allow these wastes to be recovered on site. With batch distillation, the waste is placed in the unit and volatile components are vaporized by applying heat through a steam jacket or boiler. The vapor stream is collected overhead, cooled, and condensed. As the waste's more volatile, high-vapor-pressure components are driven off, the boiling point temperature of the remaining material increases. Less volatile components will begin to vaporize, and once their concentration in the overhead vapors becomes excessive, the batch process is terminated. Alternatively, the process can be terminated when the boiling point temperature reaches a certain level. The residual materials that are not vaporized are called *still bottoms*. A schematic of an example batch distillation unit is shown in Figure 12.4.

Fractional Distillation. If a waste contains a mixture of volatile components that have similar vapor pressures, it is more difficult to separate these components, and continuous fractional distillation is required. In this type of distillation unit, a packed tower or tray column is used. An example of such a unit is shown in Figure 12.5. Steam is introduced at the bottom of the column while the waste stream is introduced above and flows downward, countercurrent to the steam. As the steam vaporizes the volatile components and rises, it passes through a rectification section above the waste feed. In this section, vapors that have been condensed from the process are refluxed to the column, contacting the rising vapors and enriching them with the more volatile components. The vapors are then collected and condensed. Organics in the condensate may be gravity separated from the aqueous stream, after which the aqueous stream, can be recycled to the column.

Factors affecting distillation include the vapor pressures of the volatile components; column temperature, pressure, and internal structure (packing or

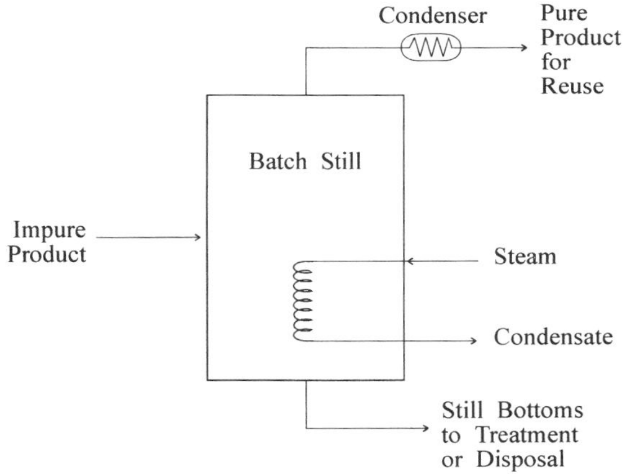

Figure 12.4 Batch distillation unit. (Adapted from EPA, 1987.)

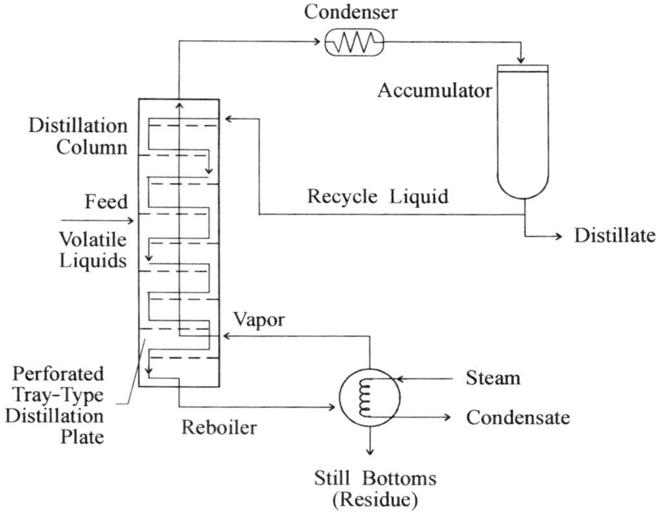

Figure 12.5 Continuous fractional distillation system. (From EPA, 1987.)

trays); and concentrations of oil and grease, total dissolved solids, and total dissolved volatile solids. The higher the vapor pressure of a volatile component, the more easily it can be distilled or stripped from solution. Column pressure influences the boiling point of the liquid stream. Column temperature can be reduced by operating the column under a vacuum. Trays used in columns (bubble cap, sieve, valve, and turbogrid) do not plug as easily with solids as do packed media and can be used with a wider range of liquid and vapor flow rates. Packed columns have lower pressure drops per stage, are more corrosion resistant, and reduce foam by distributing the liquid flow more uniformly. Oil and grease, total dissolved solids, and total dissolved volatile solids affect the partial pressures and solubilities of volatile components.

Evaporation

Evaporation can be used to separate volatile compounds from nonvolatile components and often is used to remove residual moisture or solvents from solids or semisolids. Thin-film evaporators and dryers are examples of evaporation equipment used for this type of application. Some evaporators are also appropriate for aqueous solutions.

Thin-Film Evaporators. Two types of thin-film evaporators are commonly used in industrial applications. The first type introduces feed material into the center of a rotating heated conical receiver. Centrifugal force causes the feed to travel to the outer edge of the conical receiver, where it is collected and drawn off as residue. During the process, the heat causes the volatile components to be driven from the feed. These volatile components are condensed on a chilled surface of the evaporator and collected as distillate.

The second type of thin-film evaporator, termed a *wiped-film evaporator,* introduces feed material on a heated wall of a cylinder. Rotating wiper blades continuously spread the feed along the inner wall of the cylinder to maintain uniformity of thickness and to ensure contact with the heated surface. The volatile components are driven off and collected on an internal chilled condenser surface. The condensate or distillate is removed continuously. At the end of the process, the residual becomes dry and heavy and drops to the bottom of the unit for removal. The wiped-film evaporator is best suited for treatment of viscous or high-solids content feed.

Dryers. Drying, another type of evaporation technique, is suited for waste streams of very high solids content. Several common types of dryers are vacuum rotary dryers, drum dryers, tray and compartment dryers, and pneumatic conveying dryers.

Vacuum Rotary Dryer. The vacuum rotary dryer is a batch system that uses a cylindrical rotating unit and agitator blades to mix or agitate the waste stream during the drying process. A vacuum is applied and maintained during the process to remove liquids during agitation. The process is terminated when the solids are dried to a specified moisture content.

Drum Dryer. The drum dryer is a continuously operated unit that uses rotating heated drums for the evaporation contact area. This method of drying can be used with materials high in volatiles. In several designs, the volatiles can be withdrawn and collected for reuse, as appropriate.

Tray and Compartment Dryer. A tray and compartment dryer is a batch unit that uses a stationary tray or compartment to dry the waste, generally before transport for disposal or further treatment. Some units can be mounted on removable trucks.

Continuously Operating Pneumatic Conveying Dryer. A continuously operating pneumatic conveying dryer is used for applications similar to the tray and compartment dryer. In this case, however, drying is performed in conjunction with grinding, as the solids are conveyed and dried within the unit.

Aqueous Evaporation. Evaporation also can be used in the treatment of wastewaters containing salts and dissolved solids. Generally, evaporation used in this way is considered a concentration technique instead of a treatment method. In these cases, evaporation minimizes the volume of waste requiring disposal.

One type of aqueous evaporation occurs outdoors and utilizes a lined pond or open tank with spray nozzles designed to spray the wastewater into the air as a fine mist. Mechanical aerators are also used for this purpose, especially in ponds. The spraying effectively increases the surface area of the wastewater, which enhances evaporation. As the wastewater is recycled through the spray system, the salt or solids content is increased as the wastewater volume is reduced. The brine that forms can be withdrawn for proper disposal when a certain concentration or volume reduction is achieved.

Another type of aqueous evaporation can be accomplished in a closed process vessel that uses steam to evaporate the liquid into a water vapor, which is ultimately condensed and may be reused. The concentrated liquid is collected for further treatment or disposal. An example of this type of evaporation unit is shown in Figure 12.6.

Extraction

Extraction is a technology that can be used when distillation proves ineffective. Mixtures that form azeotropes or contain substances of similar vapor pressures or boiling points may be candidates for extraction rather than distillation. Two types of extraction include solvent extraction, also termed *liquid extraction,* and supercritical fluid extraction.

Solvent Extraction. Solvent extraction involves the mixing of a waste stream, which contains a hazardous constituent to be extracted, with an extraction fluid. Extraction occurs when the hazardous constituent, termed the *solute,* has greater solubility in the extraction fluid, termed the *solvent,* than in the waste stream. Mixing of the waste stream with the solvent allows mass transfer of the solute from the waste stream to the solvent. The solvent selected for the extraction procedure must be immiscible in the waste stream. As such,

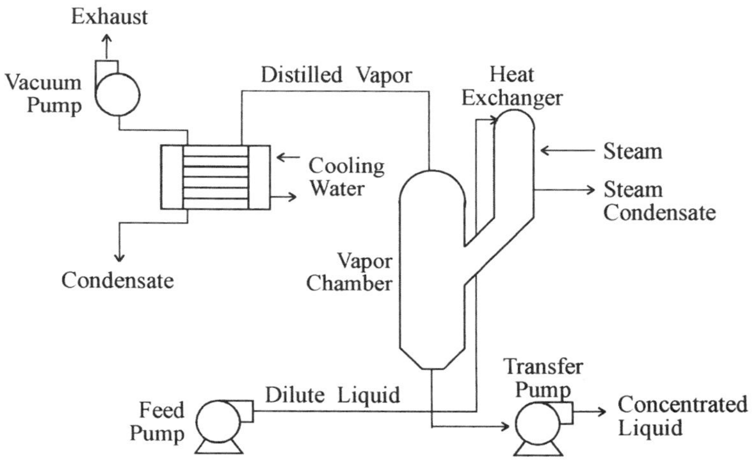

Figure 12.6 Aqueous evaporation system. (From EPA, 1987.)

the waste stream and the solute-enriched solvent, or extract, will form distinct phases. A continuous gravity decanter can be used to separate and decant the extract.

The extract is sent to a recovery unit such as a distillation unit to remove the concentrated hazardous constituent from the solvent. The solvent can then be reused. The remaining waste stream, now devoid of the hazardous constituent, is either treated or recycled to the originating process, depending on the application.

Extraction is accomplished in two steps, with solvent–waste contact occurring first and phase separation of the liquids occurring second. The process can be operated as a batch or continuous process. Examples of types of equipment used for solvent–waste contact include mixing tanks, spray columns, perforated plate and baffle towers, and centrifugal contactors. Phase separation typically is accomplished using decanters, gravity settling tanks, or other equipment. Waste streams suitable for treatment using solvent extraction include hydrocarbon-bearing waste streams generated by the petroleum and petrochemical industries.

Supercritical Fluid Extraction. Supercritical fluid extraction (SFE), an innovative extraction method, uses a pressurized or supercritical gas such as carbon dioxide as the solvent. Supercritical fluids make excellent solvents because of their comparatively low viscosities and high diffusivities. Thus they greatly increase the efficiency of the extraction process through improved mass transfer rates. Additionally, the separation process of the hazardous constituent from the extract is simplified since supercritical fluids are gaseous at room temperature; thus the solute can easily be concentrated for reclamation, reuse, or disposal, as appropriate. Compounds that have been extracted successfully using SFE include aliphatic hydrocarbons, alkenes, simple aromatics such as benzene and toluene, polynuclear aromatics, and phenols (EPA 1991).

Ion Exchange

Ion exchange is an adsorption process where ionic species are adsorbed from solution by exchanging places with a similarly charged ion on the exchange media. Ion exchange is used primarily to remove metals, although nonmetallic inorganic and organic ions can also be removed. Metals that can be removed by ion exchange include barium, cadmium, hexavalent chromium, copper, lead, mercury, nickel, selenium, silver, uranium, and zinc (Krishnan et al. 1993). Nonmetallic ions that may be removed by ion exchange include nitrate, sulfate, cyanide (when complexed with iron), and some organic acids, phenols, and amines under special conditions.

The adsorptive materials used for ion exchange are called *zeolites*. Naturally occurring zeolites belong to a group of hydrous aluminum silicate minerals. Synthetic zeolites or resins are based on organic polymers and are used more commonly today. Ion exchange resins that adsorb positively charged ions are cationic and those that adsorb negatively charged ions are anionic. These resins can be designed to remove certain ions with greater specificity than other ions; however, ions with the same charge (positive, negative) can still compete for exchange sites and adversely affect the removal efficiency of the targeted ion if they are present in high enough concentrations.

Ion exchange units can be operated in parallel or series and in downflow or upflow (fluidized) mode. A unit is regenerated when the target ion appears in the effluent (breakthrough). Ion exchange resins are regenerated by washing with a concentrated solution containing the original exchangeable ion. This results in the desorption of the target ion from the resin with replacement by the original ion. Hydrogen-based cationic resins are regenerated with acid solutions such as sulfuric, nitric, or hydrochloric acid. Anionic resins that are hydroxide-based are regenerated with caustic solutions such as sodium hydroxide. Sodium chloride solution is used to regenerate sodium-based cationic and chloride-based anionic resins. Concurrent regeneration is when the regenerant solution flows through the unit in the same direction as does the wastewater; countercurrent regeneration flows in the opposite direction and requires less solution.

Pretreatment of aqueous streams may be required prior to using ion exchange. Suspended solids that can plug an ion exchange unit should be reduced to the 10-μm level. Organics that can foul resins can be removed by carbon adsorption. Iron and manganese, commonly present in groundwaters, should be removed because they precipitate on the resin.

Membrane Filtration

Membrane filtration describes a number of well-known processes, including reverse osmosis, ultrafiltration, nanofiltration, microfiltration, and electrodialysis. The basic principle behind this technology is the use of a driving force (electricity or pressure) to filter particles, ions, and organic molecules through a membrane, producing a "clean" stream on one side and a concentrated stream on the other. Although membrane filtration is discussed in this section in relation to waste treatment, this technology also has many industrial applications, such as water desalination and softening, product concentration and purification, and metal recovery. Applications in waste treatment include metals removal and recovery, oil–water separation, and removal of toxic organic compounds.

The individual membrane filtration processes are defined chiefly by pore size, although there is some overlap. The smallest membrane pore size is used in reverse osmosis [0.0005 to 0.002 μm], followed by nanofiltration (0.001 to 0.01 μm), ultrafiltration (0.002 to 0.1 μm), and microfiltration (0.1 to 1.0 μm). Electrodialysis uses electric current to transport ionic species across a membrane. Micro- and ultrafiltration rely on pore size for material separation, reverse osmosis on pore size and diffusion, and electrodialysis on diffusion. Separation efficiency does not reach 100% for any of these membrane processes. For example, when used to desalinate or soften water for industrial processes, the concentrated salt stream (reject) from reverse osmosis can be 20% of the total flow. These concentrated, yet still dilute streams may require additional treatment or special disposal methods.

Reverse osmosis and electrodialysis are used to separate metals from electroplating bath rinse waters in the metal finishing industry. Ultrafiltration is used for oil–solids–water separation, metals recovery, and paint recovery from anodic paint processes. It has also been tested as an innovative technology for removal of solids and residual organics following biological treatment of contaminated groundwater (EPA 1995b). Microfiltration is used for oil–solids–water separation and metals removal and recovery. Although nanofiltration has been used successfully for water treatment (removal of hardness, organic compounds, and viruses), it has not been used much for waste treatment. Testing has also been conducted with another type of membrane filtration process, pervaporation, that uses a vacuum to induce transfer of volatile organics from a liquid across the membrane (EPA 1995a). Membrane filtration of gaseous streams, which is used on industrial process streams to recover volatile organics, has been tested as posttreatment for air stripping and soil venting gases.

Membranes are subject to fouling, which can be caused by metal oxides, precipitating salts, colloids, and biological growth. Cleaning agents include acids, alkalines, oxidizers, detergents, and organic solvents. Pretreatment prior to membrane filtration is required to reduce heavy solids and remove free oil.

Neutralization

Neutralization is pH adjustment of an acidic or caustic waste to a more neutral range. Neutralization is a very common treatment step for wastewaters and gases; it is used less frequently with solid wastes because the pH is less of a problem with these wastes. The most commonly used neutralizing agents are lime or calcium hydroxide for acidic wastewaters and sulfuric acid for caustic wastewaters. Other chemicals used for wastewater neutralization include sodium hydroxide and magnesium hydroxide for acidic streams and hydrochloric acid and carbon dioxide for caustic streams. At facilities where both acidic and caustic streams are generated, commingling these streams helps reduce the amount of neutralizing chemicals that would otherwise be required. These acid–caustic waste streams can only be commingled, however, where they are compatible, that is, where the mixture will not generate unwanted precipitants or chemical reactions. Neutralization of acid gases with liquid caustic solutions (wet scrubbing) is very common. For example, the flue gas from an incinerator may be scrubbed with a soda ash (sodium carbonate) solution to remove acid gases. Wet scrubbing is normally done in a packed tower; the packing referred to is usually small plastic forms that provide a high degree of surface area per unit volume in order to maximize the contact of the gas with the scrubbing solution. There are also dry scrubbing systems for acid gases where the neutralizing agent, for example, lime, is sprayed into the gas stream.

Oil–Water Separation

A type of decantation, gravity separation of oil and water is a very simple process, often used as a pretreatment step in an overall wastewater treatment system. Under quiescent conditions, free oil floats to the surface, where it can be skimmed off. Oil–water emulsions may be broken for separation by using coagulants, acids, pH adjustment, heat, centrifuging, or high-potential alternating current (Noyes 1994). The API separator and corrugated plate interceptor (CPI) are common types of oil–water separators. API separator sludge from petroleum refining wastewaters is listed as a hazardous waste, K051. Like the DAF process discussed earlier in this section, it is not the API separator itself that makes K051 hazardous, but the hazardous materials the waste contains.

Oxidation and Reduction

Oxidation and reduction (redox) reactions are used for both partial and complete degradation of many organic and inorganic compounds. A substance is oxidized when its oxidation state is increased; similarly, it is reduced when its oxidation state is reduced. For inorganic compounds, oxidation involves the loss of electrons, reduction is a gain in electrons. Redox reactions for organics are more complicated and may include the transfer of electrons, a hydrogen atom, an oxygen atom, a hydroxyl radical, a chlorine atom, a chlorinium ion (Cl^+), or similar species (Weber 1972). Examples of compounds that are treated by redox processes include alcohols, phenols, and cyanide. Common chemical oxidants for waste treatment include chlorine, ozone, and hydrogen peroxide.

Oxidizers do not discriminate among compounds and are capable of reacting with any oxidizable compounds in a waste stream. Oxidation is used either to completely degrade a compound or, quite often, to degrade a compound partially to a less toxic form or intermediate that can be discharged or, if needed, treated further by another process.

Factors affecting oxidation processes include pH, the type and quantity of oxidizable compounds in the waste, and metals that can react with the oxidizing agent. The pH affects the oxidation rate by changing the free energy of the overall reaction, changing the reactivity of the reactants, and/or affecting specific OH^- ion or H_3O^+ ion catalysis (Weber 1972). The presence of oxidizable compounds in addition to the target compound will increase the amount of oxidant required. Metal salts, especially those of lead and silver, can also react with the oxidizer and increase the required dosage or interfere with the treatment process (Noyes 1994).

Chlorine. Chlorine is a well-known disinfectant for water and wastewater treatment; however, it can react with organics to form toxic chlorinated compounds such as the trihalomethanes bromodichloromethane, dibromochloromethane, and chloroform. Chlorine dioxide may be used instead since it does not produce the troublesome chlorinated by-products such as chlorine. In addition, by-products formed by chlorine dioxide oxidation tend to be more readily biodegradable than those of chlorine; however, chlorine dioxide is not suitable for waste streams containing cyanide.

Cyanide destruction by alkaline chlorination is a widely used process. With alkaline chlorination, cyanide is first converted to cyanate with hypochlorite at a pH greater than 10. A high pH value is required to prevent the formation of cyanogen chloride, which is toxic and may evolve in gaseous form at a lower pH. With additional hypochlorite, cyanate is then oxidized to bicarbonate, nitrogen gas, and chloride. The pH for this second stage is 7 to 9.5 (Weber 1972).

Ozone. A primary advantage of oxidation with ozone is the avoidance of chlorinated by-products of the reaction. Excess ozone that is not consumed in the reaction decomposes to oxygen. The main disadvantage is the high electrical cost of producing ozone, which is generated on site from dry air or oxygen by high-voltage electric discharge.

Ozone can be used to oxidize low concentrations of organics completely in aqueous streams or partially degrade compounds that are refractory or difficult to treat by other methods. Compounds that can be treated with ozone include alkanes, alcohols, ketones, aldehydes, phenols, benzene and its derivatives, and cyanide. Ozone readily oxidizes cyanide to cyanate; however, further oxidation of the cyanate by ozone proceeds rather slowly and may require other oxidation treatment, such as alkaline chlorination to complete the degradation process.

Ozone is only slightly soluble in water. Thus factors that affect the mass transfer between the gas and liquid phases are important and include temperature, pressure, contact time, contact surface area (bubble size), and pH.

Ozonation can be enhanced by the addition of ultraviolet (UV) radiation. This combination can be effective in degrading chlorinated organic compounds and pesticides. In addition, metal ions such as iron, nickel, chromium, and titanium can act as catalysts, as can ultrasonic mixing.

Hydrogen Peroxide. Hydrogen peroxide is typically used as an oxidizer in combination with UV light, ozone, and/or metal catalysts. Fenton's reagent is hydrogen peroxide with iron as a catalyst. Hydrogen peroxide/UV light has been shown to be effective in oxidizing benzene; chlorobenzene, chloroform, chlorophenol, 1,1-dichloroethane, dichloroethene, phenol, tetrachloroethene, 1,1,1-trichloroethane, trichloroethene, toluene, xylenes (EPA 1993a–c, 1994), and many other organic compounds.

Precipitation

Precipitation processes have been used for many years to remove metals from aqueous streams. Metals precipitation is accomplished by pH adjustment and the addition of a chemical reagent that forms a precipitant with the metal that can be settled out and separated from the aqueous stream. An example of a chemical precipitation system is shown in Figure 12.7. Chemical reagents such as calcium hydroxide (lime), sodium hydroxide (caustic), and magnesium oxide/hydroxide can be used to precipitate arsenic, cadmium, trivalent chromium, copper, iron, lead, manganese, nickel, and zinc. Hydroxide precipitation, particularly with lime, is the most common precipitation process because the chemical reagents are readily available, relatively inexpensive, easy to handle, and form sludges that dewater easily. Sulfide precipitation with reagents such as sodium sulfide and ferrous sulfide can be used with cadmium, cobalt, copper, iron, mercury, manganese, nickel, silver, tin, and zinc. Carbonate precipitation with reagents such as sodium carbonate (soda ash) and calcium carbonate can be used with cadmium, lead, nickel, and zinc. Because the sulfide forms of metals are much less soluble than their hydroxide counterparts, sulfide precipitation can treat metals to lower concentrations.

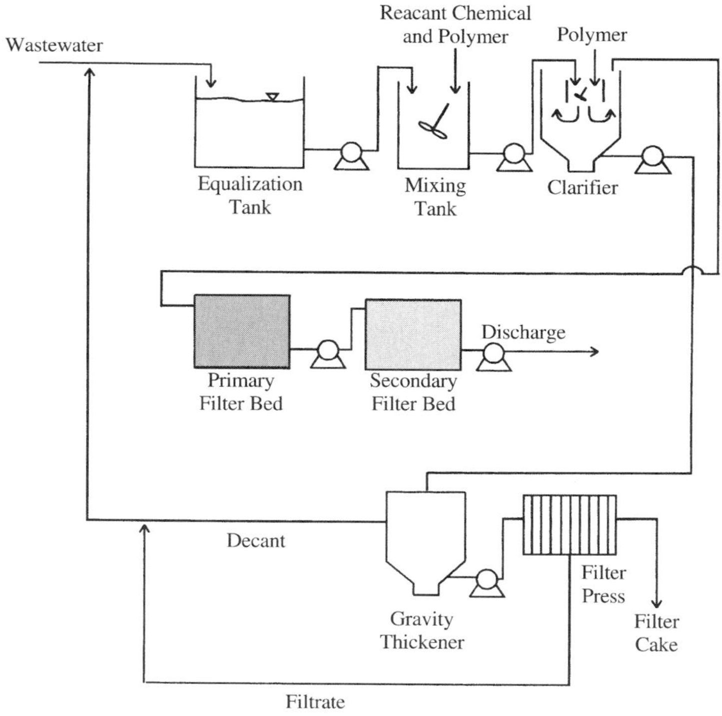

Figure 12.7 Chemical precipitation system.

Although very insoluble, metal sulfides form very fine, hard-to-settle particles and generate large quantities of sludge.

The typical precipitation process takes place in a series of tanks beginning with chemical addition, followed by flocculation and coagulation of precipitated solids, sludge thickening, and finally, sludge dewatering. Sand filtration of the wastewater effluent to remove residual solids may be used as a final polishing step. Chemicals may be added to aid flocculation and coagulation and to produce a more dense and easily dewatered sludge. Typical chemical aids are alum (aluminum sulfate), ferric chloride, and proprietary polyelectrolytes. Precipitation of arsenic first requires oxidation of the arsenite form to arsenate. Hexavalent chromium will require reduction to the trivalent form.

Precipitation is affected by pH, solubility product of the precipitant, ionic strength and temperature of the aqueous stream, and the presence of metal complexes. For each metal precipitant, there is an optimum pH where its solubility is lowest and hence the highest removals may be achieved. When an aqueous stream contains various metals, the precipitation process cannot be optimized for each metal, sometimes making it difficult to achieve effluent targets for each. Solubility products depend on the form of the metal compound and are lowest for metal sulfides, reflecting the relative insolubility of these compounds. For example, the solubility product for lead sulfide is on the order of 10^{-45} compared to 10^{-13} for lead carbonate. Metal sulfides and hydroxides are more soluble at higher temperatures, whereas carbonates are less soluble. When metals are complexed or chelated with other ions or molecules [e.g., cyanide and ethylenediaminetetraacetic acid (EDTA)], they are more soluble, may not precipitate easily, and will require additional or alternate forms of treatment.

Sedimentation and Clarification

Sedimentation–clarification is a process by which hazardous and nonhazardous grits, fines, and other suspended solids are removed from the waste stream through gravity settling. The process typically is used as a pretreatment step or in conjunction with another treatment process, such as chemical or biological treatment. Usually, sedimentation and clarification are performed in a gravity settling clarifier, pond, or basin. Examples of a gravity settling pond and a clarifier are shown in Figure 12.8. A gravitational settling system provides enough residence time to allow the heavier suspended solids to settle gravitationally to the bottom of the system. These solids are removed and thickened periodically prior to disposal.

Flocculation, commonly used in clarifiers to enhance sedimentation, is a physical–chemical process in which small particles agglomerate to form larger particles. The large particles, because they are heavier than the smaller particles, settle more effectively in the clarifier or sedimentation basin. Flocculation of certain types of particles can be induced by slow agitation of the wastewater, without the addition of a flocculating agent. However, a floccu-

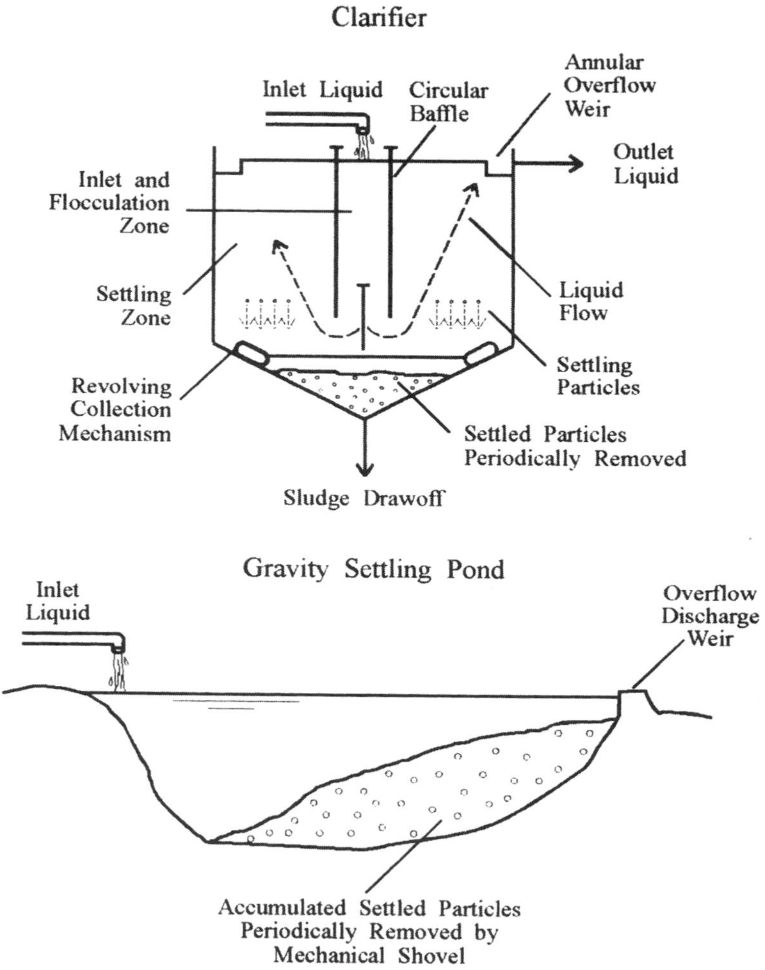

Figure 12.8 Sedimentation/clarification systems. (Adapted from EPA, 1987.)

lating chemical is usually added to the wastewater to promote flocculation of the smaller particles. The flocculants adhere readily to suspended solids and to each other to form larger particles with greater density, which settle better. The settled solids thicken at the bottom of the clarifier and in some clarifier designs are used to form a sludge blanket that becomes a filtration mechanism within the clarifier.

Clarifiers typically are used in chemical precipitation and biological treatment processes to remove precipitated metal solids and suspended biological solids. To prevent the sludge blanket from becoming too thick or heavy, part of the sludge blanket is removed continuously or intermittently from the system and thickened prior to disposal.

In settling basins or impoundments, the solids removed from the liquid typically are allowed to accumulate until the effective treatment volume of the basin and the solids removal efficiency is reduced. The accumulated solids then are removed from the impoundment by dredging, with the liquid in the basin, or by dewatering the impoundment and removing the solids with some type of excavating equipment, such as a dragline with a clamshell. It is usually necessary to take the basin out of service while the solids are being removed.

Stabilization and Solidification

Stabilization and solidification are technologies that are often used together to improve or strengthen the physical nature of a waste and to bind, immobilize, or otherwise prevent the migration of toxic constituents contained in the waste. Stabilization generally refers to processes that reduce the toxicity, leaching, or mobility of toxic constituents, perhaps through chemical reactions, but the physical form of the waste is not necessarily changed or improved. Solidification generally refers to processes that change the physical nature of the waste to make it more manageable, increase its structural strength and load-bearing capacity, or reduce its permeability, but not necessarily by chemical reaction. Many treatment technologies involve characteristics of both stabilization and solidification, making distinctions between the two processes unclear at times so that the two terms are often used together. Applying the correct definition is not as important as understanding the changes in the waste characteristics: Are toxic constituents rendered less toxic or mobile, and can the waste be handled easily and withstand the physical stresses during and after disposal?

Some of the most common stabilization and solidification processes are those using cement, lime, and pozzolanic materials. These materials are popular because they are very effective, plentiful, and relatively inexpensive. Other stabilization and solidification technologies include thermoplastics, thermosetting reactive polymers, polymerization, and vitrification. Vitrification is discussed in the section "Thermal Processess" later in this chapter; the other stabilization and solidification processes are discussed below.

Cement, Lime, and Pozzolanic Materials. Cement is the chemical binder used in making concrete, a chemically cured mixture of cement, water, and aggregate (usually sand and gravel). With sufficient water for hydration, cement forms a calcium aluminosilicate crystalline structure that can physically and/or chemically bind with toxic constituents such as metal hydroxides and carbonates. Use of lime (calcium hydroxide) involves similar reactions where aluminosilicates are supplied by the waste or treatment additives. Pozzolans are siliceous materials that can combine with lime to form cementitious materials. Common pozzolanic materials are fly ash, blast furnace slags, and cement kiln dust. The high pH value of cementitious binders protects against acid conditions that can cause metal leaching. Organics interfere with hydration and the formation of the crystalline structure binding the waste. This interference can be minimized by treatment additives such as clays (natural and organically modified), vermiculite, or organically modified silicates. Other waste constituents that can reduce the effectiveness of cementitious processes include sulfates, chlorides, borates, and silt.

Thermoplastics. Wastes containing heavy metals and/or radioactive materials may be mixed with thermoplastics. In this process the waste is mixed with a molten thermoplastic material such as bitumen, asphalt, polyethylene, or polypropylene. Wastes are normally dried, heated, mixed with the thermoplastic, cooled, and finally placed in containers for disposal. The process is not suitable for organic wastes unless air emission controls are installed because the high temperatures will drive off volatile components. It is also not suitable for hygroscopic wastes that absorb water, expanding and cracking the thermoplastic. Strongly oxidizing constituents, anhydrous inorganic salts, and aluminum salts are also unsuitable for thermoplastics.

Thermosetting Reactive Polymers. Materials used as thermosetting polymers include reactive monomers urea-formaldehyde, phenolics, polyesters, epoxides, and vinyls, which form a polymerized material when mixed with a catalyst. The treated waste forms a spongelike material that traps the solid particles but not the liquid fraction; the waste must usually be dried and placed in containers for disposal. Because the urea-formaldehyde catalysts are strongly acidic, urea-based materials are not generally suitable for metals that can leach in the untrapped liquid fractions. Thermosetting processes have greater utility for radioactive materials and acid wastes.

Polymerization. Spills of chemicals that are monomers or low-order polymers can be polymerized by adding a catalyst. Compounds that may be treated by polymerization include aromatics, aliphatics, and other oxygenated monomers, such as vinyl chloride and acrylonitrile.

Supercritical Water Oxidation

Supercritical water oxidation is like wet air oxidation (discussed below) in that it uses high-temperature and high-pressure conditions to oxidize organics. Temperature and pressure conditions, however, are higher (greater than 218 atm and 374°C) such that water is at supercritical conditions, where it has properties between those of a liquid and a gas. Organics are easily solubilized in supercritical water, and once solubilized, they are completely oxidized at the high-temperature, high-pressure conditions. Reaction times are very short (a few seconds to less than a minute) and treatment efficiencies very high (essentially 100%). Supercritical water oxidation has the capability of treating many different kinds of organics, including polycyclic aromatic hydrocarbons, chlorinated hydrocarbons, PCBs, paint, oil, dyes, pulp and paper wastes, chemical warfare agents, and missile propellants (Jensen 1994). Wastes must be either aqueous or in slurry form.

Reaction vessels for supercritical water oxidation must be highly corrosion resistant because of the aggressive nature of supercritical water and oxidation reaction products at extreme temperatures and pressures. Supercritical oxidation of PCBs and some chlorinated hydrocarbons can be difficult because acids are formed that are highly corrosive to almost any materials used for reactor components (Jensen 1994). Inorganic salts can be a problem because they are sparingly soluble under these conditions and will precipitate on the vessel walls and components.

Stripping

Stripping is a common method for removing low concentrations of volatile organics and inorganics from wastewater. Stripping is accomplished by passing air or another stripping gas such as heated nitrogen through a liquid stream. Additionally, steam can be used as the stripping agent.

Air–Gas Stripping. Compounds that have relatively high volatilities and low water solubilities can be transferred (stripped) from aqueous streams into the air or an air–gas stream. Examples of compounds that can easily be air stripped include gasoline and jet fuels [with benzene, toluene, ethylbenzene, and xylenes (BTEX) as primary components of interest], solvents such as trichloroethene and tetrachloroethene, and ammonia. The strippability of a compound is related to its Henry's law constant, which is the ratio of the vapor-phase concentration of the chemical in equilibrium with its concentration in water. In general, the higher the Henry's law constant, the more strippable is the compound. Similarly, since Henry's law constant increases with temperature, higher temperatures generally improve air stripping efficiencies. Air–gas stripping can be accomplished using one of two methods, sparging or countercurrent stripping. Factors important in the removal of organics from wastewater using this technique include temperature, pressure, air/water ratio, and surface area available for mass transfer. The airstream exiting a stripper will generally require some type of emissions control, depending on local and regulatory requirements. Carbon adsorption is often used; catalytic oxidation is another option.

Sparging. Sparging is the simplest form of air stripping and typically is used to remove insoluble organics from water. Sparging is accomplished in a tank equipped with an air supply header in the bottom. Air is forced through small holes or nozzles to create bubbles that rise to the top of the tank. The volatile organics are stripped from the liquid phase and are transferred to the gas phase. The tank must be sized to provide adequate residence time to strip the organics from the wastewater. If necessary, finer bubbles can be generated to increase the mass transfer efficiency of the system. In general, this type of system is not as efficient as a countercurrent stripping system, which is described next. Additionally, off-gas typically is not captured during this type of stripping, and air emissions requirements may not permit the use of this type of stripping for some applications.

Countercurrent Stripping. Stripping using a countercurrent stripping system is accomplished in a column, with countercurrent gas and liquid contact. Wastewater is introduced at the top of the column and withdrawn at the bottom. Air or another stripping gas is introduced at the bottom and exhausted at the top. The column height and diameter are sized to provide sufficient contact time for adequate contaminant removal. If necessary, the wastewater can be recirculated through the column several times to achieve a treatment standard.

When stripping insoluble organics from wastewater, the liquid phase controls the rate of mass transfer. To achieve the desired removal rates, the stripping column is flooded and operates as a bubbler tower. When stripping relatively soluble organics from wastewater, the gas phase controls the rate of mass transfer. For these cases the column is operated like a spray chamber and the liquid is sprayed into the column. At very dilute concentrations, both phases of mass transfer rates are important. To increase the efficiency of a spray column, the gas–liquid surface area can be increased by packing the column. Packing media—polyethylene, stainless steel, polyvinyl chloride (PVC), or ceramic that provide large surface area/volume ratios—provides a surface for the liquid droplets to travel downward, which increases the gas–liquid contact time. Usually, a demister is used on spray columns to filter water droplets from the exhaust. The diameter of the spray stripper is sized to minimize column flooding during operation. Design criteria for packed towers include surface area provided by the packing media, column height and diameter, and air/water flow rates. An example of a countercurrent packed-tower system is shown in Figure 12.9.

Steam Stripping. Steam stripping typically is used to strip more concentrated volatile organic compounds from aqueous wastewaters. Steam stripping is similar to continuous fractional distillation (see description below), except that steam, rather than reboiled bottoms, provides the direct heat to volatilize organic vapors. The steam containing the concentrated volatile constituents is collected as the overhead from the stripper column and is then condensed by cooling it in a condenser drum. Depending on the characteristics and concentration of the volatile constituent in the condensate, it can be phase separated and reused, reclaimed, or disposed of. In some cases the entire condensate stream may require additional treatment before disposal. When it is practical to separate the volatile constituent from the condensed steam, the condensate may be recycled to the steam generation process. An example of this process is presented in Figure 12.10.

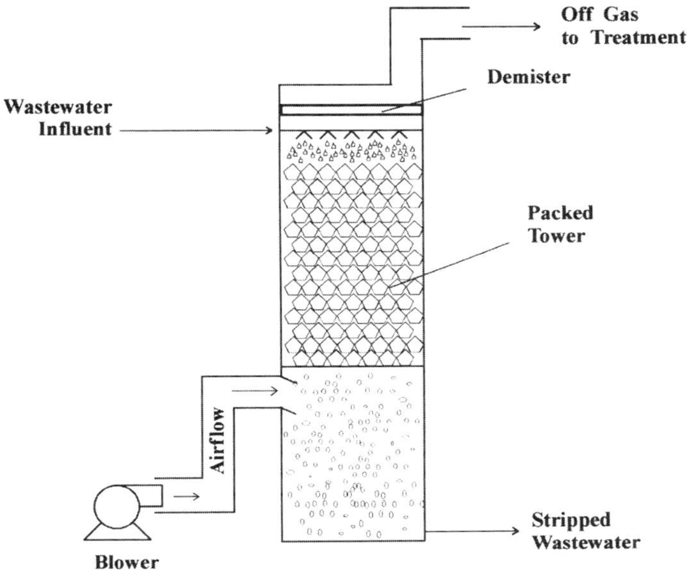

Figure 12.9 Packed tower.

Waste streams suitable for steam stripping include aqueous wastes contaminated with chlorinated hydrocarbons, aromatics, ketones, alcohols, and high-boiling-point chlorinated aromatics (EPA 1987). The steam stripping process is performed in a closed system and has no continuous air emission potential until the point at which the condensed steam–volatile constituent is managed. At this point, air emission controls are required for the volatile constituent.

Wet Air Oxidation

With wet air oxidation, increased temperature and pressure are used to oxidize dilute concentrations of organics and some inorganics, such as cyanide, in aqueous wastes that contain too much water to be incinerated but are too toxic to be treated biologically. In general, wet air oxidation provides primary treatment for wastewaters that are subsequently treated by conventional methods. This technology can be used with wastes that are pumpable such as slurries and liquids.

Waste streams that are treated by wet air oxidation generally are those having dissolved or suspended organic concentrations from 500 to 50,000 mg/L. Below 500 mg/L, oxidation rates are too slow, and above 50,000 mg/L, incineration may be more feasible (EPA 1991).

Operating parameters include temperature, pressure, oxygen concentration, and residence time. Materials of construction include stainless steel, nickel, and titanium alloys (the latter for extremely corrosive wastes containing heavy

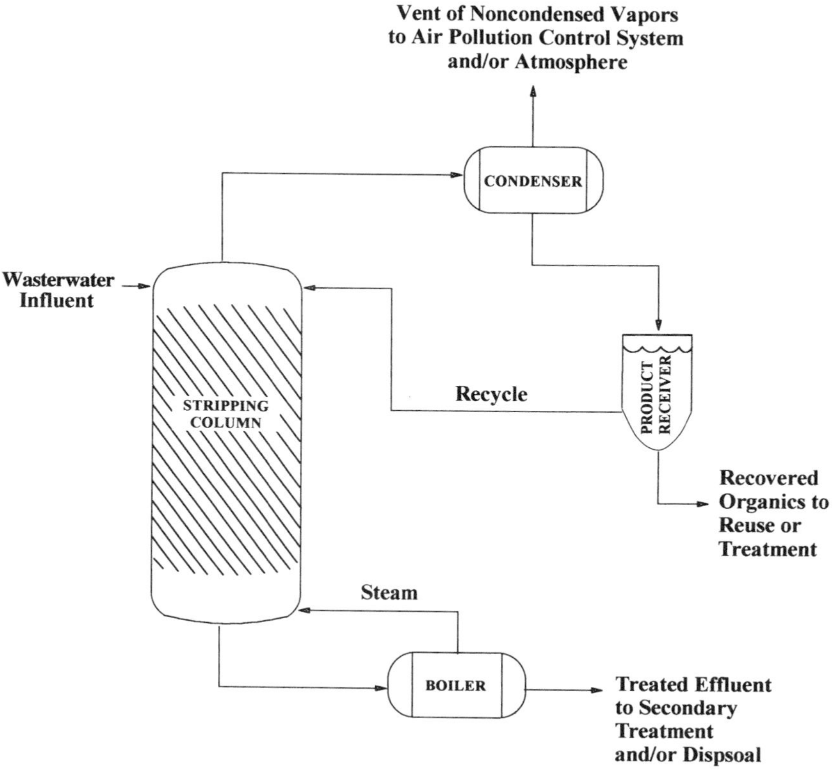

Figure 12.10 Steam stripping system.

metals). Vented gases from the process may require scrubbing or other emission controls.

Because of high energy input, wet air oxidation is relatively expensive. To reduce energy demand, the incoming wastewater can be preheated by heat exchange with the treated effluent. Operating temperatures may be lowered by using a catalyst. Typically, removal of the catalyst from the effluent is needed to avoid contamination and to minimize operating costs. An example of a continuous wet air oxidation system is presented in Figure. 12.11.

BIOLOGICAL PROCESSES

Biological processes are effective in treating a wide variety of organic and inorganic compounds. Most biological processes are aerobic and use the oxygen available in supplied air or pure oxygen. The chief products of aerobic biodegradation of carbonaceous and nitrogenous compounds are carbon dioxide, ammonia, water, and biomass (cell growth of the microorganisms).

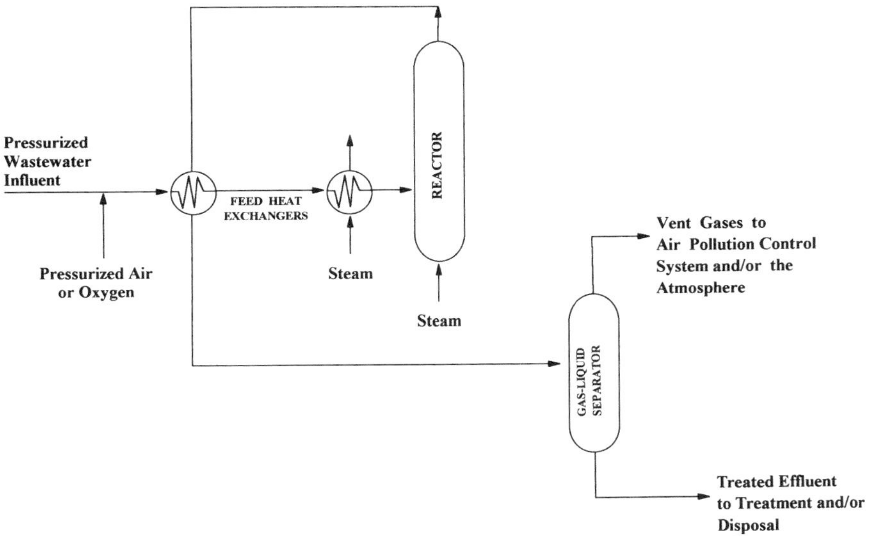

Figure 12.11 Continuous wet air oxidation system.

The ease with which a compound is degraded is related to its chemical structure. Some compounds are not degraded completely but are, instead, broken down into more simple compounds.

The biodegradability of hazardous compounds according to general structure has been summarized by Pitter and Chuboda (1990) from a number of studies. They have found that the biodegradability of hydrocarbons generally decreases in the following order: alkanes and alkenes with straight alkyl chains, mono- and dicycloalkanes, and aromatic hydrocarbons, including polycyclic ones. Mono- and dicyclic aromatic hydrocarbons are usually more biodegradable than cycloalkanes. Chain branching in alkanes and alkenes decreases biodegradability, although biodegradability of alkanes and alkenes are essentially the same if they have the identical chain configurations. Monocycloalkanes are usually readily biodegradable, although exceptions have been noted with specific compounds. Polycycloalkanes are less biodegradable, especially tetra- and higher polycycloalkanes. Naphthalene, lower alkylbenzenes, and phenanthrene are aromatic hydrocarbons that are biodegraded relatively easily. The biodegradability of alkylbenzenes depends on the number and position of the alkyl groups. Polyalkylation of aromatic hydrocarbons usually decreases biodegradability. With phenylalkanes, biodegradability depends on the branching of long alkyl chains and those phenylalkanes with a quaternary carbon atom, particularly in the terminal position, are quite resistant to biodegradation. Polycyclic aromatic hydrocarbons (except for naphthalene) are relatively slow to biodegrade, particularly tetra- and higher aromatics.

General guidelines for biodegradability have been developed by others. Eckenfelder and Musterman (1995) give the following general guidelines for the biological treatment of industrial wastewaters:

- Nontoxic aliphatic compounds containing carboxyl, ester, or hydroxyl groups are readily biodegradable. Those with dicarboxylic groups require longer acclimation times than those with a single carboxyl group.
- Compounds with carbonyl groups or double bonds are moderately degradable and slow to acclimate.
- Compounds with amino or hydroxyl groups decrease in biodegradability relative to degree of saturation, in the following order: primary, secondary, then tertiary carbon atom of attachment.
- Biodegradability decreases with increasing degree of halogenation.

Noyes (1994) presents a more general and simplistic order of decreasing biodegradability: straight-chain compounds, aromatic compounds, chlorinated straight-chain compounds, and chlorinated aromatic compounds.

Many different factors influence the performance of biological treatment systems. Although each specific biological process has special requirements, factors common to biological processes, besides biodegradability of the waste constituents, include: organic concentration, temperature, pH, nutrients, and oxygen (aerobic or anaerobic).

Activated Sludge

Activated sludge is widely used to treat both municipal and industrial wastewater. Used in a variety of process configurations and operating modes, the activated sludge process contains three basic elements—an aeration basin, clarification, and sludge recycle—as shown in Figure 12.12. The aeration

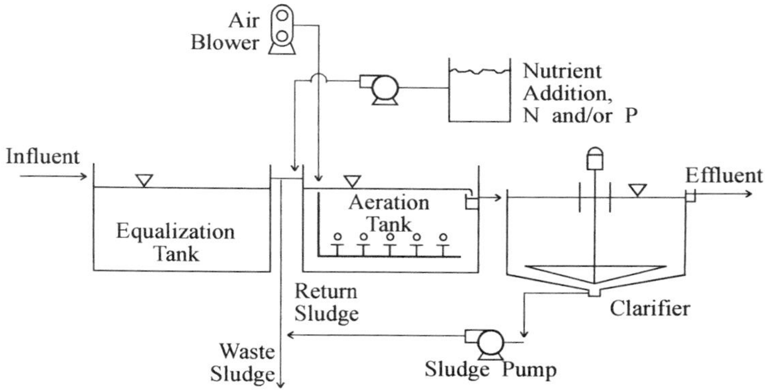

Figure 12.12 Activated sludge process. (From EPA, 1987.)

basin contains a suspension of *activated* microorganisms, primarily bacteria and protozoa, that biodegrade the wastewater constituents under aerobic conditions. Aerators supply oxygen for the microorganisms and mixing energy to keep them in suspension and in contact with the waste. The microorganisms are separated from the treated wastewater by clarification and a portion of this biological sludge is recycled back to the aeration basin to maintain the microbial population and begin a new treatment cycle.

Although the activated sludge process is relatively robust and able to handle variable wasteloads and operating conditions, it must be protected from shock loadings that are toxic or excessively high in concentration. Influent variability, both in flow and quality, can be reduced through equalization prior to the aeration basin. Other common types of pretreatment that may be necessary to protect the aeration basin are neutralization, oil–water separation, and grit or solids removal.

The microorganisms grow in response to the food source supplied in the wastewater and produce more biological sludge than is needed to maintain the process. This excess sludge must be wasted from the process and is usually treated by dewatering and aerobic or anaerobic digestion. The digested sludge may be disposed of by incineration or landfilling or may be used beneficially in land application for its nutrient (nitrogen, phosphorus) and organic value.

Aeration basins can be constructed as concrete or steel tanks or earthen impoundments, although tanks are more common now because of groundwater problems with leakage from impoundments and stringent regulation of impoundments for the treatment of hazardous waste.

Process Variations. Many variations of the activated sludge process have been developed in response to different wastewater characteristics and process requirements. The mixing regime of the process may be plug flow, completely mixed, or more often, something between the two. In a plug flow system, influent wastewater and return activated sludge are mixed at the head of the aeration basin and travel as a *plug* along the length of the basin. Therefore, the organic concentration is greatest at the head of the basin and diminishes as the wastewater is treated traveling toward the basin outlet. In a completely mixed activated sludge system, the influent wastewater and return activated sludge are uniformly or completely mixed in the aeration basin. Mixing the influent wastewater with the larger volume of the aeration basin allows high organic loads and toxic or inhibitory compounds to be diluted to protect the system from shock, an advantage over plug flow systems. In general, complete mix systems have higher mixed-liquor concentrations than plug flow systems and consequently can treat higher organic loads. However, complete mix systems tend to produce a sludge that is less dense and may be more difficult to settle. In reality, pure plug flow and complete mixing do not exist as the process operates somewhere in between. Some of the most common process variations of activated sludge systems are described in the following sections.

Tapered Aeration. In this process, aeration is provided in a tapered or stepped fashion to match oxygen requirements in a plug flow system. Oxygen requirements are highest at the head of the aeration basin, where the organic-laden influent enters. Aeration is decreased along the basin as the organics are degraded. Toward the end of the basin, it may be necessary to maintain a certain level of aeration in excess of oxygen requirements in order to provide enough mixing energy to keep the microorganisms in suspension.

Contact Stabilization. This process takes advantage of the ability of the activated sludge to adsorb organics quickly. Contact stabilization is operated either as a two-unit system or as two zones within a single aeration basin. In the two-unit system, the influent wastewater is first mixed with return activated sludge in a contact basin. Contact time is short, 30 to 90 minutes, but sufficient for the sludge to adsorb the organics. The mixture then flows to a clarifier where the sludge and adsorbed material are settled out. The treated effluent is discharged from the clarifier. A portion of the settled sludge is wasted as excess biomass and the remainder enters the stabilization basin, where the adsorbed organics are biodegraded over a period of 3 to 6 hours, thereby reactivating the sludge. In the two-zone system, the aeration basin is operated as plug flow. The contact zone for the influent wastewater is near the end of the aeration basin, where organic concentrations are lowest and the sludge is ready to adsorb additional organics. The stabilization zone begins at the head of the aeration basin, where the settled sludge with its adsorbed organic load enters. Contact stabilization requires about 50% less aeration basin capacity than conventional systems and produces less biomass, which settles better. On the down side, effluents are poorly nitrified and the process is better suited to wastewaters that contain high proportions of BOD as colloidal or suspended matter.

Extended Aeration. As its name implies, extended aeration operates with a long aeration period. With a long sludge age and low food/microorganism ratio, sludge production is low and the sludge settles well. Because aeration and basin volume requirements are high, extended aeration treating domestic sewage is normally used for small flows and small communities. With industrial wastewaters, extended aeration is common when there is a high concentration of organics and low biodegradation rates of some of the organic constituents. The oxidation ditch is a common form of the extended aeration process. Viewed from above, an oxidation ditch looks like a racetrack around which the wastewater flows. Aeration is provided by rotors or brushes, which also provide the energy to keep the water flowing around the track. Influent wastewater usually enters upstream of the aerator. Anoxic zones can be set up in oxidation ditches, allowing nitrogen removal through denitrification. In addition to the racetrack design, oxidation ditches can be of the carousel type, which resembles a long racetrack folded in half. The carousel has two aerators that allow larger flows to be treated.

Pure Oxygen. Activated sludge systems can be designed to use pure oxygen instead of air to increase oxygen transfer rates. Advantages of pure oxygen systems are increased BOD removal rates, smaller aeration basin volumes, lower sludge production, and improved sludge settling. The aeration basin is covered and divided into a series of stages. Influent wastewater, return activated sludge, and oxygen are introduced into the first stage. In each progressive stage, both the BOD and oxygen content decrease, acting like a tapered aeration plug-flow system. Operating parameters for pure oxygen systems [mixed liquor suspended solids (MLSS), mixed liquor oxygen levels, food-to-microorganism (F/M) loading rate] are higher than for conventional activated sludge. Disadvantages of pure oxygen systems include the high cost of oxygen production and the need for safety precautions and monitoring with combustible hydrocarbon in high-oxygen environments.

PACT System. Powdered activated carbon treatment, or PACT, is the simple addition of powdered activated carbon to an activated sludge system. This process is used to prevent inhibitory effects of toxic compounds, remove refractory organics or color, or remove effluent toxicity. Other benefits of the process are reduced emissions of volatile compounds from wastewaters and improved sludge settling. Carbon dosage rates depend on the given situation, but usually range from 50 to 3000 mg/L (Eckenfelder and Musterman 1995).

Process Control. There are many different process control variables that determine the efficiency of the activated sludge process. Effluent quality depends primarily on wastewater characteristics, plant loading (organics and flow), residence time in the system, temperature, type and concentration of microorganisms degrading the waste, oxygen supply, and degree of mixing. Major process control variables are described in the following sections.

Hydraulic Retention Time. The hydraulic retention time (HRT) is a measure of the average time the wastewater spends in the aeration basin. It is calculated by dividing the volume of the aeration basin by the influent wastewater flow rate. Since the flow rate of recycle sludge is not included in the calculation, the actual retention time would be less. The HRT depends on the type of activated sludge process being used. Conventional systems for domestic sewage and some readily degradable industrial wastewater have HRTs around 4 to 8 hours. Extended aeration systems can have much longer HRTs, around 18 hours to as much as 72 hours.

MLSS/MLVSS. The concentration of microorganisms in the aeration basin determines the organic loading and the efficiency of the process. The contents of the aeration basin, the mixture of wastewater and microorganisms, is referred to as *mixed liquor*. The mixed liquor suspended solids (MLSS) and mixed liquor volatile suspended solids (MLVSS) are both used as measures of the microorganism concentration in the aeration basin. The MLVSS more closely approximates the biologically active portion of the solids in the mixed liquor because microbial cellular material is organic and volatilizes or burns at 500°C (the temperature for volatile solids analysis). The MLSS includes both the volatile and inert solids in the mixed liquor. It is often used or at least monitored more frequently than the MLVSS because the analytical test for suspended solids is quicker than for volatile suspended solids. The MLSS concentration for conventional systems ranges from 1500 to 3500 mg/L. The volatile fraction representing the MLVSS varies, but a typical value is 0.8. MLSS concentrations will be higher in other process configurations. For example, pure oxygen systems may have MLSS concentrations from 5000 to 8000 mg/L. Specially acclimated systems may have MLSS concentrations of 10,000 mg/L and higher. MLSS concentrations are limited by the availability of oxygen in the aeration basin and recycle sludge flow rates.

F/M Ratio. The food/microorganism ratio (F/M) is a measure of the organic loading on the biological population. The F/M ratio is calculated as the mass of BOD per day divided by the mass of MLSS or MLVSS in the aeration basin. Thus the F/M ratio has units of day^{-1}. The F/M ratio can be controlled by adjusting both the flow rate and concentration of the sludge recycled to the aeration basin. When based on MLVSS, typical values for the F/M ratio for conventional systems range from 0.2 to 0.4 day^{-1}. The F/M ratio is higher for contact stabilization (0.2 to 0.6 day^{-1}) and pure oxygen systems (0.4 to 1.0 day^{-1}), and lower for extended aeration systems (0.03 to 0.15 day^{-1}).

Mean Cell Residence Time. The mean cell residence time (MCRT) is a measure of the average length of time (in days) spent in the system by the microorganisms and relates to the overall growth phase of the microbial population. It is calculated as the mass of microorganisms in the aeration basin, as represented by the MLSS or MLVSS, divided by the mass of microorganisms removed from the system through sludge wastage and in the effluent discharge. The MCRT may also be referred to as the *solids retention time* or *sludge residence time* (SRT) or *sludge age*. The MCRT in conventional systems typically ranges from 3 to 5 days. Pure oxygen systems have higher MCRTs (8 to 20 days), as do extended aeration systems (6 to 40 days). Systems used for nitrification operate at MCRTs greater than 10 days.

Sludge Recirculation. The concentration of microorganisms in the aeration basin is achieved by recirculating settled sludge from the clarifier. Both the flow rate and concentration of the return activated sludge determine the solids in the aeration basin. The recirculation rate is calculated as the flow rate of the return activated sludge divided by the flow rate of the influent wastewater. Typical values for conventional systems are 0.25 to 0.5. Pure oxygen systems have similar rates, while the recirculation rates are higher for both contact stabilization (0.25 to 1.0) and extended aeration (0.75 to 1.5).

Aeration. Aeration not only delivers oxygen to the activated sludge system but provides mixing energy to keep the microorganisms in suspension. Aeration is provided by mechanical aeration using surface aerators or by diffusers on the bottom of the aeration basin. Both types of aeration equipment are common in conventional and pure oxygen systems, while horizontal surface aerator brushes are used in oxidation ditches. In activated sludge systems using air, oxygen concentrations in the aeration basin are usually maintained between 1.5 and 2.0 mg/L. In pure oxygen systems, the concentrations are much higher (4 to 8 mg/L).

Aerated Lagoons and Polishing Ponds

Surface impoundments (ponds) can be used as activated sludge basins. Other uses of ponds in biological wastewater treatment systems are aerated lagoons and polishing ponds. As in an activated sludge system, surface aerators in an aerated lagoon provide oxygen and mix the wastewater and microorganisms together. A settling pond follows the aerated lagoon to separate the microorganisms from the treated wastewater. Alternatively, a section of the aerated lagoon downstream of the aerators can be used for settling under quiscient conditions. In either case, the settled sludge is not recirculated back to the aeration basin, which distinguishes the aerated lagoon from the activated sludge system. A polishing pond following an activated sludge system or aerated lagoon has no mechanical aeration and is used to "polish" the wastewater by providing additional biodegradation and solids settling.

Rotating Biological Contactors

A rotating biological contactor (RBC) is physically quite different from an activated sludge system. With an RBC, the biological solids are attached to a solid surface as a fixed film rather than being suspended in the wastewater as in the activated sludge system. An RBC is a series of large closely spaced plastic disks on a rotating horizontal shaft placed in a rectangular basin containing the wastewater. The disks, which are several feet in diameter, are partially submerged in the wastewater. The shaft rotates the disks slowly through the wastewater and out again, contacting the biological film growing on the disks with the wastewater and allowing reaeration. The biological film continously grows and sloughs off the disks into the wastewater. Downstream of the RBC, the treated wastewater and biological solids are separated in a clarifier. RBCs require less space and energy than an activated sludge plant; however, they are used much less frequently. They are not well suited for high organic strength industrial wastewaters because of insufficient oxygenation and plugging of the space between disks with excessive biomass that does not slough off.

Trickling Filters

Trickling filters are similar to RBCs in that the microorganisms grow as a film on support media. The support media in a trickling filter can be rock or other inert natural material or a synthetic plastic. The support media is usually contained with a circular basin. A central shaft rotates a set of distributor arms that "trickle" the wastewater over the media. The wastewater is continuously aerated because air can pass freely through the void spaces in the bed. Excess biological growth sloughs off and is separated from the treated wastewater in a clarifier. Trickling filters can be operated as low- or high-rate systems. Low-rate systems are highly dependable, perform consistently, and nitrify (remove ammonia) well; however, odor and filter flies can be nuisance problems. High-rate systems eliminate the odor and fly problems by recirculating part of the clarified wastewater back to the filter. The primary reason for recirculation, however, is to increase treatment efficiency so that higher organic loadings can be achieved.

Sludge Digestion

Organic sludges from biological wastewater treatment systems are normally treated by aerobic or anaerobic digestion to reduce the organic content and thereby stabilize the material before land application or landfilling. Waste activated sludge and primary sludge (if a solids clarifier precedes the aeration basin) are combined for digestion. Anaerobic digestors that are not mixed form an aqueous layer (supernatant) above the digesting solids. Because it is rich in soluble biodegradation by-products, the supernatant is returned to the wastewater treatment system. The methane generated by anaerobic digestion can be used on site to heat both the sludge digestor and facility buildings. Advantages of anaerobic digestion are no oxygen requirements, which makes treatment of high-strength wastes economical; low residual-sludge quantities, methane production; and easily dewatered sludges. Disadvantages of anaerobic digestion are larger digestor volumes required to maintain a longer MCRT, heating requirements, and sensitivity to upsets and environmental changes. Aerobic digestors are mixed to distribute the supplied oxygen and are usually unheated. Digested solids are separated from the supernatant in a sludge thickener tank.

THERMAL PROCESSES

Thermal treatment is used to destroy, break down, or aid in the desorption of contaminants in gases, vapors, liquids, sludges, and solids. There are a variety of thermal processes that destroy contaminants, most of which are classified as incineration. *Incineration* literally means to become ash. With respect to the incineration of hazardous wastes regulated under the Resource Conservation and Recovery Act (RCRA), however, there is a strict legal definition of what constitutes an incinerator. A more simplified description is a unit that combusts materials in the presence of oxygen at temperatures normally ranging from 800 to 1650°C. Typical types of incineration units that are discussed in this section are liquid injection, rotary kiln, multiple hearth, fluidized beds, and catalytic oxidation. Thermal desorption is also discussed in this section. Prior to discussing these individual technologies, an overview of the main factors affecting incinerator performance is given in the following section.

Factors Affecting Performance

There are many factors that affect both the choice of a particular thermal treatment and its performance. Chief among these are waste characteristics, temperature, residence time, mixing or turbulence, and air supply.

Waste Characteristics. The physical form of the waste restricts the applicability of some thermal technologies. A convenient summary of physical characteristics and thermal processes has been presented by Cheremisinoff (1994). Those processes that can treat the widest range in physical form are incineration, pyrolysis, calcination, and microwave discharge. These processes are capable of treating wastes in essentially any form: containerized, solid, sludge, slurry, liquid, or fume. Molten salt reactors and plasma technology can be used with any of these types of wastes, except fumes. Evaporative processes are amenable to solids, sludges, slurries, and liquids. Wet air oxidation, sometimes considered thermal treatment but also described as a physical–chemical process, can be used for slurries and liquids. Distillation (with and without steam) and steam stripping, also considered physical–chemical treatment, are limited to liquid treatment. Catalytic incineration is limited to fume treatment.

Other waste characteristics that affect the choice of thermal treatment and its operation include heating value, contaminant boiling points, metal content, heterogeneity, moisture content, decomposition products and by-products, and the presence of explosive constituents. The heating value and moisture content of the waste determine if combustion is self-supporting or if auxiliary fuel will be required. Elemental constituents in the waste may be carried in the exhaust gases and must be removed. Problem metals are antimony, arsenic, cadmium, lead, and mercury. Acid gases containing chlorine, sulfur, fluorine, or bromine that are generated must be neutralized and scrubbed from the flue gases. The heterogeneity of the waste is determined by the distribution and concentration of combustible organics; localized concentrated materials can produce hot spots in the combustion chamber. Heterogeneity is also defined by the type and amount of debris that may be included in excavated wastes. These materials can interfere with the process by shielding the waste, fouling equipment, or slagging in the reactor. Undesirable decomposition products and by-products include nitrogen oxides and dioxins.

Temperature. The temperature for combustion processes must be balanced between the minimum temperature required to combust completely the original contaminants and any intermediate by-products and the maximum temperature at which the ash becomes molten. Typical operating temperatures for thermal processes are: incineration (750 to 1650°C), catalytic incineration (315 to 550°C), pyrolysis (475 to 815°C), and wet air oxidation (150 to 260°C at 1500 psig) (Cheremisinoff 1994). Pyrolysis is thermal decomposition in the absence of oxygen or with less than the stoichiometric amount of oxygen required. Because exhaust gases from pyrolytic operations are somewhat "dirty" with particulate matter and organics, it is not often used for hazardous wastes.

Residence Time. For cost efficiency, residence time in the reactor should be minimized but must be long enough to achieve complete combustion. Typical residence times for various thermal processes are incineration (0.1 second to 1.5 hours), catalytic incineration (1 second), pyrolysis (12 to 15 minutes), and wet air oxidation (10 to 30 minutes) (Cheremisinoff 1994).

Turbulence. Turbulence is important to achieve efficient mixing of the waste, oxygen, and heat. Effective turbulence is achieved by liquid atomization (in liquid injection incinerators), solids agitation, gas velocity, physical configuration of the reactor interior (baffles, mixing chambers), and cyclonic flow (by design and location of waste and fuel burners) (Noyes 1994, WEF 1990).

Air Supply. Oxygen in excess of stoichiometric requirements for complete combustion is needed because incineration processes are not 100% efficient, and excess air is needed to absorb a portion of the combustion heat to control the operating temperature. In general, units that have higher degrees of turbulence such as liquid injection incinerators require less excess air (20 to 60%), while units with less mixing such as hearth incinerators require more (30 to 100%).

Catalytic Oxidation

Catalytic oxidation is used only for gaseous streams because combustion reactions take place on the surface of the catalyst, which otherwise would be covered by liquid or solid material. Common catalysts are palladium and platinum. Because of the catalytic boost, operating temperatures and residence times are much lower, which reduces operating costs. Catalysts in any treatment system are susceptible to poisoning (masking of or interference with the active sites). Catalysts can be poisoned or deactivated by sulfur, bismuth, phosphorus, arsenic, antimony, mercury, lead, zinc, tin, or halogens (notably chlorine); platinum catalysts can tolerate sulfur compounds but can be poisoned by chlorine.

Fluidized Bed

In this type of incinerator, the waste and an inert bed material (sand normally, or alumina) are fluidized by blowing heated air through a distributor plate at the bottom of the bed, as shown in Figure 12.13. Fluidization results in good mixing and uniform distribution of materials within the bed, which results in lower operating temperatures (450 to 710°C) and lower excess air requirements (Noyes 1994). Fluidized beds can be used to incinerate gases, liquids, or solids, although each type of waste is introduced into the bed differently. Fluidized beds are susceptible to seizing or binding up or agglomeration on individual bed particles, which can lead to seizing up. Bed seizure can occur with wastes containing clays, inorganics (salts), or high concentrations of lime (Noyes 1994).

Liquid Injection

Liquid injection units are the most common type of incinerator today for the destruction of liquid hazardous wastes such as solvents. Atomizers break the liquid into fine droplets (100 to 150 μm) (LaGrega et al. 1994), which allows the residence time to be extremely short (0.5 to 2.5 seconds). The viscosity of the waste is very important; the waste must be both pumpable and capable of being atomized into fine droplets. Both gases and liquids can be incinerated

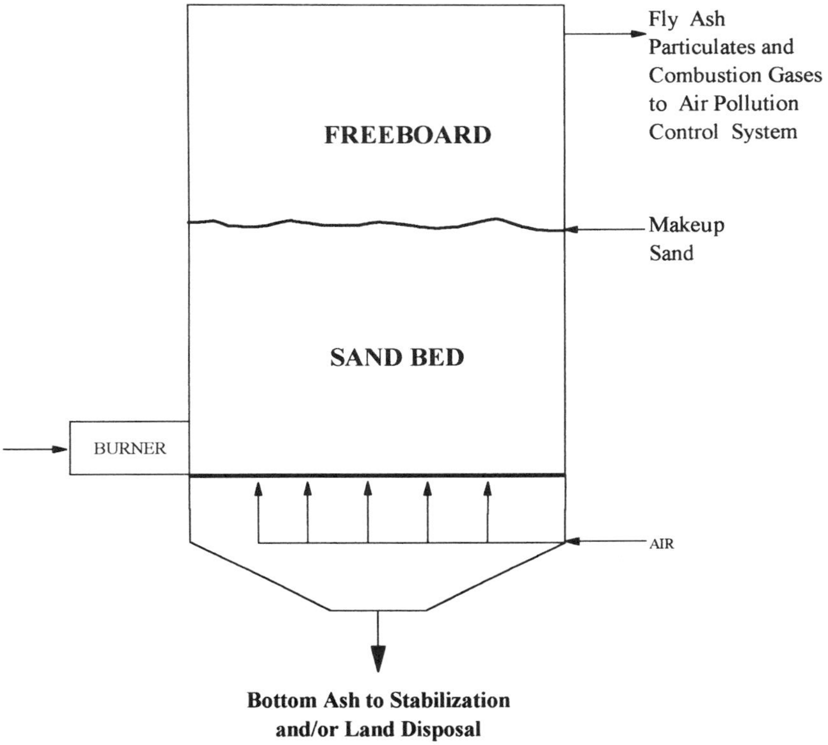

Figure 12.13 Fluidized-bed incineration system.

in liquid injection units. Gases include organic streams from process vents and those from other thermal processes; in the latter case, the liquid injection incinerator operates as an afterburner (WEF 1990). Aqueous wastes containing less than 75% water can be incinerated in liquid injection units (LaGrega et al. 1994).

Multiple Hearth

Although well suited for combustion of wet sludges, particularly municipal biological wastewater sludges, multiple hearth incinerators are not often used for hazardous wastes because operating temperatures (800 to 1000°C) of this type of incinerator are too low for many hazardous compounds (WEF 1990). The multiple hearths are stacked vertically and sludges are introduced into the top hearth. A series of "rabble" arms extending from a center rotating shaft mix and move the waste along the circular hearth until the waste falls through holes unto the hearth below. For incineration, usually a minimum of six hearths is required; more hearths are required for pyrolysis (Noyes 1994). Liquid wastes may be incinerated along with sludges when injected through burner nozzles.

Rotary Kiln

A rotary kiln is a long, cylindrical incinerator that is sloped a few degrees from horizontal. Waste is introduced at the upper end. The gentle slope and slow rotation of the kiln continually mix and reexpose the waste to the hot refractory walls, moving the waste toward the exit point. An example of a rotary kiln is presented in Figure 12.14.

Rotary kilns operate in continuous or batch mode, although continuous operation is more cost-efficient and less wearing on the refractory material. Operating temperatures range from about 800 to 1650°C (WEF 1990). They can be operated in either ashing or slagging mode; the former is common in the United States. Rotary kilns are relatively large incinerators, but because of their versatility in treating a wide range of wastes and their treatment efficiency, they are often used for incinerating hazardous wastes. Rotary kilns can be used to incinerate gases, liquids, sludges, and solids. However, highly aqueous organic sludges tend to form a ring on the walls of the kiln that prevent the discharge of ash and they can also form sludge balls that are not completely incinerated (WEF 1990).

Thermal Desorption

Thermal desorption is an innovative treatment that has been applied primarily to soils (see Chapter 13). Wastes are heated to temperatures of 200 to 600°C to increase the volatilization of organic contaminants. Volatilized organics in the gas stream are removed by a variety of methods, including incineration, carbon adsorption, and chemical reduction.

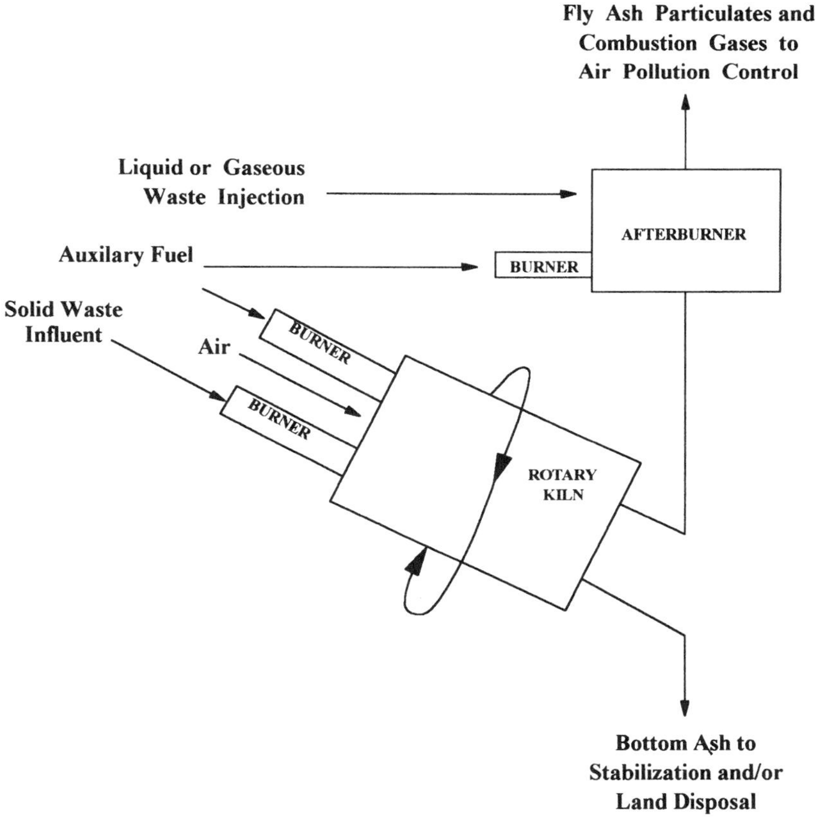

Figure 12.14 Rotary kiln incineration system.

LAND-BASED DISPOSAL SYSTEMS

Concern about soil and groundwater contamination and inadequate waste treatment has given rise to stringent control of land-based systems such as landfills, surface impoundments, land treatment units, and underground injection wells. Federal regulations have been developed under RCRA and its amendments covering the design, operation, and closure of land-based systems.

Land disposal restriction regulations, promulgated as a result of HSWA and defined under 40 CFR Part 268, placed significant restrictions on land disposal of certain wastes, including sludges, solvents, toxic chemicals, and wastewaters. The definition of land disposal includes placement of any hazardous waste in landfills, land treatment facilities, injection wells, salt domes or salt bed formations, underground mines or caves, and other waste management units, such as surface impoundments and waste piles intended for disposal purposes.

Maximum concentration limits or treatment technology-based standards have been established for land restricted wastes. If a waste meets the established concentration limits (in the waste extract or in the waste itself, depending on the waste), or if it meets the treatment technology-based standard, the waste may be land disposed. Dilution of a waste stream as a substitute for treatment is prohibited.

Wastes regulated under the land disposal restrictions include wastewaters and nonwastewaters. Wastewaters contain less than 1% by weight of total suspended solids and total organic carbon, with some exceptions. Concentration limits and acceptable technologies to treat the same constituent in the two forms will vary. Selected examples of concentration limits for land restricted wastes in the form of wastewaters and nonwastewaters are presented in Table 12.1. Selected examples of technologies that have been established as acceptable for treatment of restricted wastes are presented in Table 12.2.

Acceptable concentrations in the waste extract are defined under 40 CFR §268.41, Table CCWE, and acceptable concentrations in the waste itself are defined under 40 CFR §268.43, Table CCW. A complete list of technologies acceptable for treatment of restricted wastes is provided in 40 CFR §268.42, Table 2.

Landfills

Major federal regulations covering landfill design and operation are the Subpart N standards of 40 CFR Parts 264 and 265 for hazardous waste and the standards for municipal landfills at 40 CFR Part 258. The standards for hazardous waste landfills cover:

- Design and operation of liner, leachate collection, and stormwater systems
- Action leakage rates and response plan for leak detection systems
- Monitoring and inspection on liners, covers, and leachate collection systems
- Surveying and record keeping for each landfill cell
- Closure of the landfill or individual landfill cells and postclosure care
- Requirements for ignitable, reactive, and otherwise incompatible wastes
- Requirements for bulk and containerized liquids, overpacked drums (lab packs), containers, and certain dioxin-contaminated listed wastes

The standards for municipal waste landfills under Part 258 cover:

- Location restrictions regarding airports, floodplains, wetlands, fault areas, seismic impact zones, and unstable areas
- Operating criteria for hazardous waste prohibition, cover material, disease vector control, explosive gas control, air polution, access, storm-

TABLE 12.1 Examples of Concentration Limits for Land Restricted Chemicals in the Form of Wastewaters and Nonwastewaters

Constituent	Wastewaters (mg/L)	Nonwastewaters (mg/kg)
Acetone (from waste codes F001–F005)	0.05[a]	0.59[a]
Chlordane: alpha and gamma (waste code U036)	0.0033	0.13
Chloroform (from waste codes K009 and K010)	0.10	6.0
Total cyanides (from waste codes F011 and F012)	1.9	110
Cyclohexane (waste code U057)	0.36	NA[b]
Dimethyl phthalate (waste code U102)	0.54	28
Mercury (from waste code K071)	0.030	0.025[a]
Phenol (waste code U188)	0.039	6.2
Toluene (waste code U220)	0.080	28
Xylenes (waste code U239)	0.32	28

[a]Treatment standard expressed as concentration in the waste extract.
[b]Technology-based standard defined under 40 CFR §268.42, Table 2 applies.

Source: Information from 40 CFR §268.41, Table CCWE and §268.43, Table CCW.

water, surface water discharges, liquid waste restrictions, and record keeping;

- Design criteria for liners and leachate collection
- Groundwater monitoring and corrective action for groundwater contamination
- Closure and postclosure
- Financial assurance

The following sections deal primarily with landfill construction and design. The reader is referred to Parts 258, 264, and 265 of 40 CFR for detail on specific regulatory requirements.

TABLE 12.2 Examples of Technologies Acceptable for Treatment of Restricted Wastes

EPA Code	Technology Description	Waste Applications
CHOXD fb CRBN	Chemical or electrolytic oxidation followed by carbon adsorption	Wastewaters containing allyl alcohol, fluoroacetamide, acrylic acid, cumene, formaldehyde, others
INCIN	Incineration	Concentrated wastes (nonwastewaters) containing cyclohexane, phosphine, allyl alcohol, fluoroacetamide, acrolein, strychnine and salts, pentachloroethane, others
FSUBS	Fuel substitution	Non wastewaters of high-TOC ignitable liquids, ignitable regulated residues and heavy ends, nitroglycerin, acrylic acid, cyclohexane, allyl alcohol, cumene, others
HLVIT	High-level mixed radioactive waste vitrification	Radioactived high-level wastes generated during the reprocessing of fuel rods

Source: Information from 40 CFR §268.42, Table 2.

Basic Landfill Components. The basic components of a landfill are a liner, a leachate collection system, and a final cover. Many different types and combinations of materials are used for these components. In addition, since both liners and covers are designed to prevent liquid migration, many of the same materials are used for both.

Common landfill materials include clay or low-permeable soil materials, geosynthetics (geogrids, geonets, geotextiles, geomembranes, flexible membrane liners), and synthetic clay liners. To aid in the discussion, brief descriptions of these materials are as follows:

- *Clay or soil material:* normally, natural clays with a hydraulic conductivity of at least 1×10^{-7} centimeter per second (cm/s). More permeable soils may be mixed with clay or other additives to create a "soil material" with sufficient impermeability.
- *Geogrid:* a geotextile used for additional structural stability for steep slopes and liner support under differential settlement.

- *Geonet:* a geotextile used as a drainage net.
- *Geosynthetic:* a general term referring to a group of synthetic materials that includes geogrids, geonets, geotextiles, and geomembranes.
- *Geotextile:* a thin, fabric-type material that is used as a particle filter, drainage net, or geocell containment. Also referred to as *geofabric*. When used for drainage, the term *geonet* is often used.
- *Synthetic clay liner:* a manufactured material made of a thin layer of sodium montmorillonitic clay sandwiched between two geotextile layers.
- *Synthetic membrane:* a thin, flexible, plastic-type material used as an impermeable layer. Also called *geomembrane* or *flexible membrane liner* (FML).

The main purpose of synthetic membranes is to prevent liquids from migrating into or out of a landfill. In final covers, synthetic membranes prevent precipitation from entering a landfill. In landfill liners, these membranes prevent leachate from migrating through to secondary landfill liners or outside the landfill into the soil and groundwater.

Synthetic membranes are made from common polymers such as butyl rubber; chlorinated polyethylene; chlorosulfonated polyethylene; ethylene–propylene rubber; high-density polyethylene; medium- to very-low-density polyethylene, and polyvinyl chloride. Other polymers that may be used in synthetic membranes are nylon, polyester, neoprene, ethylene propylene diene monomer, and urethane.

Key factors in choosing synthetic membranes are compatibility with the waste leachate, permeability, and mechanical properties. Compatibility with the waste leachate is important because the membrane can be chemically weakened or damaged and fail to contain the leachate. Permeability is important because a membrane must be a barrier not only to water but also to other leachate constituents, such as hydrocarbons. Mechanical properties are important because they determine how easy the membrane is to install, how well it resists tears and punctures, and how well it holds up under temperature extremes and sunlight.

Liners. Landfill liners are meant to prevent waste leachate from migrating into the soil and groundwater outside the landfill. They range in complexity from natural liners (where the existing soil acts as a barrier) to single, double, and multiple liners. Natural liners were common many years ago before soil and groundwater contamination problems became widely known. Today, new landfill construction includes at least a single liner; hazardous waste landfills must have at least a double liner.

The materials used for landfill liners include synthetic membranes, clay, soils that have been mixed with additives to improve their characteristics (amended soils), and alternate barrier materials such as synthetic clay liners.

Leachate Collection. Leachate originates from a mixture of precipitation falling on a landfill and the liquids present in landfilled wastes. As this mixture percolates through the landfill, it picks up or leaches contaminants from the wastes. Leachate is removed from a landfill to reduce the hydraulic head on liners and the amount of leachate that could breach the liner and contaminate soil and groundwater.

The leachate collection system usually consists of a gridwork of perforated pipe placed above the landfill liner. The piping grid is covered with a layer of sand and gravel that allows easy drainage of the leachate into the pipe and prevents the pipe from being crushed by the weight of the waste above it. As an alternative to a piping grid, a leachate collection system may also be constructed with geonets. An example of a primary leachate collection system is shown in Figure 12.15.

Covers. A final cap or cover is placed over the waste when a landfill is closed. The primary purpose of this final cover is to minimize the infiltration of precipitation and thus minimize the generation of leachate. A cover also helps to maintain the integrity of the landfill while improving the final appearance of the disposal site.

Final covers are actually multiple layers. Typically, the order in which these layers are placed over the waste are a grading layer, a gas venting layer, a barrier layer, a drainage layer, a protective layer, and a top soil–vegetative layer.

The grading layer is placed directly over the waste to even out the contours of the landfill surface, providing a stable and smooth base for the next layer. The grading layer may also serve as the gas venting layer if a coarse-grained material such as sand or gravel is used.

Gases evolve from a landfill as wastes decompose. Gas vents or extraction wells may be installed for various reasons: to prevent rupture of the final

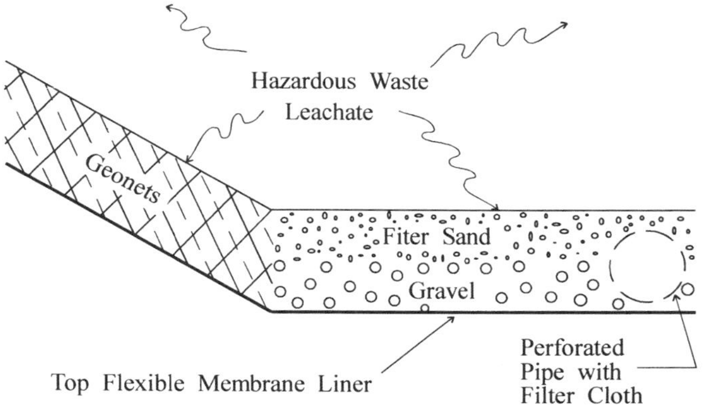

Figure 12.15 Cross-sectional view of a primary leachate collection system. (Adapted from EPA, 1989.)

cover; to prevent stress on the vegetation cover, which can lead to erosion; to meet air quality criteria; to minimize health risks to the local population; or to burn for heating value. The type and quantity of gases generated depend on the type of waste. When putrescible (biodegradable) wastes are landfilled, gas venting is considered essential (Bagchi 1994). Municipal landfills, which contain a high percentage of biodegradable wastes, can produce significant quantities of methane and carbon dioxide. Unless gas extraction wells are installed in the landfill, a venting layer is needed as part of the final cover (LaGrega et al. 1994). Coarse-grained materials such as sand and gravel are used to allow these gases to pass freely to the vents. Vents may be isolated or connected together in a grid system.

A barrier layer is placed over the grading and gas venting layers. The barrier layer minimizes the infiltration of precipitation and stormwater. Barriers are typically constructed of clay or a synthetic membrane over a low-permeable material such as clay. Landfills that are lined with synthetic membranes typically use synthetic membranes in the barrier layer. This prevents the buildup of excessive hydraulic head on the liner since the permeabilities of the barrier and liner will be the same. Synthetic clay liners may also be used as barriers.

A drainage layer placed over the barrier layer diverts water away laterally from the barrier, preventing the buildup of hydraulic head. When synthetic membranes are used in the barrier layer, a drainage layer over the membrane prevents the saturation of the upper soil cover, which can lead to failure and erosion. The drainage layer also protects synthetic membranes from damage when final soil covers are placed. Drainage layers may be constructed of sand and gravel, geonets, or geocomposites.

Topsoil is placed over the drainage layer. The type of soil used must be cohesive enough to maintain the contours of the landfill, minimize erosion, and support the vegetative cover. Together with the drainage layer, topsoil protects the barrier layer from cracks that can develop from drying and freeze–thaw cycles. If the top soil is sandy and has a permeability or 1×10^{-3} cm/s or more, a separate drainage layer may not be needed (Bagchi 1994). Many different types of vegetative (grass) covers can be used, but it is important to choose a vegetative cover that grows well in the area and is easy to maintain. Natural vegetation may be allowed to grow as the final cover; however, until the natural vegetation is fully established, additional seeding or covering will be needed to prevent the final cover from erosion.

Hazardous Waste Landfills. Hazardous waste landfills constructed after January 29, 1992 must have a double liner and a leachate collection system above each liner. An example of a double-liner leachate collection system is shown in Figure 12.16. The double-liner requirements may be waived, however, for landfills containing only wastes from foundry furnace emission controls or metal casting molding sand (monofills) and certain replacement landfill units. The top liner must be designed to prevent hazardous constituents from migrating into the liner, not only while the landfill is in operation, but throughout

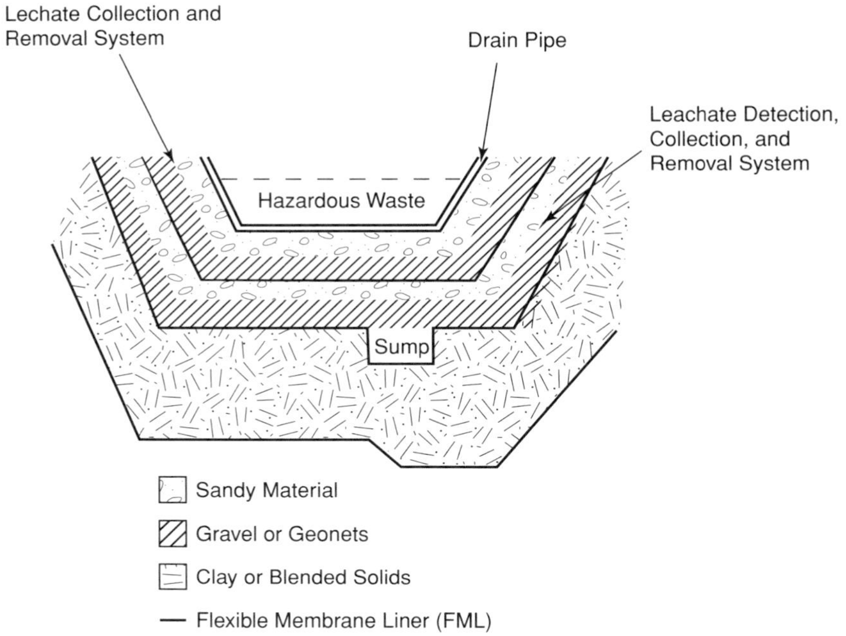

Figure 12.16 Cross section of a hazardous waste landfill. (Adapted from EPA, 1989.)

the postclosure care period, which may be for 30 years or longer. The bottom liner must be a composite liner of at least two components. Like the top liner, the upper component must be designed to prevent the migration of hazardous constituents. The lower component of the bottom liner must have at least 3 feet of compacted soil material with a hydraulic conductivity of 1×10^{-7} cm/s.

The top liner and upper component of the bottom liner are usually geomembranes. The compacted soil material for the lower component of the bottom liner is usually a clayey soil or one that has been amended to meet the hydraulic conductivity criterion.

There must be a leachate collection system above the top liner and one above the bottom liner. This system must be constructed either with a granular drainage material such as gravel and sand or a synthetic material such as a geonet. If a granular layer is used, it must be at least 12 inches thick and have a hydraulic conductivity of at least 1×10^{-2} cm/s. If a geonet is used, it must have a transmissivity of at least 3×10^{-5} m^2/s. The bottom slope of the leachate collection system must be at least 1%.

The final cover for a hazardous waste landfill must provide long-term minimization of liquid migration; require minimum maintenance; promote drainage while minimizing erosion or abrasion; maintain its integrity with settling and subsidence of the landfill contents; and have a premeability less than or equal to that of any bottom liner or natural subsoils.

Surface Impoundments

Surface impoundments or ponds are commonly used for biological wastewater treatment and for impounding stormwater. Impoundments are used more often for nonhazardous wastewaters because the hazardous waste regulations for this type of unit are very strict. Impoundments provide temporary storage for stormwater prior to discharge or treatment. In arid areas, stormwater basins may be used as evaporation ponds in lieu of creating a discharge that must have a wastewater permit.

Surface impoundments used to manage hazardous wastewaters must meet technical standards at 40 CFR Parts 265 and 264 as well as land disposal restrictions in Part 268. Although there are some exceptions, hazardous waste impoundments are subject to minimum technology requirements (MTRs). MTRs include two or more liners and a leachate collection and removal system between liners. Requirements for liner components and the leachate system are similar to those for hazardous waste landfills except that granular drainage materials must have a hydraulic conductivity of at least 1×10^{-1} cm/s, and geonets must have a transmissivity of at least 3×10^{-4} cm/s. LDRs require that treatment residues generated in hazardous waste impoundments be removed at least once a year unless the residues meet applicable treatment or prohibition levels.

Land Application

In land application or treatment, waste is mixed into the upper soil layer, where microorganisms degrade organic compounds and metals are adsorbed. Land application of municipal wastewater biosolids is common because such materials have value as fertilizers (nitrogen, phosphorus) and as soil amendments. Land treatment of hazardous wastes has been significantly curtailed because of land disposal restrictions on hazardous wastes.

Land application of biosolids are regulated under 40 CFR Part 503. These regulations contain limits for arsenic, cadmium, chromium, copper, lead, mercury, molybdenum, nickel, selenium, and zinc. There are limits for both the metal content in the biosolids (ceiling and high-quality limits), as well as maximum loading rates on the land (annual and total cumulative). Metal concentrations in biosolids must not exceed the ceiling or maximum concentration limits. Cumulative loading rates apply to biosolids that do not meet the high-quality limits. Biosolids meeting the high-quality limits are exempted from certain requirements.

Land treatment of hazardous waste is regulated under 40 CFR Parts 264, 265, and 268. The land treatment standards of Parts 264 and 265 cover treatment demonstrations; design and operating requirements; restrictions regarding food-chain crops; monitoring of the unsaturated zone; record keeping; closure and postclosure; and special requirements for ignitable, reactive, incompatible, and dioxin-contaminated wastes.

Design Features

General. The specific design and layout of a land treatment facility will be based on terrain, soil type, expected loading rates, and the types of wastes being treated. The main feature of the treatment facility is the treatment zone. Other components include a monitoring system, runoff management system, waste staging area, road system, and security. An example of a layout of a land treatment unit is shown in Figure 12.17. As shown in the figure, the treatment unit is divided into sections. If more than one waste type is applied at the treatment facility, the use of a separate section for each waste is common to simplify loading rate calculations and record keeping.

Waste Characteristics. Waste characteristics such as chemical, physical, and biological properties should be evaluated to determine appropriateness of land treatment technology for a given waste. Information from agency documents and technical literature can provide useful case studies that can serve as a

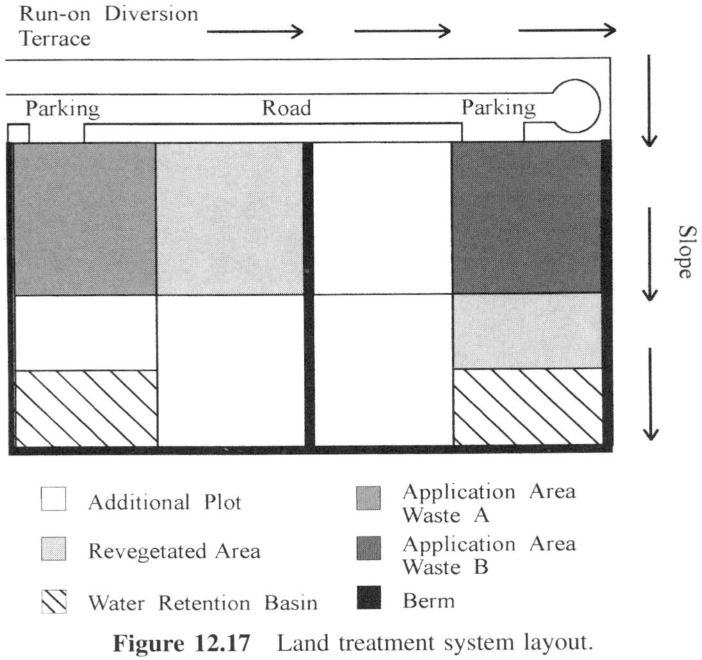

Figure 12.17 Land treatment system layout.

prescreening method for determining expected treatment efficiencies for elected wastes.

Site Characteristics. Site characteristics need to be evaluated carefully during the design phase of the facility to ensure that the waste does not migrate to underlying groundwater and to optimize the facility. Characteriscs that are pertinent include the site's hydrogeology, groundwater table characteristics, and climate. Hydrogeology information should include data on geological strata, aquifers, soil type, and other data. Groundwater table data should include items such as the location of perched zones and groundwater flow patterns. Climate data should include freeze–thaw and wet–dry cycles, temperature variations, and other climate-related data.

Waste–Soil Interaction. Testing must be performed to determine adequacy of waste–soil interaction prior to selecting land treatment as a disposal technology. Bench and field testing can be used to determine final fate of contaminants in soil as well as assimilative capacity of soils for waste constituents, degradation factors, immobilization, and potential pathways of contaminants to air, surface water, and groundwater.

Treatment Zone. The treatment zone consists of two sections: a zone of incorporation and the lower treatment zone. The maximum depth of the entire treatment zone, as specified in the regulations, is 5 feet. The lower boundary of the lower treatment zone should be at least 3 feet above the highest seasonal level of the water table. The depth of the lower boundary should also be well above any perched water table, and it should be adjusted to accommodate water table variations in a particular section, if necessary (DOE 1987). An example of a profile of a land treatment unit incorporating these features is shown in Figure 12.18.

Other Design Components. Monitoring of the system is an integral part of the design and must include methods for assessing leachate–contamination migration as well as treatment efficiency. Monitoring methods used to make this assessment are discussed in the next section.

Run-on and runoff management is an important part of the design. Run-on control can be achieved using terracing and other diversion mechanisms. The design should also include features that eliminate or minimize runoff from the facility. If runoff does occur, it should be diverted to water retention basins or water filtration units. An NPDES permit for discharge of the runoff typically is required, and associated discharge monitoring and treatment requirements, as necessary, are incorporated into the permit.

Roads must be designed to withstand the vehicular loads of heavy equipment. In addition, they should be placed to allow easy access to the treatment areas and waste staging areas. Security should include, at a minimum, a fence or other barrier that protects the facility from entry by unauthorized persons.

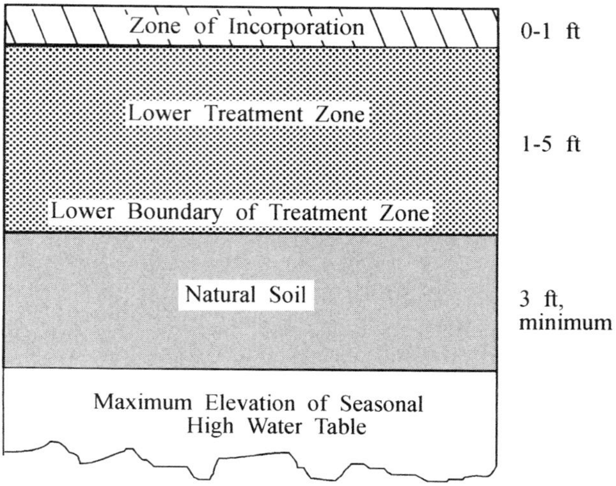

Figure 12.18 Profile of a land treatment facility.

There must be a means to control entry into the facility, and "Danger—Unauthorized Personnel Keep Out" signs must be posted at each entrance to the facility.

Monitoring Program. The monitoring program for a land treatment facility is integral to its operation. Elements of a comprehensive monitoring program include waste monitoring, unsaturated zone and groundwater monitoring, run-off monitoring, treatment zone monitoring, vegetation monitoring, and air quality monitoring.

Waste Monitoring. Waste streams must be checked periodically for changes in composition. A waste monitoring plan should be developed to ensure that allowable cumulative applications are met and that limiting waste constituents stay at uniform levels during application.

Unsaturated Zone Monitoring. Two aspects of the unsaturated zone must be monitored in order to determine whether hazardous constituents have migrated out of the treatment zone. The soil itself should be analyzed from soil cores and soil-pore liquid should be analyzed using monitoring devices such as lysimeters. The samples are collected at a frequency defined in the permit and are compared to background samples to determine whether there is a statically significant change over background values for constituents monitored.

Groundwater Monitoring. All land treatment facilities are required by regulation to have a groundwater monitoring program. Details for groundwater monitoring system design and installation are presented in chapter 13. Groundwater monitoring supplements unsaturated zone monitoring.

Runoff Monitoring. When NPDES permits are issued for discharge of runoff from the land treatment facility, monitoring requirements outlined in the permit must be met. Normally, a flow measuring device and an automated sampler will be required to acquire flow-proportional samples. For runoff management systems that contain and treat the runoff before discharge, composite (grab) sampling and analysis prior to discharge normally are required.

Treatment Zone Monitoring. Monitoring the treatment zone is necessary to evaluate soil pH, moisture content, and degradation rates of organic and other waste constituents. Additionally, rates of accumulation are evaluated to assess loading rates and to estimate the active life of the facility.

Vegetation Monitoring. Where food chain crops are grown, analysis of the vegetation at the land treatment facility will ensure that harmful quantities of metals or other toxic constituents are not accumulating in the plants. This is normally a part of the postclosure requirements.

Air Monitoring. Air monitoring may be required to assure that waste constituents are not leaving the site through wind dispersal. Additionally, personal sampling or another means of evaluating worker exposure to volatile constituents may be part of the overall land treatment facility monitoring program.

Operating Records. As part of a comprehensive operating record and as a good management practice, numerous records pertaining to the facility's operation should be maintained. Examples of these records are presented in Table 12.3.

A good management practice is to maintain all records pertaining to land treatment at the facility for the life of the facility. A record of waste disposal locations and quantities must be submitted to the regional administrator and local land authority upon closure of the facility. In addition, records pertaining to corrective action, as applicable, should be kept and will be required upon closure.

Applications. Use of land treatment has been significantly curtailed for many hazardous waste sludges and other wastes as a result of land disposal restrictions. (Land disposal restrictions are discussed later in the chapter.) These restrictions included listed metal-bearing wastes from petroleum refineries, which once were primary candidates for disposal using this technology. In addition, the owner or operator must demonstrate prior to application of a waste, that the hazardous constitutents can be completely degraded, transformed, or immobilized in the treatment zone. Approvals for adequate dem-

TABLE 12.3 Examples of Records Maintained at a Land Treatment Facility

Initial Site and Waste Assessment Records

- U.S. Geological Survey maps and other topography information
- Hydrogeological surveys
- Initial waste and soil analyses
- Climate information

Inspections of Facility (Signed and Dated)

- Inspections of the treatment area for excessive moisture
- Inspections of the security of the facility
- Inspection of berms and runoff and run-on management systems
- Inspection of road conditions

Waste Analysis Records

- Waste analysis plan
- Waste analysis records
- Documentation of changes in waste composition

Monitoring Data

- Data from ground water monitoring
- Data from soil monitoring
- Data from retention pond monitoring
- Data from soil pore monitoring
- Data from other monitoring
- Documentation of statistically significant changes below the treatment zone

Records of All Waste Applications

- Description of wastes applied to the facility
- Records of dates, amounts, wastetypes, and rates of application

Records of Climate Parameters During Operation

- Rainfall records
- Records of unusual climatic events during operation

Records of Vegetation Planting

- Records of vegetative planting including dates, crop types, and area covered
- Records of fertilization of crops, including fertilizer type and amounts
- Records of crop growth

Maintenance Records

- Records of maintenance schedules and activities for levees, berms, roads, and other areas and items
- Records of decontamination of equipment

TABLE 12.3 (*Continued*)

Records of Runoff Management

- NPDES permit and monitoring reports
- Documentation of breaches of runoff retention
- Documentation of water depth in retention basins after rainfall events
- Records of dates and times when the retention basins were emptied
- Other documents required in the permit

Records of Treatment Zone Activities

- Records of tilling activities
- Documentation of control methods to demonstrate that wastes were not applied below treatment zone lower boundary
- Records of measures to control soil pH
- Records of measures to control the moisture content of the soil

Manifest Documents and Records

- For wastes accompanied by a manifest, records of where the waste was applied, cross-referenced with the manifest number.
- For wastes accompanied by amanifest, records of quantities and application dates.
- Copies of the manifest and certifications by the generator that the waste is not land restricted.

Accidents or Incidents

- Records of spills during the operation of the facility
- Records of personnel injuries
- Records of breaches of security
- Records of other unplanned events during the operation of the facility

Source: Information from DOE (1987).

onstration have been difficult to obtain. Land treatment is now typically used for nonhazardous waste.

Closure. The cumulative amount of any waste that can be applied to a given parcel of land usually is predetermined by the concentration of metals in the waste and the acceptable contamination levels for the metals. For many wastes, metals are present only in trace amounts, which allows the facility to have long-term use. Once a limit for a specific metal is reached, however, the land treatment facility must be closed.

During the closure period, the owner or operator of the facility must continue all operations necessary to maximize degradation, transformation, or immobilization of the hazardous constituents within the treatment zone. When planted, the vegetative cover cannot substantially impede continued degradation of the waste. The owner or operator must maintain the runoff management system, control wind dispersal of hazardous waste, and comply with food-chain crop regulations. In addition, the owner or operator must continue to monitor the unsaturated zone, except that soil-pore liquid monitoring may be terminated 90 days after the last application of waste to the treatment zone. These requirements do not apply if the owner or operator can demonstrate that the level of hazardous constituents in the treatment zone soil does not exceed the background value of those constituents by an amount that is statistically significant. Postclosure requirements are similar.

Underground Injection

Underground injection wells are used for the disposal of aqueous solutions containing acids, heavy metals, organics, or inorganics that are either more difficult or expensive to treat by other methods. Typically, wastewaters are pretreated prior to injection to remove oils or solids that could plug the injection zone.

Injection wells are regulated under the Safe Drinking Water Act (SDWA) and RCRA. Hazardous waste injection wells are subject to land disposal restrictions at 40 CFR Part 268. Pretreatment systems for hazardous waste are also subject to RCRA technical standards at 40 CFR Parts 264 and 265.

Hazardous waste subject to land disposal restrictions cannot be disposed in an injection well unless the waste meets the treatment standards of Part 268 or is exempted by the regulatory authority. In obtain this exemption, the injection system must meet stringent criteria defined at 40 CFR Part 148, which includes a no migration standard. Under this standard, injected wastes cannot migrate upward out of the injection zone or laterally within the zone to a point of discharge or interface with an underground source of drinking water for 10,000 years. In addition, before the waste can migrate out of the zone or to a point of discharge or interface with a drinking water source, it must be rendered nonhazardous by attenuation, transformation, or immobilization of its hazardous constituents. Information required to support the exemption petition must include a description of the waste treatment methods and a monitoring plan.

DISPOSAL OF RADIOACTIVE WASTES

Low-Level Radioactive Wastes

Disposal of low-level waste (LLW) is regulated under 10 CFR Part 61. Requirements pertaining to near-surface disposal, including site characterization, site selection, and construction of the site are documented in 10 CFR §§61.50. 61.51, and 61.53. Site closure and stabilization requirements are found in 10 CFR §§61.27-28, 61.52, and 61.53.

Site Selection and Design. Suitability of the site for LLW disposal is determined by the site's natural characteristics such as hydrogeology, climate, soil chemistry, and other factors (NRC 1989). Technical evaluation of the site includes:

- Evaluation of ability of the site to maximize long-term (over 500 years) stability and isolation of waste
- Evaluation of ability of the site to be characterized, modeled, and monitored
- Evaluation of the location of the site to determine sufficient distance from projected population growth areas, other future developments, or known natural resources
- Evaluation of the topography to ascertain if it is free from flooding or frequent ponding
- Evaluation of the location of the site to determine if there is sufficient distance above groundwater to prevent any possible groundwater intrusion into the bottom of the disposal unit
- Evaluation of the site's potential for faulting, folding, or seismic and/or volcanic activity
- Evaluation of the ability of the site to maintain a comprehensive environmental monitoring program and to meet all performance criteria

After the selection criteria are met, the LLW disposal facility must be designed to meet basic criteria. First, the facility should be designed to minimize water contact with waste during storage, disposal, and after disposal. In addition, the design should minimize the need for any active maintenance after closure to ensure long-term isolation of the waste. Finally, the design should complement the site's natural characteristics and provide for erosion control and run-on and runoff management.

Environmental Monitoring and Other Management Practices. As with other disposal facilities discussed in this chapter, LLW disposal facilities must be monitored continually throughout the life of the facility and after closure to ensure that the facility poses no threat to human health and the environment. As specified in the regulations, monitoring systems must be capable of providing early warning of releases of radionuclides before they leave the site boundaries. Typically, upgradient and downgradient groundwater monitoring wells are used to collect the required data.

Other management practices include intruder protection for some Class B and all Class C facilities. For facilities that accept Class B waste, intruder protection in the form of waste stabilization is sufficient. For facilities accepting Class C wastes, waste stabilization and deeper disposal or barriers are required. Class A waste does not require stabilization since it is cost prohibitive and impractical to require small generators, such as medical and university research centers to meet these requirements. NRC has ruled that segregation of Class A from Classes B and C is sufficient (NRC 1989).

Closure of LLW Disposal Facilities. The following standards must be met at the time of LLW disposal site closure and during postclosure:

- The site must be closed according to an approved plan.
- All Class A wastes must be segregated from Class B and C wastes.
- Class C wastes must be disposed of so that waste containers are not less than 5 meters below the top of the disposal unit covers.
- All waste containers must be backfilled properly with soil or other materials to reduce the possibility of cracking of the cap or barrier due to subsidence.
- The boundaries of the disposal unit must be surveyed, marked, and recorded on a map.
- Buffer zones around the disposal unit and beneath the disposal zone must be provided and maintained.

Postclosure observation and maintenance must be performed for a period of time to assure that the site is performing as expected. Transfer of control to a governmental agency can occur after the period of observation if no problems are encountered.

High-Level Radioactive Wastes

Disposal of high-level radioactive waste (HLW) is regulated under 10 CFR Part 60. This type of waste requires permanent isolation in a geologic repository. NRC issues licenses to DOE to receive and possess source, spent nuclear, and by-product materials at a geologic repository operations area sited, constructed, or operated in accordance with the Nuclear Waste Policy Act of 1982.

Technical Performance Criteria. Technical performance criteria for HLW repositories are defined in the regulations. All geologic repository operations areas must be designed so that until permanent closure has been completed, radiation exposure, radiation levels, and release of radioactive materials will meet exposure standards defined by NRC and EPA. In addition, waste must be retrievable for up to 50 years. Upon permanent closure, engineered barriers must be designed to effect substantially complete containment. Any release of radionuclides must be small and occur over long periods of time. Containment of a HLW within waste packages must be substantially complete for a period of 300 to 1000 years after permanent closure of the facility.

Design Criteria. The following design criteria for the geologic repository operations area must be met before a license is issued to the facility (ORAU 1988):

- A means to limit concentrations of radioactive material in air must be established.
- A means to limit time required to perform work in the vicinity of radioactive materials must be developed.
- Suitable shielding must be designed.
- A means to control access to high-radiation areas or airborne radioactivity areas must be established.
- Surface facilities must be designed to control the release of radioactive materials in effluents within the limits specified by NRC and EPA.
- HLW waste packages must be designed so that the chemical, physical, and nuclear properties of the waste package and its interaction with the site environment do not compromise the performance of the waste package or the underground facility or geologic setting.
- A geologic setting must exhibit an appropriate combination of favorable conditions to reasonably assure that the performance objectives relating to isolation of the waste will be met.

REFERENCES

Bagchi, A. (1994). *Design, Construction, and Monitoring of Landfills,* 2nd ed., Wiley, New York.

Cheremisinoff, P. N. (1994). "Selecting Hazardous Waste Treatment Systems: Thermal Classification Is Key," *National Environmental Journal,* vol. 4, no. 1.

Corbin, M. H., N. A. Metzer, and M. F. Kress (1994). *Project Summary: Field Investigation of Effectiveness of Soil Vapor Extraction Technology,* EPA/600/SR-94/142, U.S. EPA, Cincinnati, OH, 1994.

Department of Energy (1987). *Hazardous Waste Land Treatment: A Technology and Regulatory Assessment,* DE88005571, prepared by M. Overcash, K. W. Brown, and G. B. Evans, Jr. for Energy and Environmental Systems Division, Argonne National Laboratory, Argonne, IL.

Eckenfelder, W. W., Jr., and J. L. Musterman (1995). *Activated Sludge Treatment of Industrial Wastewater,* Technomic Publishing Company, Lancaster, PA.

Environmental Protection Agency (1987). *Compendium of Technologies Used in the Treatment of Hazardous Waste,* EPA/625/8-87/014, Center for Environmental Research Information, U.S. EPA, Cincinnati OH.

——— (1988). *Best Demonstrated Available Technology (BDAT) Background Document for Aniline Production Treatability Group (K103, K104), Volume 7, Proposed,* EPA/530/SW-88/0009g, U.S. EPA, Washington, DC.

——— (1991). *Treatment Technology Background Document,* PB91-160556, U.S. EPA, Washington, DC.

——— (1993a). *Demonstration Bulletin: CAV-OX® Ultraviolet Oxidation Process, Magnum Water Technology,* EPA/540/MR-93/520, U.S. EPA, Washington, DC.

——— (1993b). *Emerging Technology Summary: Laser Induced Photochemical Oxidative Destruction of Toxic Organics in Leachates and Groundwaters,* EPA/540/SR-92/080, U.S. EPA, Washington, DC.

——— (1993c). *perox-pure Chemical Oxidation Technology Peroxidation Systems, Inc., Applications Analysis Report,* EPA/540/AR-93/501, U.S. EPA, Washington, DC.

——— (1994). *CAV-OX Cavitation Oxidation Process, Magnum Water Technology, Inc., Application Analysis Report,* EPA/540/SR-93/520, U.S. EPA, Washington, DC.

——— (1995a). *SITE Technology Capsule: KAI Radio Frequency Heating Technology,* EPA/540/R-94/528a, U.S. EPA, Cincinnati, OH.

——— (1995b). *SITE Demonstration Bulletin: ZenoGem Wastewater Treatment Process, ZENON Environmental Systems,* EPA/540/MR-95/503, U.S. EPA, Cincinnati, OH.

Hutchins, et al. (1991). *Nitrate for Biorestoration of an Aquifer Contaminated with Jet Fuel; Report for June 1988–Sept. 1990,* U.S. Environmental Protection Agency in cooperation with Solar Universal Technologies, Inc., NSI Technology Services Corp., Traverse Group, Inc., and 9th Coast Guard District.

Jensen, R. (1994). "Successful Treatment with Supercritical Water Oxidation," *Environmental Protection,* vol. 5, no. 6.

Krishnan, E. R. et al. (1993). *Recovery of Metals from Sludges and Wastewaters,* Noyes Data Corporation, Park Ridge, NJ.

LaGrega, M. D., P. L. Buckingham, and J. C. Evans (1994). *Hazardous Waste Management,* McGraw-Hill, New York.

Musterman, J. L. and V. J. Boero (1995). "Granular Activated Carbon vs. Macroreticular Resins," *Industrial Wastewater*, vol. 3, no. 2.

Noyes, R. (1994). *Unit Operations in Environmental Engineering,* Noyes Data Corporation, Park Ridge, NJ.

Nuclear Regulatory Commission (1989). "Regulating the Disposal of Radioactive Waste," NUREG/BR-0121, Washington, DC.

Oak Ridge Associated Universities (1988). "A Compendium of Major U.S. Radiation Protection Standards and Guides," Oak Ridge, TN.

Pitter, P., and J. Chuboda (1990). *Biodegradability of Organic Substances in the Aquatic Environment,* CRC Press, Boca Raton, FL.

Roeck, D. R., and A. Sigg (1996). "Carbon Injection Proves Effective in Removing Dioxins," *Environmental Protection,* vol. 7, no. 1.

Water Environment Federation (formerly Water Pollution Control Federation) (1990). *Hazardous Waste Treatment Processes,* Manual of Practice FD-18, prepared by Task Force on Hazardous Waste Treatment, WEF, Alexandria, VA.

Weber, Walter J., Jr. (1972). *Physiochemical Processes for Water Quality Control,* Wiley, New York.

BIBLIOGRAPHY

Adams, C. E., Jr., D. L. Ford, and W. W. Eckenfelder, Jr. (1981). *Development of Design and Operational Criteria for Wastewater Treatment,* Enviro Press, Nashville, TN.

American Petroleum Institute (1988). *Evaluation of the Treatment Technologies for Listed Petroleum Refinery Wastes,* API Publication 4465, API, Washington, DC.

——— (1995). *Modeling Aerobic Biodegradation of Dissolved Hydrocarbons in Heterogenous Geologic Formations,* Publication 848-00200, API, Washington, DC.

American Society for Testing and Materials (1989). *Environmental Aspects of Stabilization and Solidification of Hazardous and Radioactive Wastes,* Special Technical Publication 1033, Cote and Gilliam, eds., ASTM, Philadelphia.

American Society of Civil Engineers and American Water Works Association (1990). *Water Treatment Plant Design,* 2nd ed., McGraw-Hill, New York.

Arozarena, M. M., et al. (1990). *Stabilization and Solidification of Hazardous Wastes,* Noyes Data Corporation, Park Ridge, NJ.

Bouley, J. (1993). "Liners & Lining Systems," *Pollution Engineering,* vol. 25, no. 4.

Breton, M., et al. (1988). *Treatment Technologies for Solvent Containing Wastes,* Noyes Data Corporation, Park Ridge, NJ.

Burton, D. J., and K. Ravishankar (1989). *Treatment of Hazardous Petrochemical and Petroleum Wastes,* Noyes Data Corporation, Park Ridge, NJ.

Clifton, C. E. (1967). *Introduction to Bacterial Physiology,* McGraw-Hill, New York.

Conner, J. R. (1989). *Chemical Fixation and Solidification of Hazardous Wastes,* Van Nostrand Reinhold, New York.

Costa, Michael J. (1995). "Case Study: Vacuum Extraction Harnessed for Emergency Cleanup," *Environmental Protection,* vol. 6, no. 4.

Department of Energy (1990). *Decontamination of Low-Level Wastewaters by Continuous Countercurrent Ion Exchange,* DE90 011077/XAB, Oak Ridge National Laboratory, Oak Ridge, TN.

——— (1991). *Application of High Level Waste-Glass Technology to the Volume Reduction and Immobilization of TRU, Low Level, and Mixed Wastes,* DE91 009319/XAD, prepared by D. F. Bickford et al., Westinghouse Savannah River Co., Aiken, SC, for U.S. DOE and American Nuclear Society, Washington, DC.

——— (1991). *Long-Term Durability of Polyethylene for Encapsulation of Low-Level Radioactive, Hazardous, and Mixed Wastes,* DE92 000486, Brookhaven National Laboratory, Upton, NY.

Dietrich, J. A. (1995). "Membrane Technology Comes of Age," *Pollution Engineering,* vol. 27, no. 7.

Environmental Protection Agency (1989). *Incineration of Solid Waste,* EPA/600/J-89/531, Risk Reduction Engineering Laboratory, U.S. EPA, Cincinnati, OH.

——— (1990). *Incinerability Ranking Systems for RCRA Hazardous Constituents,* EPA/600/J-90/496, Risk Reduction Engineering Laboratory, U.S. EPA, Cincinnati, OH.

——— (1990). *Overview of Metals Recovery Technologies for Hazardous Waste,* EPA/600/D-91/026, prepared by PEI Associates, Inc., for Risk Reduction Engineering Laboratory, U.S. EPA, Cincinnati, OH.

——— (1990). *Solvent Waste Reduction,* prepared by ICF Consulting Associates for U.S. EPA, published by Noyes Data Corporation, Park Ridge, NJ.

——— (1991). *Treatment Technology Background Document,* PB91-160556, Office of Solid Waste, U.S. EPA, Washington, DC.

——— (1992). *Low Temperature Thermal Treatment (LT^3) Technology, Roy F. Weston, Inc., Applications Analysis Report,* EPA/540/AR-92/019, U.S. EPA, Washington, DC.

——— (1992). *Mercury and Arsenic Wastes: Removal, Recovery, Treatment, and Disposal,* Noyes Data Corporation, Park Ridge, NJ.

——— (1993). *Resources Conservation Company B.E.S.T. Solvent Extraction Technology: Applications Analysis Report,* EPA/540/AR-92/079, Washington, DC.

——— (1993). *SITE Demonstration Bulletin: Microfiltration Technology, EPOC Water, Inc.,* EPA/540/MR-93/513, U.S. EPA, Cincinnati, OH.

——— (1994). *SITE Demonstration Bulletin, Thermal Desorption System: Clean Berkshires, Inc.,* EPA/540/MR-94/507, U.S. EPA, Cincinnati, OH.

——— (1994). *SITE Emerging Technology Summary: Cross-Flow Pervaporation for Removal of VOCs from Contaminated Wastewater,* EPA/540/SR-94/512, U.S. EPA, Cincinnati, OH, 1994d.

——— (1994). *SITE Emerging Technology Bulletin: Volatile Organic Compound Removal from Air Streams by Membranes Separation, Membrane Technology and Research, Inc.,* EPA/540/F-94/503, U.S. EPA, Cincinnati, OH.

——— (1994) *SITE Technology Capsule: Clean Berkshires, Inc. Thermal Desorption System,* EPA/540/R-94/507a, U.S. EPA, Cincinnati, OH.

——— (1995). *SITE Technology Capsule: Terra-Kleen Solvent Extraction Technology,* EPA/540/R-94/521a, U.S. EPA, Cincinnati, OH.

——— (1995). *Development Document for Proposed Effluent Limitations Guidelines and Standards for the Pharmaceutical Manufacturing Point Source Category,* EPA/821/R-95/019, U.S. EPA, Washington, DC.

——— (1995). *Geosafe Corporation In Situ Vitrification: Innovative Technology Evaluation Report,* EPA/540/R-94/520, U.S. EPA, Washington, DC.

Fagan, M. R. (1994). "Peroxygens Enhance Biological Treatment," *Environmental Protection,* vol. 5, no. 9.

Flood, D. R. (1994). "Synthetic Linings for Hazardous Wastes", *National Environmental Journal,* vol. 4, no. 3.

Gray, N. F. (1990). *Activated Sludge: Theory and Practice,* Oxford Science Publications, New York.

Huling, S. G., et al. (1990). *Enhanced Bioremediation Utilizing Hydrogen Peroxide as a Supplemental Source of Oxygen: A Laboratory and Field Study,* EPA/600/2-90/006, U.S. EPA, Washington, DC.

Jackman, A. P., and R. L. Powell (1991). *Hazardous Waste Treatment Technologies,* Noyes Data Corporation, Park Ridge, NJ.

Jenkins, D., M. G. Richard, and G. T. Daigger (1993). *Manual on the Causes and Control of Activated Sludge Bulking and Foaming,* Lewis Publishers, Boca Raton, FL.

Metcalf & Eddy, Inc. (1972). *Wastewater Engineering: Collection, Treatment, Disposal,* McGraw-Hill, New York.

Nelson, C. H. (1995). "Using Ozone to Speed Up Air Sparging," *Environmental Protection,* vol. 6, no. 4.

Reynolds, J. P., R. R. Dupont, and L. Theodore (1991). *Hazardous Waste Incineration Calculations: Problems and Software,* Wiley, New York.

Russell, D. L. (1992). *Remediation Manual for Petroleum-Contaminated Sites,* Technomic Publishing Co., Lancaster, PA.

Santos, E. (1995). "Standardized and Customized pH Neutralization Systems," *Environmental Protection*, vol. 6, no. 12.

Water Environment Federation (formerly Water Pollution Control Federation) and American Society of Civil Engineers (1988). *Aeration: A Wastewater Treatment Process,* Manual of Practice FD-13, WEF, Alexandria, VA, and ASCE, New York.

Weymann, David (1995). "Biosparging and Subsurface pH Adjustment: A Realistic Assessment," *National Environmental Journal,* vol. 5, no. 1.

13

GROUNDWATER AND SOILS ASSESSMENT AND REMEDIATION

A basic understanding of pollution problems at an industrial or waste management facility can be obtained from an assessment of the groundwater. Sometimes, activities from past operations have caused contamination, which is evident only years later when an environmental investigation takes place. In this chapter we discuss methodologies for determining pollution in groundwater and soils. In addition, the topic of groundwater and soils remediation is discussed.

GROUNDWATER ASSESSMENT

Designing Groundwater Monitoring Systems

The primary reason for installing a groundwater monitoring system is to provide access to groundwater to determine in situ groundwater conditions. If the system is installed properly, samples collected and analyzed from the wells will provide a means for evaluating groundwater contamination. An excellent guidance document on the subject of design and installation of groundwater monitoring systems is the *Handbook of Suggested Practices for the Design and Installation of Ground-Water Monitoring Wells* (EPA 1990).

Factors to Consider. Several factors influence the design and installation of an effective groundwater monitoring system. These include:

- *Geologic and hydrogeologic conditions.* Geologic and hydrogeologic conditions directly affect the occurrence and movement of groundwater and contaminant transport. Considerations include soil type, soil permeability, water table height and seasonal fluctuations, characteristics of aquifers, groundwater direction, and others. In addition, the presence of consolidated or unconsolidated formations must be understood to evaluate and design water intake methods properly.

- *Facility characteristics.* Different types of facilities (i.e., land treatment facilities, hazardous waste landfills, underground injection control systems, waste treatment facilities, underground storage facilities, etc.) will require different monitoring system designs. Waste characteristics, such as solubility and miscibility of wastes managed at the facility, should also be considered.
- *Other site-specific information.* Other information about the site should be taken into account, such as age of tanks, activities at the site, location of previously existing pumping wells, exact property boundaries, and other data that may be available.
- *Equipment to be used in the well.* The well must be designed to accommodate certain equipment, such as water-level measuring devices and groundwater sampling devices.

Drilling. Drilling methods should be adequate for the geologic conditions. Examples of common drilling methods and their applications and limitations are presented in the section "Sample Collection Methods" later in this chapter.

Design Components of Monitoring Wells. Several components of the monitoring well must be designed properly to provide an adequate sampling well. These include the well casing, well intake, annular seals, and surface completion. An example of a completed monitoring well is presented in Figure 13.1.

Well Casing. The well casing should be of adequate strength to resist borehole pressures (collapse forces) and compressive and tensile stresses exerted during installation. In addition, the casing should be made of materials that will resist chemical attack and do not themselves have the potential to leach constituents into the groundwater. Selected examples of casing materials commonly used for groundwater monitoring installations are presented in Table 13.1.

Compatibility of the material with the expected groundwater constituents should be investigated thoroughly before well installation, since some materials have limited use. For instance, stainless steel can be used only in noncorrosive conditions, and thermoplastics are not recommended for use where potentially sorbing organics are of concern. In addition, studies have indicated that cuts and abrasions on stainless steel and, to a lesser degree, PVC casing during installation are readily susceptible to surface oxidation and provide active sites for sorption of metal impurities such as lead, cadmium, and other heavy metals (U.S. Army Corps of Engineers 1989).

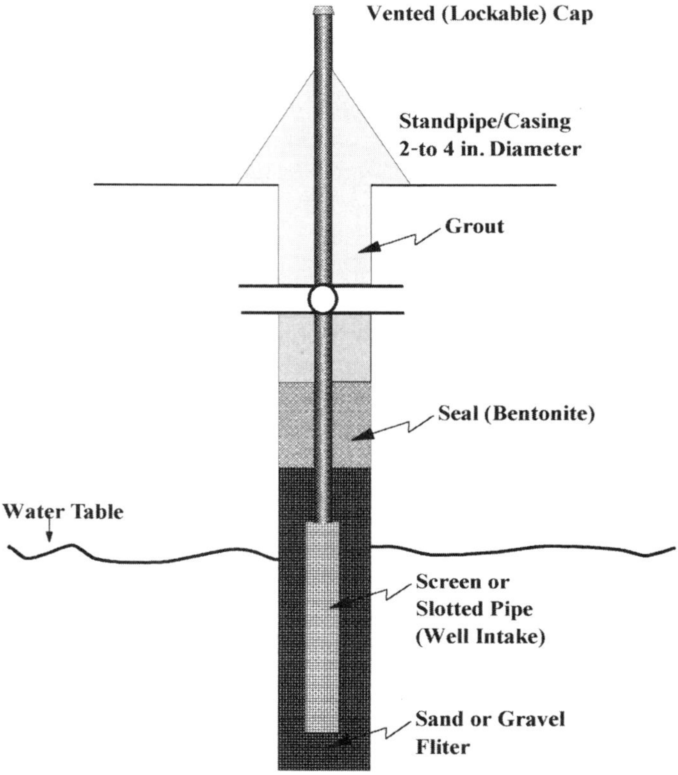

Figure 13.1 Schematic of a groundwater monitoring well.

TABLE 13.1 Examples of Materials Used for Well Casing

Material Category	Material Type
Fluoropolymer materials	Polytetrafluoroethylene (PTFE), tetrafluoroethylene (TFE), fluorinated ethylene propylene (FEP), perfluoroalkoxy (PFA), polyvinylidene fluoride (PVDF)
Metallic materials	Carbon steel, low-carbon steel, galvanized steel, AISI 304 and 316 stainless steels
Thermoplastic materials	Polyvinyl chloride (PVC), acrylonitrile–butadiene–styrene (ABS)
Fiberglass-reinforced materials[a]	Fiberglass-reinforced epoxy (FRE), fiberglass-reinforced plastic (FRP)

[a]These materials have not yet been used in general application across the country, and very few data are available on characteristics and performances.

Source: Information from EPA (1990).

Materials used in joining sections of casing should be of adequate strength and should be chemically compatible with the application. When possible, mechanical joints, such as threaded joints, are recommended. A fluoropolymer tape can be wrapped around the threads of the joint for watertightness. Other joints commonly found in the field include welded joints for metal applications and solvent-cemented joints for thermoplastic casings. Thermoplastic casings with solvent joints generally are not acceptable on new groundwater monitoring system installations. Solvent-cemented joints are never appropriate for fluoropolymer applications since these materials are inert and a bond cannot be formed.

Well Intake. The design and installation of the monitoring well intake is critical to proper functioning of the well. In unconsolidated or poorly consolidated materials, the intake design includes a natural or artificial filter pack of granular materials to maximize well development and prevent plugging. Materials in natural filter packs that are effective include gravels and coarse sands. Materials normally used in artificial filter packs include quartz sand or glass beads. The filter pack should extend from the bottom of the well intake to approximately 2 to 5 feet above the well intake, unless this extension would provide a cross connection into an overlying water-bearing zone.

Another key aspect of well intake design is the opening or slot size. The slot size is selected based on the uniformity coefficient of the formation material for naturally packed wells or the size of the materials in the filter pack for artificially packed wells. Examples of types of well intake slots are presented in Figure 13.2.

Well intakes are typically 2 to 10 feet in length (EPA 1990). The intake should be placed in the aquifer of interest for evaluation of that specific aquifer. If several aquifers are to be evaluated, separate wells should be con-

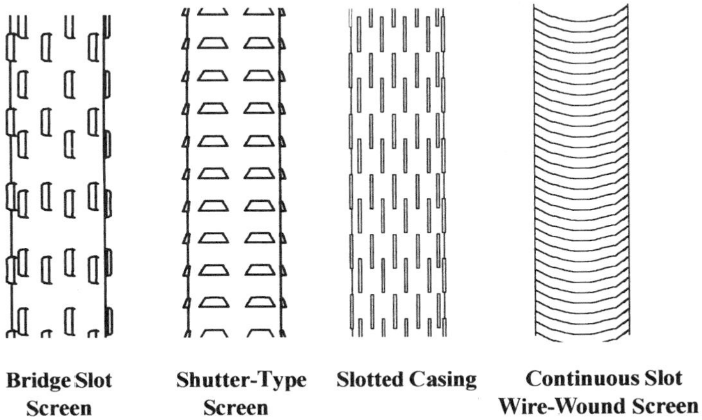

Figure 13.2 Types of well intakes.

structed with the intakes placed at proper depths for sampling discrete water-bearing zones.

Annular Seals. Once the casing is placed in the borehole and the well intake is protected with filter packs, the annular space above the filter pack must be sealed. This will prevent vertical migration of the groundwater and will provide protection from surface water runoff into the well. In addition, the seal allows for the sampling of discrete zones. The annular seal is made with a stable, low-permeable material such as bentonite or cement.

Surface Completion. There are two types of surface completions for monitoring wells: aboveground completion and flush-to-the-ground completion. In both types of completions, there is a surface seal of cement or concrete placed around the well casing for added protection to the well. In the aboveground completion, a protective casing 18 to 24 inches high is installed in the cement or concrete surface seal while it is wet and uncured. This casing should be fitted with a lockable cap for well protection and should be vented and have a drain hole. In the flush-to-the-ground completion, a lockable protective cover such as a meter box should be installed around the well casing. As in the aboveground completion, the structure is placed in the surface seal while it is still wet. This type of completion is susceptible to invasion by surface runoff if watertight seals are not maintained.

Monitoring Releases

General groundwater monitoring requirements for detection of releases from hazardous waste management units, including detection monitoring, compliance monitoring, and corrective action, are detailed in 40 CFR §§264.90-264.101. The key to a valid monitoring program is proper placement of the monitoring wells for statistical evaluation of contamination in the uppermost aquifer. Guidance for installing a groundwater monitoring system and evaluating potential contamination are presented in several EPA guidance documents, including:

- SW-611—*Procedures Manual for Groundwater Monitoring at Solid Waste Disposal Facilities* (EPA 1980)
- OSWER 9950.1—*RCRA Ground Water Monitoring Technical Enforcement Guidance Document* (EPA 1986a)
- EPA/600/2-85/104—*A Practical Guide for Ground Water Sampling* (EPA 1985)

The groundwater monitoring system must consist of a sufficient number of wells, installed at appropriate locations and depths, to yield representative background and point-of-compliance groundwater samples from the uppermost aquifer. Upgradient wells should be installed in positions that will yield samples of groundwater not affected by the hazardous waste management unit. Downgradient well locations should be selected to ensure that potential pathways of contaminant migration are sampled. If several water-bearing zones are to monitored, this can be accomplished through well clusters, multiport well systems, nested piezometers (for shallow depths), and other groundwater profile systems.

As required under RCRA regulations, upgradient wells and downgradient wells must be compared statistically to demonstrate that waste management units have not contaminated the groundwater. For permitted facilities, statistical evaluation of groundwater contamination can be performed using one of four statistical methodologies. These include parametric analysis of variance (ANOVA), nonparametric ANOVA based on ranks, tolerance or prediction interval analysis, or control chart analysis. RCRA-permitted facilities may use an alternative method if it meets the same performance standards required of the four methods delineated. Interim status facilities (those that have applied for, but have not received a RCRA permit) must make their statistical demonstration using Student's *t*-test. EPA-approved statistical methods are delineated in EPA's guidance document *Statistical Analysis of Ground-Water Monitoring Data at RCRA Facilities: Interim Final Guidance* (EPA, 1989a).

ANALYTICAL METHODS

Acceptable analytical methods for evaluating groundwater are documented in EPA's SW-846, *Test Methods for Evaluating Solid Waste: Physical/Chemical Methods* (EPA 1986n). Two quantitation limits are defined in the manual: the method quantitation limit (MQL), which is the minimum concentration of a substance that can be measured using the method, and the practical quantitation limit (PQL), which is the lowest level that can reliably be achieved within specified limits of precision and accuracy during routine laboratory operating conditions. The PQL generally is accepted as adequate for groundwater evaluation.

Laboratory Analysis

Most commonly, sample analysis for groundwater is performed in a laboratory. Typical analytical methods used in groundwater analysis of organics and pesticides include gas chromatographic (GC) methods and gas chromatographic/mass spectroscopic (GC/MS) methods. Hydrocarbons are analyzed using GC and high-performance liquid chromatographic methods. Metals (except for hexavalent chromium and mercury) generally are evaluated using atomic absorption (AA) methods. Other constituents are evaluated using colorimetric, spectrophotometric, and other test methods.

Selected examples of types contaminants commonly found in groundwater, appropriate method number, and description of method are listed in Table 13.2. Analytical methods and PQLs for RCRA-regulated groundwater con-

TABLE 13.2 Examples of Contaminants Commonly Found Groundwater and Appropriate Analytical Methods

Contaminant Type	Method Number	Method Description
Metals: aluminum, antimony, arsenic, barium, beryllium, cadmium, calcium, chromium	7020–7191	Atomic absorption (AA), direct aspiration, or furnace technique (method 7061 for arsenic uses AA, gaseous hydride)
Hexavalent chromium	7195	Coprecipitation
	7196	Colorimetric
	7197	Chelation/extraction
	7198	Differential pulse polarography
Metals: cobalt, copper, iron, lead, lithium, magnesium, manganese	7200–7461	AA, direct aspiration, or furnace technique
Mercury	7470–7471	Manual cold-vapor technique
Metals: molybdenum, nickel, osmium, potassium, selenium, silver, sodium, strontium, thallium, tin, vanadium, zinc	7480–7950	AA, direct aspiration, or furnace technique
Halogenated volatile organics	8010	Gas chromatography (GC)
Nonhalogenated volatile organics	8015	GC
Aromatic volatile organics	8020	GC
Volatile organics	8240	Gas chromatography/mass spectrometry (GC/MS)
	8260	GC/MS, capillary column technique
	8021	GC with photoionization and electrolytic conductivity detectors in series
Semivolatile organics	8250	GC/MS, packed column technique
	8270	GC/MS, capillary column technique
Polynuclear aromatic hydrocarbons	8100	GC
	8310	High-performance liquid chromatographic methods
Chlorinated hydrocarbons	8120	GC

TABLE 13.2 (*Continued*)

Contaminant Type	Method Number	Method Description
Organophosphorus pesticides	8140	GC
	8141	GC, capillary column technique
Total cyanide	9010	Colorimetric, manual
	9012	Colorimetric, automated UV
Sulfate	9035	Colorimetric, automated chloranilate
	9036	Colorimetric, automated methylthymol blue, AA II
Phenolics	9065	Spectrophotometric
	9066	Colorimetric
Total recoverable oil and grease	9070	Gravimetric, separatory funnel extraction
Radium-228	9320	Coprecipitation, beta counter

Source: Information from EPA (1986b).

stituents are identified in 40 CFR Part 264, Appendix IX. Chemicals species (contaminants) that are most frequently detected in groundwater at hazardous waste sites are presented in Table 13.3.

In Situ Analytical Methods

As a result of Superfund and other remediation activities, analytical methods for groundwater evaluation that can be used directly in the field are being investigated and, when possible, employed. Some of these methods that have proved successful in pilot or field tests include:

- *Optical detection of organic nitro compounds (experimental).* This method uses a membrane for in situ optical detection of organic nitro compounds based on fluorescence quenching. A fluorophor is incorporated into a plasticized membrane, which preconcentrates organic nitro compounds from the groundwater. This leads to fluorescence quenching. The membrane can be used to sense explosives that are present at the parts per million (ppm) level or higher (U.S. Army Corps of Engineers 1991).
- *Optical detection of polynitroaromatic hydrocarbons.* This method uses a membrane that reacts with polynitroaromatic hydrocarbons to form a brown product when placed in groundwater contaminated with these compounds. This brown product can be remotely detected in the parts per billion (ppb) range with fiber optics, as a one-time reading (U.S. Army Corps of Engineers 1991).
- *Synchronous excitation (SE) fluorescence spectroscopy.* This method is an advanced type of ultraviolet fluorescence spectroscopy. The method

TABLE 13.3 Most Frequently Detected Groundwater Contaminants at Hazardous Waste Sites

Rank	Compound	Common Sources
1	Trichloroethylene	Dry cleaning, metal degreasing
2	Lead	Gasoline (prior to 1975), mining, construction material (pipes), manufacturing
3	Tetrachloroethylene	Dry cleaning, metal degreasing
4	Benzene	Gasoline, manufacturing
5	Toluene	Gasoline, manufacturing
6	Chromium	Metal plating
7	Methylene chloride	Degreasing, solvents, paint removal
8	Zinc	Manufacturing, mining
9	1,1,1-Trichloroethane	Metal and plastic cleaning
10	Arsenic	Mining, manufacturing
11	Chloroform	Solvents
12	1,1-Dichloroethane	Degreasing, solvents
13	1,2-Dichloroethane, *trans-*	Transformation product of 1,1,1-trichloroethane
14	Cadmium	Mining, plating
15	Manganese	Manufacturing, mining, occurs in nature as oxide
16	Copper	Manufacturing, mining
17	1,1-Dichloroethane	Manufacturing
18	Vinyl chloride	Plastic and record manufacturing
19	Barium	Manufacturing, energy production
20	1,2-Dichloroethane	Metal degreasing, paint removal
21	Ethylbenzene	Styrene and asphalt manufacturing, gasoline
22	Nickel	Manufacturing, mining
23	Di(2-ethylhexyl)phthalate	Plastics manufacturing
24	Xylenes	Solvents, gasoline
25	Phenol	Wood treating, medicines

Source: National Research Council (1994).

is different from the conventional single-wavelength excitation (SWE) technique in that the SE procedure can avoid scatter distortion in situ that the SWE cannot. Examples of contaminants that can be quantified in the ppb (and sometimes parts per trillion) range include aniline, *o*-cresol, naphthalene, phenol, toluene, and xylene (EPA 1988).

- *Fiber-optic sensor.* This method uses a differential-absorption fiber-optic sensor to monitor certain volatile organochlorides such as trichloroethylene and chloroform. Optical fibers are sealed into a capillary tube containing a reagent. A porous Teflon membrane at one end allows the target molecules to enter the tube and mix with the reagent, which turns color in the presence of these constituents. This color results in a decreased transmission of light, which the optical fibers measure (DOE 1991).

Other methods for monitoring changes in the groundwater are being investigated by Lawrence Livermore National Laboratory's Environmental Technology Program and include high-frequency electromagnetic measurements and electrical resistance tomography (DOE 1990a).

SOILS ASSESSMENT

A soils assessment typically is required during the closure of a regulated waste storage, treatment, or disposal facility to demonstrate that the area has been decontaminated to an extent that it poses no threat to human health or the environment. A soils assessment may also be performed to assure adequate cleanup of a spill or to verify that activities involving hazardous materials and wastes have not affected the surrounding area adversely. If the assessment reveals that there is contamination, some type of remediation normally is required. In this chapter we discuss a methodology for performing a soils assessment and describe remediation technologies used for cleanup of contaminated soils.

Facility History

To plan a soils assessment strategy properly, a facility history must be developed. The history is used to determine areas of potential contamination so that the number of samples and the money spent on sampling and analysis can be limited as much as possible. Examples of items that might be included in the facility history are shown in Table 13.4. Using the facility history, areas that are potentially contaminated can be defined.

Determination of Sampling Points

Once the areas of potential contamination are defined, sampling points and sampling techniques can be determined. A useful guidance document for this process is EPA's SW-846 *Test Methods for Evaluating Solid Waste* (EPA 1986b). A three-dimensional grid can be superimposed over the area of investigation. Samples can be taken at varying locations and depths within the grid to assess the extent and depth of contamination, if any. Typically, a circular pattern of sampling is used, with samples being concentrated near expected contamination.

To effectively interpret data gathered from the analysis of the samples, background samples must be collected. These background samples will indicate what constituents are naturally occurring in the soil. The background

TABLE 13.4 Examples of Items to Include in a Facility History

Facility age	Description of major upgrades
Materials/wastes stored at the facility over time	List of spills, location, quantities, and cleanup techniques
Description of facility's chemical operations, present and past	List of chemical operations performed by vendors, present and past
RCRA permit	RCRA and other closures
Air emissions permit and monitoring data	Air modeling data (plume assessments)
Wastewater discharge permit and reports	Inspection logs for docks, tanks, and containment structures

samples should come from areas of the same or similar soil classification as the area under investigation. This determination can be made using the visual–manual procedure outlined by the American Society of Testing and Materials (ASTM) in ASTM D2488-90. In addition to having the same or similar classification, the areas must have been secluded from chemical and waste traffic, manufacturing, and/or other chemical-use activities to provide effective background data.

Sample Collection Methods

Several methods are used for soil sample collection that are recognized methods throughout industry. Numerous ASTM standards for soil collection and classification are documented in Volume 4.08 of the *Annual Book of Standards* (ASTM 1991). Standards and practices pertaining to this topic are delineated in Appendix C. A brief description of the most frequently used drilling/sample collection methods is presented in Table 13.5.

Whenever possible, liners should be used with sampling devices designed to incorporate liners to help prevent cross-contamination during sampling. Liners should be made of materials such as stainless steel, brass, or another material that prevents leaching of contaminants into the sample from the liner. In particular, flexible PVC and polyethylene liners are not recommended when sampling for certain volatile organics due to sorption. Diagrams of a split-spoon sampler and a thin-walled sampler are shown in Figures 13.3 and 13.4.

In some applications, preparation methods are required, such as coring of concrete for sampling under a slab or boring of the soil in order to collect samples at progressively deeper locations. In these instances, drilling equipment typically is used, and drilling operations occur in tandem with sample collection.

In most applications, the most efficient preservation method for soil samples is refrigeration. Sample integrity can be maintained for several days and

TABLE 13.5 Drilling Methods

Hollow-Stem Augers

- *Description:* uses "giant screw" and continuous flighting; as augers advance downward, the cutting move upward; hollow stem or core allows drill rods and samplers to be inserted through the center of the augers
- *Applications:* all types of soil investigations, including unconsolidated formations and stable formations; allows for soil sampling with split-spoon or thin-walled samplers; water quality sampling
- *Limitations:* preserving sample integrity difficult in unconsolidated formations; possible cross contamination of aquifers where annular space is not positively controlled by water, drilling mud, or surface casing; sand and gravel heaving may be difficult to control; limited diameter of augers limits casing size; smearing of clays may seal off aquifer to be monitored

Solid-Stem Augers

- *Description:* method similar to that of hollow-stem augers; auger made of solid steel, and therefore must be removed from the borehole to collect "undisturbed" split-spoon or thin-walled samples and to install casing
- *Applications:* shallow soils investigations; soil samples; vadose zone monitoring wells (lysimeters); monitoring wells in saturated, stable soils; identification of depth to bedrock
- *Limitations:* split-spoon or thin-walled samplers must be used for sample integrity; soil sample data limited to areas and depths where stable soils are predominant; unsuitable for most unconsolidated aquifers because of bore caving upon auger removal; depth capacity decreases as diameter of auger increases; monitoring well diameter limited by auger diameter

Cable Tool Drilling

- *Description:* utilizes a drilling rig and interlocking steel hammers (jars) that slide independently of each other; the hammering action drives of the jars drives the sampling/well drilling barrel into the ground
- *Applications:* drilling in all types of geologic formations; accommodates almost any depth and diameter range; excellent samples can be obtained from coarse-grained materials
- *Limitations:* heaving of unconsolidated materials must be controlled; equipment availability limited in some parts of the United States; drilling process is relatively slow

Air Rotary Drilling

- *Description:* involves the use of circulating fluids such as mud, water, or air to remove the drill cutting and to maintain an open hole as drilling progresses; forces air down the drill pipe and back up the borehole to remove the drill cuttings
- *Applications:* easily used for drilling of semiconsolidated and consolidated rock; provides good-quality formation samples; allows for easy and quick identification of lithologic changes; allows for identification of water-bearing zones and estimate of water yields

TABLE 13.5 ***(Continued)***

- *Limitations:* not suitable for unconsolidated formations; air may alter chemical or biological conditions

Air Rotary with Casing Driver Drilling

- *Description:* uses air rotary drilling technique with the addition of a casing driver
- *Applications:* commonly used in unconsolidated sands, silts, and clays; also used for drilling in alluvial material; interaquifer cross contamination minimized from casing supports; provides good-quality formation samples; minimal formation damage
- *Limitations:* thin, low-pressure water-bearing zones easily missed; samples may be pulverized; air may modify chemical or biological conditions

Mud Rotary and Water Drilling

- *Description:* involves the introduction of drilling fluids (either drilling muds or water) into the borehole through the drill pipe; drilling fluids maintain open hole, provide lubrication to the drill bit, and remove drill cuttings
- *Applications:* commonly used in clay, silt, and reasonably compacted sand and gravel; allows for split-spoon and thin-walled sampling in unconsolidated materials; allows for core sampling in consolidated rock; flexible range of tool sizes and depth capabilities
- *Limitations:* bentonite or other drilling fluid additives may influence quality of groundwater samples; aquifer identification difficult; samples inadequate for monitoring well screen selection; drilling fluid invasion of permeable zones may compromise validity of subsequent monitoring well samples

Dual-Wall Reverse-Circulation

- *Description:* utilizes a double-walled drill pipe and has the reverse circulation of other conventional rotary drilling methods; air or water is forced down the outer casing and is circulated up the dinner drill pipe, with the cuttings lifted up to the surface through the inner drill pipe
- *Applications:* used in both unconsolidated and consolidated formations; minimal risk of contamination of sample and/or water-bearing zone from sample collection; in stable formations, wells with diameters as large as 6 inches can be installed in open hole completions; in unstable formations, well diameters are limited to approximately 4 inches
- *Limitations:* air may modify chemical or biological conditions; recovery time is uncertain; filter pack cannot be installed unless the well is a completed open hole

Driven Wells

- *Description:* consist of a steel well screen that is either welded or attached with drive couplings to a steel casing; well screen and attached casing are forced into the ground by hand using a weighted drive sleeve or a heavy drive head mounted on a hoist; as well is driven, new sections of casing are attached to the well
- *Applications:* water-level monitoring in shallow formations; can be used for water supply wells

TABLE 13.5 (*Continued*)

• *Limitations:* depth limited to approximately 50 feet; small-diameter casing; steel casing interferes with some chemical analysis; not suitable for dense and/or dry materials
Jet Percussion
• *Description:* uses a wedge-shaped drill bit attached to the end of the drill pipe; water is forced under pressure down the drill pipe and is discharged through ports on the sides of the drill bit; the drill is lifted and dropped while rotating and the water is forced up the annular space between the drill pipe and the borehole wall, carrying a cutting to the surface
• *Applications:* used primarily in unconsolidated formations, but may be used in some softer consolidated rock; best application is a 4-inch borehole with a 2-inch casing and screen installed, sealed, and grouted
• *Limitations:* diameter limited to 4 inches; disturbance of formation possible if borehole is not cased immediately

up to several weeks, depending on soil type and preservation temperature, with the optimal temperature range being 0 to 4°C (EPA 1991). If the sample has to be transferred to a container, the time the sample is exposed to the atmosphere should be minimized and measures should be taken to ensure that the sample strata is not disturbed.

Field Considerations

Field sampling is not always performed under optimum conditions, so adequate field equipment must be on hand to ensure proper sampling for varying sets of circumstances. Items that might be included in a field vehicle are presented Table 13.6.

The field log is an integral part of the sampling event and should be completed as accurately as possible. Items to be included in the field log are presented in Table 13.7. Once the log for an event is completed, it should be kept as part of the sampling record and should be filed with sampling data and other information about the project.

Analtyical Methods

Standard Methods. Standard methods for soils analysis are documented in EPA's manual SW-846 (EPA 1986b). The methods presented for analysis of soils are the same methods used for groundwater, described earlier in this chapter. Preparation procedures typically will be different, however, since soils will require digestion and extraction before the analysis is performed. Because of this, practical quantitation limits (PQLs) of soil and groundwater

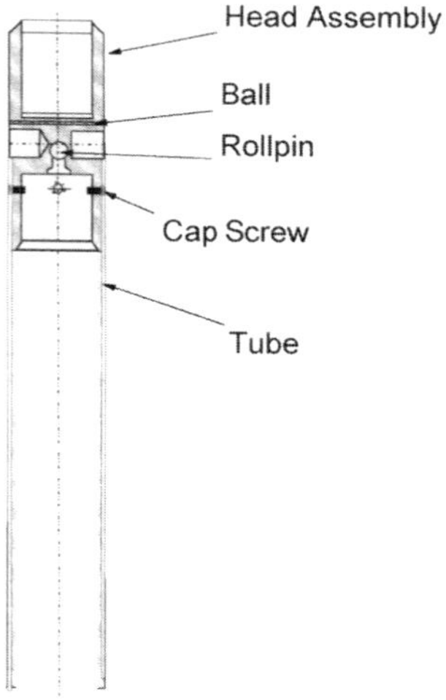

Figure 13.3 Diagram of a thin-wall sampler.

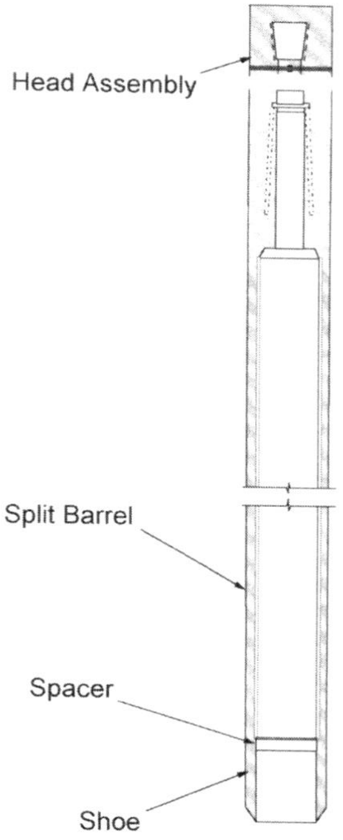

Figure 13.4 Diagram of a split-spoon sampler.

TABLE 13.6 Items That Might Be Included in a Field Vehicle During Sampling Events

Category	Equipment
Safety and personal equipment	Pylons, tape, or rope for marking off area, "Do Not Enter" signs, oxygen meter and LEL meter for sampling in confined spaces, dust mask, water bottle, earplugs or other hearing protection, safety glasses, hard hat, sunscreen, coveralls, other appropriate clothing
Record-keeping items	Clipboard, all-weather paper, field notebook with all-weather paper, indelible ink pens and markers, other pens, cellophane tape, watch, camera and film, compass, thermometer
Sampling items	Appropriate noncontaminated jars, plastic containers, lids, sample labels, indelible ink pens, airtight tape, thin-walled tubes with plastic end caps, hand augur, trowel, ice chest, ice, sampling equipment decontamination solutions, clean rags, chain of custody forms

may vary significantly from contaminant to contaminant, with quantitation limits of groundwater contaminants generally being lower.

In Situ Analysis

Soil-Gas Analysis. Several analytical devices currently are marketed that can perform in situ soil-gas analysis, an example of which is presented in Figure 13.5. In situ analysis cuts down on sample handling and exposure of the sample to the atmosphere. In addition, in situ soils analysis allows for better selection of sampling locations, since concentrations can be mapped during the sampling event. Many of the procedures used in soil-gas analysis were originally developed for oil and gas exploration and have been modified in recent years for investigations of hazardous waste sites. At present, none of the procedures currently used in the field have been standardized by EPA (DOE 1990b).

Active in situ soil-gas analysis utilizes soil-gas cores or probes, which are 1 to 3 meters in depth, and a vacuum system such as a syringe to pull the vapor to the surface for immediate analysis. Probes can be inserted into the ground with hammers or hydraulic rams to reach depths of 3 to 5 feet. Numerous chemical detectors can be used in the field to give a quantitative measure of soil-gas contamination. These detectors are listed in Table 13.8.

TABLE 13.7 Items to Include in a Field Log

Personnel Data

- Name, company, and telephone number of person collecting the sample and recording the event
- Names of other field crew members, company affiliations, and phone numbers
- Name of driller, company affiliation, and phone number
- Name of other persons involved in the sampling activity
- Signature of person recording information in the log

Site Data

- Date and time of the start of the sampling event
- Time of various events during sampling, such as drilling activities, sample collection and preservation, and transport to lab
- Location of sampling point (include map or sketch, if possible)
- Weather conditions, including temperature, cloud conditions, humidity (if known), and other conditions such as mist, rain, or sleet
- Out of the ordinary aspects, such as confined space, enclosed building, or other

Sample Data

- Depth to sample point
- Soil color, soil type or classification, and unusual characteristics of soil
- Sample equipment used and method of collection
- Preservation techniques utilized
- Other observations

In addition to active soil-gas analysis using detectors listed in the table, initial screening for contamination can be performed using an organic vapor analyzer (OVA). This type of device does not distinguish between hydrocarbons and chlorinated solvents. Thus, once contamination is identified, more sophisticated equipment must be used to speciate and quantify the contaminants. Other hand-held instruments that can be used as screening devices include portable infrared spectrometers, ion mobility spectrometers, and fiber optical sensors. If budgets and trained personnel do not allow for active sampling, passive samplers made of an adsorbent such as activated charcoal can be buried in the soil and collected later for analysis in the laboratory.

Surface Radiation Survey. Portable beta/gamma detectors are used widely for screening contamination at a hazardous waste site containing radioactive wastes. Most of these detectors are easy to operate, with probes that are used to scan the surface and provide continuous readings while the operator walks in a predetermined pattern. When areas contaminated above background levels are found, they can be marked with a stake, and measured values can be recorded on the stake and in a log book. Using this method, the extent of contamination easily can be charted (DOE 1990b).

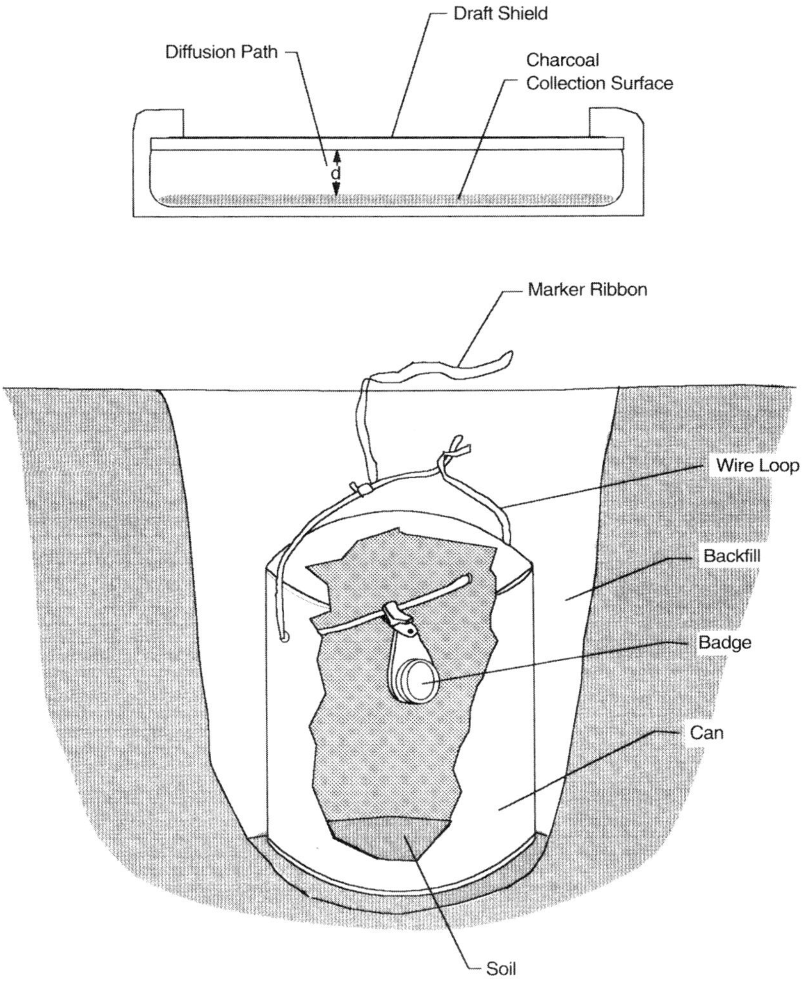

Figure 13.5 Passive soil-gas sampling badge and manifold.

SOIL AND GROUNDWATER REMEDIATION

Contaminated soil and groundwater present special challenges because the contamination is diffuse and difficult to access. In situ processes, those that leave both soil and groundwater in place during treatment, are often favored because they are less disruptive, take advantage of natural existing conditions, and preserve the existing subsurface environment. Ex situ and pump and treat processes remove soil and groundwater for treatment. They may be favored over in situ treatment where they will reduce cleanup times, their operation

TABLE 13.8 Selected Detectors That Can Be Used in Soil-Gas Analysis

Detection Method	Sensitivity	Applications
Flame ionization detector (FID)	Can detect 4 picograms/second (pg/s) or carbon	Can detect total hydrocarbons
Electron capture detection (ECD)	Can detect 0.01 pg/s of material	Can detect halogenated hydrocarbons; more sensitive to iodine than chlorine or fluorine
Hall electrolytic conductivity detector (HECD)	Can detect 0.5 pg/s of chlorine	Can detect halogenated species, nitrogen-containing organics, and sulfur-containing organics
Photoionization detector (PID)	Can detect 10 to 100 pg/s	Can detect aromatic hydrocarbons of aromatic material
Flame photometric detector (FPD)	Can detect 1 pg/s of phosphorus or 100 pg/s of sulfur	Can detect sulfur and phosphorus compounds

Source: Information from EPA (1989b).

and capabilities are considered more reliable or better understood, or they can achieve lower cleanup levels. Both in situ and ex situ treatment for soil and groundwater rely on a combination of unit processes, which often include biological degradation of organics.

Nonaqueous phase liquids (NAPLs) present special problems for soil and groundwater cleanup. Contaminant transport through groundwater depends in part on the water solubility of the compound. Because NAPLs cling to subsurface particles and are slow to dissolve in groundwater, they hinder cleanups and prolong cleanup times. Dense nonaqueous-phase liquids (DNAPLs) migrate downward in the aquifer and can collect in pools or pockets of the substructure. Examples of DNAPLs are the common solvents tetrachloroethylene [also known as perchloroethylene (PCE)] and trichloroethylene (TCE), which were used extensively at many facilities before the extent of subsurface contamination problems was realized.

Today, strict regulation of hazardous waste and material management promises to minimize future soil and groundwater contamination; however, past practices have left us with a large legacy of contaminated sites that are often difficult and costly to remediate. This is the driving force behind the development of innovative technologies for soil and groundwater cleanup. The EPA strongly supports the development of innovative technologies and maintains the vendor information system for innovative treatment technologies (VISITT) database to provide information on availability, performance, and cost of these technologies.

In this section, soil and groundwater cleanup technologies are divided into five subsections: plume containment, bioremediation, physical–chemical in situ treatment, pump-and-treat technology for groundwater, and ex situ soil treatment. Many of the unit processes for soil and groundwater treatment are general waste treatment technologies but are discussed here in the context of soil and groundwater cleanup. We discuss these technologies more generally elsewhere in the chapter.

Plume Containment

Wells can be placed at a contaminated site to prevent the contamination from spreading further or migrating off site. In the past, containment efforts often relied on physical methods such as bentonite slurry trenches, grout curtains, sheet pilings, well points, and fixative injections (Bouwer and Cobb 1987). Containment by judiciously placed wells generally costs less, takes less time to install, and is more flexible because pumping rates and locations can be varied (Sims et al. 1992).

Four typical well patterns for contaminant plume containment are described by Sims et al. (1992). The first is a pair of injection-production wells. The second is a line of downgradient pumping wells. The third is a pattern of injection-production wells around the boundary of a plume. The fourth, the double-cell system, uses an inner cell and outer recirculation cell, with four cells along a line bisecting the plume in the direction of flow. Two other methods of plume containment are biofilters and a funnel-and-gate system, which are described in the section "In Situ Bioremediation."

Bioremediation

To say that bioremediation is all the rage in groundwater and soil cleanup is not an overstatement. Bioremediation has great appeal. It is a natural process that degrades hazardous organic chemicals into innocuous carbon dioxide and water or nonhazardous by-products and it is often less expensive and more effective than pump and treat methods. Articles on bioremediation appear regularly in environmental journals and the EPA has its own regular series of reports on current activities called *Bioremediation in the Field.*

On the down side, there is a lot that is not known or completely understood about the process, which makes it difficult to design and prove in field applications. It is a relatively slow process; complete remediation of a site can take several years. The efficiency of the bioremediation cannot always meet target cleanup levels. Bioremediation has been very successful in cleaning up sites contaminated with gasoline and other fuels because the primary contaminants—benzene, toluene, ethylbenzene, and xylenes (BTEX)—are relatively easy to extract and degrade. Other compounds have been biodegraded successfully; however, many compounds are still subjects of bioremediation research. Nevertheless, what is known and has been done so far with bioremediation is very encouraging and points to the great potential that bioremediation has in the future of contaminated groundwater and soil remediation.

Biodegradation Principles. Microorganisms degrade organic compounds to synthesize cellular materials and to obtain the energy to grow and reproduce. Anabolism results in the synthesis of new cellular material, and catabolism (degradation) produces chemicals with lower free energy and in the case of aerobic biodegradation, will often result in complete decomposition of a hydrocarbon to carbon dioxide and water. The principal microorganisms involved in the degradation of organic chemicals in water and soil are bacteria and fungi.

Microorganisms can degrade compounds under aerobic or anaerobic conditions, although the types of microorganisms and compounds and degradation process and products may differ. In aerobic environments, free oxygen is present. In anaerobic conditions, free oxygen is not present, although oxygen may be present in the form of such compounds as nitrates, sulfates, and iron oxide.

Respiration, or biological oxidation, is the use of oxygen as an electron receptor in the catabolic degradation of an organic and can occur either aerobically or anaerobically. Aerobic respiration uses free oxygen as an electron receptor, whereas anaerobic respiration uses inorganic oxygen. In both cases, however, water and carbon dioxide are the principal end products, as with any oxidation reaction.

Fermentation is an anaerobic catabolic process that uses organics as electron receptors. Since fermentation produces organic products that have lower free energy than their precursors, it is useful in remediation. The lowest free-energy form of carbon produced is methane.

Although both fermentation and respiration degrade organics, microorganisms that use respiration to meet their energy requirements for synthesis and reproduction generally grow much more quickly. This and the fact that respiration oxidizes organics completely to carbon dioxide and water usually make aerobic biodegradation the treatment of choice. However, because certain synthetic organics will not or are difficult to degrade aerobically, there is much interest in anaerobic processes.

Cometabolism refers to situations where a compound cannot be biodegraded effectively unless another food source is available. The recalcitrant compound, such as TCE, does not provide the energy to allow the microorganisms to grow and thrive. When another food source is available, such as methane, energy is produced for growth and enzymes are produced to metabolize the energy-providing food source and to cometabolize the recalcitrant compounds.

Types of Treatment. Bioremediation can be conducted either ex situ or in situ. Ex situ is where groundwater is pumped to the surface or soil is excavated for treatment. *Pump-and-treat bioremediation* refers to the ex situ removal and treatment of groundwater. After treatment, the groundwater can either be reinjected or discharged; treated soils can be redeposited, landfilled, or recycled. *In situ bioremediation* takes place in the ground; neither groundwater or soil is extracted, although groundwater may be pumped to the surface for oxygen and nutrient addition. In situ bioremediation has become very popular because the process takes full advantage of natural conditions (microorganisms, soil, water) as a bioreactor, little site disturbance is necessary, and it is less equipment intensive than is ex situ treatment.

Ex situ bioremediation may use various biological wastewater treatment processes, soil piles, or land application. With in situ bioremediation, the basic process is the same—microbes, soil, and water working together as a bioreactor. Where the in situ techniques differ are in how contaminants and microbes are brought in contact and how oxygen, nutrients, and other chemical supplements are distributed in the soil–water–air matrix. Typical in situ bioremediation techniques include natural or intrinsic attenuation, air sparging, and bioventing. Treatment processes for both ex situ and in situ bioremediation are discussed in the following sections.

In Situ Bioremediation. In situ bioremediation can be an aerobic or an anaerobic process, or a combination of the two. In designing an in situ bioremediation system, one should consider the types of microorganisms available (naturally in place or added), the structural and chemical makeup of the soil matrix, types of contaminants, oxygen and nutrient addition and distribution, and temperature. These factors are discussed prior to introducing the individual techniques for in situ bioremediation.

Aerobic, Anaerobic, and Combined Systems. The vast majority of in situ bioremediations are conducted under aerobic conditions because most organics can be degraded aerobically and more rapidly than under anaerobic conditions. Some synthetic chemicals are highly resistant to aerobic biodegradation, such as highly oxidized, chlorinated hydrocarbons and polynuclear aromatic hydrocarbons (PAHs). Examples of such compounds are PCE, TCE, benzo[*a*]pyrene, PCBs, and pesticides.

Besides being slower, anaerobic treatment is more difficult to manage and can generate by-products that are more mobile or toxic that the original compound: for example, the daughter products of TCE—dichloroethenes and vinyl chloride. It requires a longer acclimation period which means slower startup times in the field. The microbial processes are less well understood and hence are less controllable than for aerobic systems.

Nevertheless, an anaerobic system may be the method of choice under certain conditions: (1) contamination with compounds that degrade only or better under anaerobic conditions; (2) low-yield aquifers that make pump-and-treat methods or oxygen and nutrient distribution impractical; (3) mixed waste contamination, where oxidizable compounds drive reductive dehalogenation of chlorinated compounds; or (4) deep aquifers that make oxygen and nutrient distribution more difficult and costly (Sewell et al. 1990).

Anaerobic respiration can degrade organics by using nitrate or sulfate as oxygen sources. Compounds that have been shown to be biodegraded by nitrate-respiring microorganisms are methanes, carbon tetrachloride, *m*-xylene, and some phenols, cresols, and PAHs. Chloroform and stable chlorinated ethanes and ethenes are not degraded by nitrate respiration. Reductive dehalogenation, not degradation, has been observed of naturally occurring aromatics where sulfate respiration was occurring. Sulfate respiration can be inhibited by the accumulation of its by-product, hydrogen sulfide (King et al. 1992).

Combined aerobic–anaerobic systems use sequenced aerobic and anaerobic conditions to degrade compounds that are resistant to aerobic biodegradation. Groundwater passes through the anaerobic zone first, which partially degrades the resistant compounds. These degradation products are then further degraded in an aerobic zone. A two-zone, plume interception approach was tested under EPA's Superfund Innovative Technology Evaluation (SITE) Emerging Technologies Program to degrade chlorinated compounds (Hasbach 1993). The first zone was anaerobic, where methanogenic (methane-producing) microorganisms were expected to promote reductive dechlorination of chlorinated solvents. The second zone was aerobic, where the partially dechlorinated products were to be biodegraded. Combined systems do not necessarily have to be separate zones in the aquifer. Aerobic and anaerobic conditions can be alternated by eliminating the oxygen source at timed intervals.

In one study, a coarse-grained sand aquifer was injected with methane and oxygen to stimulate the production of methane monooxygenase (MMO) enzyme, which is capable of degrading TCE (Semprini et al. 1987). TCE, added at 60 to 100 μg/L, was degraded by 20 to 0%. Injected concentrations of methane and oxygen were approximately 20 and 32 mg/L, respectively.

Design Considerations. The effectiveness of in situ bioremediation is influenced by many factors, including microorganisms, soils, oxygen, pH, temperature, type and quantity of contaminants, and nutrients. These factors are discussed in the following sections.

ICROORGANISMS. A large number of naturally occurring bacteria and fungi can use hydrocarbons for growth. The most commonly found genera of bacteria and fungi that are capable of hydrocarbon biodegradation are (1) bacteria—*Pseudomonas, Arthrobacter, Alcaligenes, Corynebacterium, Flavobacterium, Achromobacter, Micrococcus, Nocardia,* and *Mycobacterium;* and (2) fungi—*Thricoderma, Penicillium, Aspergillus,* and *Mortierella* (Miller 1990).

Soils contain a diverse culture of microorganisms, with the number and type present being a function of local environmental conditions. Usually, the limiting factor on the microbial population at an uncontaminated site is the amount and type of carbonaceous material available for biodegradation (Miller 1990). However, there are almost always some microorganisums present in every soil environment that will take advantage of a new or different carbon source when it becomes available and will rapidly increase their populations to metabolize it. Addition of more or specially cultured microbes has been studied and applied in the field; however, some people believe that all that is needed in most cases is simply to optimize environmental conditions for the naturally occurring microbes. But for recalcitrant compounds or for more rapid biodegradation, specialized microbes may have potential. For example, EPA has agreed to license specialized microbes that degrade chlorinated aromatic and chlorinated aliphatic compounds such as TCE under aerobic conditions, which avoids the generation of vinyl chloride as a by-product (EPA 1993).

SOILS. Soils with hydraulic conductivities greater than 10^{-4} centimeters per second (cm/s) are good candidates for in situ bioremediation because they are permeable enough to allow the transport of oxygen and nutrients through the aquifer (Sims et al. 1992). The degree of homogeneity of the soil is important in order to predict groundwater flow patterns for plume containment and distribution of oxygen and nutrients.

CONTAMINANTS. The type and concentration of contaminants in an aquifer dictate what type of in situ bioremediation system—aerobic, anaerobic, or combination—will be most effective.

OXYGEN. Many microorganisms require either free oxygen or inorganic oxygen for growth. Some bacteria can use either free oxygen or inorganic oxygen for respiration; these are called *facultative bacteria.* Many bacteria, though, require free oxygen for growth.

A rule of thumb used in wastewater treatment is that at least 1.5 mg/L of dissolved oxygen is needed to prevent oxygen from being a growth-limiting factor. Kemblowski et al. (1987) documented a threshold dissolved oxygen concentration of 0.5 mg/L in groundwater. Below 0.5 mg/L it was found that the biodegradation rates of benzene, toluene, and xylene were retarded. It was also shown that biodegradation rates increased until the groundwater dissolved-oxygen concentration reached 2 mg/L, at which point no further increase was observed. Thus it appears that a target of 1.5 to 2.0 mg/L dissolved oxygen will maximize biodegradation rates and put an upper limit on the cost to supply the oxygen.

Field data collected from groundwater in spill areas have shown that dissolved oxygen concentrations are typically reduced below 1 mg/L due to biodegradation. For example, at the site of a gasoline spill where gasoline-degrading bacteria had been identified, dissolved oxygen in the contaminated groundwater ranged from 0 to 0.5 mg/L, while in the uncontaminated areas, levels were from 2.3 to 4 mg/L (Raymond et al. 1976). In another study of an aquifer contaminated with chlorinated solvents, dissolved oxygen was always less than 0.2 mg/L (Semprini et al. 1987). These studies demonstrate that biological growth and biodegradation of hydrocarbons is oxygen limited, a fact that stimulates a continuing interest in developing effective oxygen sources and the means of distributing oxygen in groundwater and soil.

Hydrogen Peroxide. Hydrogen peroxide is an effective means of getting more oxygen into the groundwater for bioremediation and has many advantages over air or pure oxygen. Hydrogen peroxide has been demonstrated not only to boost dissolved oxygen levels, but it also stimulates microbial activity. It also has the advantage of chemically degrading contaminants partially or fully. Hydrogen peroxide mixes intimately with groundwater for better distribution in the aquifer and can prevent injection well plugging caused by bacterial growth.

On the downside, various studies (Chambers et al. 1991, EPA 1990a,b) have shown that hydrogen peroxide decomposes rapidly after soil contact. It is cytotoxic at a 3% solution, and unless stabilized, oxygen bubbles can escape prematurely through the unsaturated zone before they have a chance to disperse well in the groundwater. Catalase enzymes generated by aerobic bacteria near the injection wells appear to be the major cause of hydrogen peroxide decomposition and oxygen off-gassing. Others have noted that hydrogen peroxide will also react with dissolved iron, manganese, and other inorganics to produce oxides that can plug wells and formations (D. Russell 1992, Ware 1993). Precipitation of inorganic compounds during bioremediation has been studied in more detail by Morgan and Watkinson (1992). Concentrated solutions of hydrogen peroxide (30%) used in bioremediation must be handled carefully to avoid skin burns.

Reports of groundwater remediation studies using hydrogen peroxide are found frequently in the literature. Hydrogen peroxide reactions as an oxidant during bioremediation have been reviewed by Pardieck et al. (1992). A field-scale study of an area contaminated by an aviation gasoline fuel spill showed that hydrogen peroxide did increase the concentration of oxygen downgradient of the injection well (EPA 1990b). Hydrogen peroxide was used to remediate groundwater after a jet fuel leak at Eglin Air Force Base in Florida (EPA 1990a). Field applications using hydrogen peroxide for groundwater cleanup are continuously reported by EPA in its *Bioremediation in the Field* series. Treatment conditions common to the EPA studies are the use of hydrogen peroxide and nutrient addition to stimulate biodegradation by the indigenous microorganisms.

Most reports citing hydrogen peroxide do not specify dosage rates. Concentrations as high as 500 mg/L have been shown to be nontoxic to most microorganisms (D. Russell 1992). Other studies report dosage rates of 100 to 500 mg/L; unpublished reports of rates as high as 1000 to 10,000 mg/L are not uncommon (EPA 1990b). Research studies have shown that a 0.05%

solution (500 mg/L) was the maximum tolerated by a mixed culture of microorganisms capable of degrading gasoline. Concentrations up to 2000 mg/L could be tolerated if the dosage was stepped up incrementally; however, most toxicity studies showing no toxicity at relatively high concentrations are usually based on bacteria counts. Other studies have shown that concentrations greater than 100 mg/L decrease the oxygen utilization rate by microorganisms, which is counterproductive to biodegradation and therefore increases the time needed for remediation.

Premature decomposition of hydrogen peroxide can be inhibited by adding monobasic potassium phosphate as a stabilizer (EPA 1990b). The phosphorus in the solution has the added benefit of being a nutrient for the microorganisms. A concentration of 190 mg/L potassium phosphate was used in Huling's field study of groundwater contaminated with aviation gasoline. The phosphate is added ahead of the hydrogen peroxide until breakthrough occurs in the downgradient wells. By adding the phosphate first, the phosphate binds with the inorganics in the soils and allows the hydrogen peroxide to be distributed further. Other chemicals have been used or studied to increase the stability of hydrogen peroxide are polyphosphates, stannate or phosphate, sodium pyrophosphate, citrate, and sodium silicate (Elizardo 1993).

Hydrogen peroxide is also used as a treatment to clean wells of biological growth. One treatment is to pour a 0.5% solution into the well and allow it to sit for 4 hours, after which pumping is resumed (King et al. 1992). Some treatments use solution concentrations as high as 10% to remove biofouling from wells (Sims et al. 1992).

Pure Oxygen. By increasing the oxygen level in soils and groundwater, pure oxygen or oxygen-enriched streams stimulate bacterial activity (Chambers et al. 1991). Using bottled oxygen, 35 to 40 mg/L of dissolved oxygen can be supplied at most field temperatures (King et al. 1992).

Being less stable than hydrogen peroxide, a disadvantage with oxygen is that it may not distribute as well in the aquifer, although surfactants can promote highly stable microbubbles that are better able to travel throughout the groundwater. Unlike hydrogen peroxide, oxygen does not aid in keeping the injection well free from biological growth that can plug the well.

Nitrates and Sulfates. There is not a lot of information on actual applications of nitrate or sulfates as oxygen sources. Advantages of nitrate are that it is more soluble than oxygen, less costly, and less toxic than hydrogen peroxide (EPA et al. 1991). Regulatory agencies are less inclined to approve the use of nitrates in groundwater cleanup as an oxygen or nutrient supplement because it can potentially contaminate drinking water supplies and plug the aquifer with biological growth. If nitrate is used, dosages must be carefully calculated and controlled to avoid these problems. In EPA et al. (1991), nitrate was used as the oxygen source for microbial respiration in the cleanup of a drinking water aquifer contaminated with jet fuel, and concentrations of

BTEX were treated below 5 μg/L. There is less literature on sulfate as an oxygen source than nitrate. Most references merely state that sulfate can be used as an oxygen source for anaerobic respiration.

Solid-Phase Oxygen. Solid-phase oxygen in the form of peroxides (calcium peroxide, magnesium peroxide) has been investigated as a slow, continuous-release system that avoids problems associated with the transient nature of molecular oxygen and hydrogen peroxide (Brubaker 1995, Fagan 1994). Calcium peroxide releases oxygen into the system by first breaking down into hydrogen peroxide and calcium hydroxide; the hydrogen peroxide then decomposes to oxygen and water. Cages or socks of the material are suspended directly in the groundwater well and the material can be blended into the soil for ex situ treatment and land farming. For groundwater treatment, this form of oxygen has the most potential for containment or remediation of low concentrations of dissolved organics.

Iron Oxides. Chelated iron oxide, using nitrilotriacetic acid and EDTA, has been studied as an alternative oxygen source (WEF 1994). Iron oxide, which is often difficult for the microbes to access, is made more available by chelating agents.

NUTRIENTS. In addition to carbon and oxygen, other nutrients are needed for microbial growth. Nitrogen, phosphorus, potassium, calcium, magnesium, sodium, sulfur, chlorine, iron, manganese, cobalt, copper, boron, zinc, molybdenum, and aluminum are all present in the microbial cellular material (Clifton 1967). With the possible exceptions of nitrogen and phosphorus, which are present in the cell in relatively large amounts, these nutrients are needed at only trace levels and there is almost always enough in the natural environment to prevent growth-limiting conditions.

In the presence of large amounts of degradable carbon, the naturally available nitrogen and phosphorus could potentially limit growth and thus biodegradation rates. Based on a literature review conducted by Miller (1990), typical aerobic soil microorganisms contain 5 to 15 parts of carbon per part of nitrogen, with 10 parts being an acceptable average. This is about twice the 5:1 ratio of carbon to nitrogen that is usually assumed to represent the composition of the biomass in wastewater treatment processes (Eckenfelder 1966). If this difference is real, it may be due to the specific environmental conditions in a particular location, since such conditions are known to influence the amount of nitrogen accumulated with cellular material (Eckenfelder 1966). Phosphorus requirements are about one-fifth that of nitrogen (Eckenfelder 1966, Miller 1990).

These estimates of nitrogen and phosphorus do not consider that which is reintroduced by biological recycling. Bacterial growth and death continually recycle some fraction of the synthesized nitrogen and phosphorus. Miller cites an optimum carbon to nitrogen to phosphorus ratio of 250:10:3 for biodegradation of carbon compounds in soil, which accounts for the carbon oxidized to carbon dioxide and that assimilated into cellular material. Assuming that one part of carbon is incorporated into cellular material for every two parts of carbon used for energy and converted to carbon dioxide, the ratio of carbon to nitrogen in the cell is about 8.3:1, a reasonably conservative estimate. Miller states that a C/N/P ratio of 300:10:1 should be adequate for biodegradation, assuming no recycle.

Nutrients are usually added at concentrations ranging from 0.005 to 0.02% by weight (Sims et al. 1992). In a field application using hydrogen peroxide, nutrients were added to the injected water at the following concentrations: 380 mg/L ammonium chloride, 190 mg/L disodium phosphate, and 190 mg/L potassium phosphate, the latter used primarily to complex with iron in the formation to prevent decomposition of hydrogen peroxide (EPA 1990b).

TEMPERATURE. Ambient temperature is a key factor to successful biodegradation. In the range 15 to 35°C, bacterial growth rates double for every 10°C rise in temperature. For most soil microorganisms, the optimal temperature for maximum growth is 30 to 35°C (Miller 1990). Since vadose zone and groundwater temperatures are typically in the range 10 to 25°C (Hart and Couvillion 1986), microbial growth will not be prevented but will be at less than the maximum possible rates.

pH. A pH value between 6 and 8 should be maintained for in situ bioremediation. Since most soil pHs are within this range, pH should not be a controlling factor for biodegradation of hydrocarbons unless the soil has also been contaminated with acidic or caustic substances.

Air Sparging. With this technique, air is injected in the saturated zone to volatilize organics and to deliver oxygen for biodegradation. Soil vapor extraction may be used in conjunction with air sparging to capture the volatilized organics. Air sparging is used frequently at sites that have been contaminated with volatile compounds such as BTEX. However, performance data are lacking and difficult ro measure (EPA 1995b, API 1995).

Biosparging. Biosparging is a form of air sparging. The difference is that the primary purpose of biosparging is to deliver just enough air to meet oxygen requirements for bioremediation. Volatilization of organics may be an added benefit, but it is secondary to oxygen delivery. An enhancement is to sparge ozone in place of air (Nelson 1995). Besides providing oxygen for biodegradation, ozone aids breaking down recalcitrant compounds such as chlorinated ethenes, polycyclic aromatic hydrocarbons, and pentachlorophenol.

Biofilters. Biofilters, also known as *biobarriers* or *microbial fences,* are used to hinder migration of a contaminant plume. A biofilter is essentially a zone of biological activity that treats the contaminant as the groundwater flows through the area. The biofilter zone can be established by installing a line of air sparging wells perpendicular to the direction of groundwater flow.

Bioventing. Bioventing is soil venting that enhances biodegradation while extracting volatile compounds from the unsaturated zone.

Funnel-and-Gate. This in situ bioremediation method is so-named because groundwater is funneled through openings or gates in an impermeable sheet piling or slurry wall. Zones of biological activity are created at the gates through air sparging. Contaminants are biodegraded as groundwater is forced to pass through the gates. The funnel-and-gate method (also known as *flume-and-gate*) is used where groundwater gradients are small. Because smaller treatment zones are created, operating expenses can be reduced and there is better process control (Brubaker 1995).

Phytoremediation. Phytoremediation is a developing technology that uses plants, trees, and grasses to biodegrade, extract, or stabilize organic and metal contaminants in soil and groundwater. An engineered wetland is a well-known form of phytoremediation. Phytoremediation is likely to be used in conjunction with other treatment processes, primarily as a polishing step (Matso 1995). As reported by Matso, this method has potential in the cleanup of residual contamination in soil micropores, hydraulic control (poplars and weeping willows pump 50 to 350 gallons per day per tree), alternative landfill caps for leachate control, leachate treatment, and buffer zones for plume containment.

Ex Situ Bioremediation. Groundwater can be pumped and soils can be excavated for biological treatment ex situ. Ex situ methods include biological wastewater treatment for groundwater and slurry-phase treatment and "pile" treatment for excavated soils.

Biological Wastewater Treatment. Groundwater and leachate can be treated in a biological wastewater treatment system such as rotating biological contactors (RBCs), trickling filters, sequencing batch reactors, fluidized-bed reactors, activated sludge, or aerated impoundments. For biological treatment to be feasible, the groundwater or leachate must contain sufficient organics to support microbial growth; otherwise, dilute streams should be treated by physical–chemical means or combined with higher-strength wastewaters for biological treatment in a larger wastewater treatment system (e.g., in an on-site wastewater plant treating process wastewaters at a manufacturing facility). A sequencing batch reactor is a form of activated sludge by which treatment takes place in one tank in batch mode; that is, biodegradation with mixing, followed by quiescent settling and sludge withdrawal. A fluidized-bed reactor uses granular activated carbon or sand as the support medium for microbial growth. The groundwater to be treated is introduced through the bottom of the reactor, which serves to fluidize or suspend the support media. Pretreatment to remove solids is required to avoid plugging the reactor.

Slurry-Phase Soil Treatment. Contaminated soil can be treated biologically in slurry form, not unlike biological treatment of wastewater. However, one important difference with contaminated soil is that the contaminants preferentially adsorb unto the soil particles and must be desorbed before the microorganisms can effectively degrade the compounds. Contaminated soil may be pretreated prior to the bioreactor, for example, by soil washing to separate the fine particles, which normally contain the majority of the contamination. Contaminated soil can be slurried by mixing with water, wastewater, or contaminated groundwater. The solids concentration of the slurry can be fairly high, up to 50%, which makes mixing to keep the soil in suspension an important design parameter.

Soil Heaps, Piles, Beds, and Windrows. Contaminated soil is excavated and placed in heaps, piles, beds, or windrows. Other organic or bulking agents, such as wood chips or straw, may be added to aid in composting, mixing, and aerating the soil. In a windrow system, the soil is placed in long rows and turned periodically to mix and aerate the soil. Soil piles and heaps differ in size, heaps being larger and used where large volumes of contaminated soil must be treated. Piles and heaps may be covered or uncovered, depending on the volatility of the contaminants and air emissions requirements and whether a vacuum is drawn through or air is blown into the soil. With treatment beds, contaminated soil is placed on top of a liner system consisting of a synthetic liner, sand, and leachate drainage pipe. The bed may be covered or enclosed. In any of these treatment systems, volatilized contaminants that are removed by vacuum may be treated by incineration, carbon adsorption, or vapor-phase biodegradation. Leachate may be treated biologically in a wastewater treatment system.

Physical–Chemical–Thermal In Situ Treatment

Electroosmosis. Electroosmosis is the basis of an innovative technology dubbed the "lasagna" process, which was created to treat difficult wastes in low-permeability silt- and clay-laden soils. The lasagna process is so named because it consists of a number of layered subsurface electrodes and treatment zones. These layers can be constructed either horizontally, where contaminants are forced to move upward, or in vertical position, where lateral contaminant movement is desired.

A low-voltage electric current is applied by subsurface electrodes, which forces the migration of contaminants from low-permeable soils into high-permeable treatment zones. These treatment zones that may be created by hydrofracturing (injecting a slurry to create porous pancake-shaped zones), directional drilling, or sheet piling. Treatment in these zones may include physical, chemical, or biological processes such as adsorption by activated carbon, which would then serve as a biodegradation matrix.

In Situ Air Stripping. An innovation to conventional pump and treat air stripping is in situ air stripping (Hazen 1995). Two horizontal wells are installed, one below the water table and one in the vadose zone. Air is injected in the lower well while contaminated soil vapor is extracted by vacuum through the upper well.

Soil Flushing. Soil flushing is similar to soil washing of excavated soils except that flushing is done in situ below the ground surface. With soil flushing, the contaminated area is flooded with water. Surfactants or detergents may be added to the water to enhance removal of organics.

Soil Vapor Extraction. Volatile compounds can be extracted from subsurface soils by applying a vacuum. An added benefit is that the vacuum pulls in air from the ground surface, which stimulates biodegradation through increased oxygen transport. The groundwater below the vadose zone can also be aerated to volatilize contaminants in the groundwater (air sparging), which then move into the vadose zone, where they are removed along with the soil vapors.

An EPA study (EPA 1994a) showed that soil vapor extraction (SVE) is an effective treatment for removing volatile contaminants from the vadose zone. Sandy soils are more effectively treated than clay or soils with higher organic content because higher airflows are possible in sand and clays and organic soils tend to adsorb or retain more contaminants. Removal of volatiles is rapid in the initial phase of treatment and decreases rapidly thereafter, an important consideration in the design of air emissions control over the life of the project.

An innovative companion technology to SVE is radio-frequency heating of the soil (EPA 1995c). Heating the soil increases the volatilization of the contaminants, which are removed by SVE. Antennas are installed near the center of the contaminated area; the radio-frequency energy is applied through the antennae to heat the soil to target levels of 100 to 150°C.

Vitrification. Vitrification is an innovative treatment that turns contaminated soils into a glasslike monolithic mass. Heat is applied through electrodes placed in the ground, the soil reaching temperatures of 2900 to 3600°F (1600 to 2000°C). A layer of graphite and glass frit is first placed on the surface of the ground between the electrodes to act as the initial conductive starter path. The conductive layer and adjacent soils become a molten mass that becomes the primary electrical conductor and heat transfer medium. As heat continues to be supplied by the electrodes, the molten mass moves both outward and downward. As organic contaminants in the soil are heated, they begin to vaporize and eventually pyrolize with the rising temperature. Inorganic contaminants are immobilized in the molten material. Off-gases, which will include vaporized organics and the by-products of organic and inorganic pyrolysis, are captured above the site and treated to meet air emissions standards. Once cooled, the vitrified mass is very stable, with low leaching potential.

Vitrification is effective at destroying and immobilizing hazardous materials, but it is very energy intensive and thus expensive. Consequently, it is used primarily where wastes are difficult to treat or destruction/immobilization of contaminants is very important, such as with radionuclides.

Pump-and-Treat

In the early years of groundwater and soil remediation, pump-and-treat was the conventional technology. Contaminated groundwater is pumped to the surface, where it is treated and reinjected or discharged to surface waters or wastewater treatment plants. Reinjection may be used to stimulate in situ bioremediation, where the treated groundwater is enriched with oxygen and nutrients prior to reinjection. Many years of experience with pump-and-treat methods have brought to light certain limitations, as well as improvements and enhancements. Therefore, pump-and-treat technology must be reviewed for its applicability to a particular site.

Pump-and-treat technology is inherently slow because it depends on groundwater for transport of the contaminant to the extraction well. This characteristic is particularly troublesome when the contaminant is only slightly water soluble, adheres to the soil, or collects in pools within the aquifer.

There are many cases of contamination by DNAPLs that have frustrated pump-and-treat efforts. The general consensus is that pump-and-treat can reduce contamination or keep it from spreading, but it has failed in many cases to remediate aquifers to stringent cleanup goals.

Common technologies for surface treatment of contaminated groundwater are air stripping, carbon adsorption, and biodegradation. If cleanup goals are in the $\mu g/L$ range, then typically pump-and-treat options are limited to air stripping and carbon adsorption (D. Russell 1992). Conventional biological treatment is usually not suitable for the relatively low levels of contaminants found in groundwater (Reidy et al. 1990). Biological treatment of extracted groundwater is discussed in the section "Ex Situ Bioremediation." Air stripping, carbon adsorption, and other pump-and-treat technologies are discussed in the following sections. In addition, since the treatment technologies used for pump-and-treat operations are very common technologies that are also used for many other types of wastes, we discuss them further in the section "Physical–Chemical–Thermal In Situ Treatment."

Extraction Techniques. Groundwater can be extracted through vertical or horizontal wells, with vertical wells being the most common. A schematic of this type of system is presented in Figure 13.6.

Well systems are designed not only to extract groundwater for treatment, but may also be designed for containment of the contaminant plume and reinjection of treated groundwater. Although more costly and difficult to in-

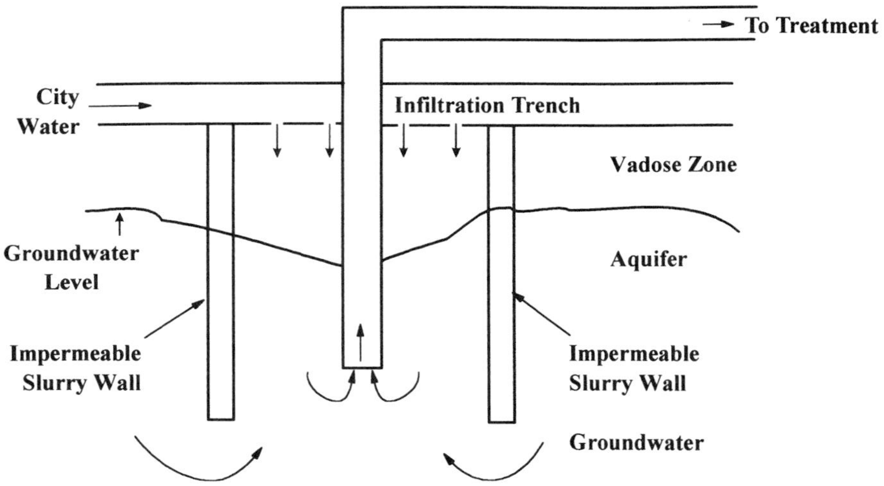

Figure 13.6 Schematic of an extraction well.

stall, horizontal wells may be the method of choice where buildings or other surface features such as roads or streams prevent direct access to the contamination from the surface. In addition, the longer lengths of horizontal wells translate into higher groundwater extraction rates.

Two-phase vacuum extraction (TPVE) is a way of pumping both contaminated air and groundwater through a single well (Costa 1995). A vacuum is applied through a well that is screened through both the vadose and groundwater zones. Both air and groundwater are pumped to the surface, where they are separated for treatment. The advantages of this system are that it does not require separate wells for air extraction and groundwater pumping and that water pumping rates can be greatly increased. A variation of this technique is *bioslurping,* used to extract light nonaqueous-phase liquids (LNAPLs) that float on the surface of the groundwater and are retained in the vadose zone (Baker 1996). Bioslurping uses a slurp tube within a well to vacuum-extract LNAPLs, groundwater, and contaminated vapors. The tube can be raised or lowered to extract these phases separately or as mixtures.

Air Stripping. Volatile contaminants can be removed from groundwater in an air stripper. Air, blown into a column or tank where groundwater is introduced, strips out the volatile compounds through volatilization. Air strippers are in widespread use in groundwater treatment and in other environmental areas (see "In Situ Air Stripping" in the "Physical–Chemical–Thermal In Situ Treatment" section) because the technology is well understood, design is rather straightforward, and the technology is proven and cost-effective. However, they are not as effective as other treatment methods, such as carbon adsorption, in removing contaminants (Reidy et al. 1990).

Four basic types of air strippers are used in groundwater treatment: packed towers, diffused aeration, spray aeration, and tray aeration (Reidy et al. 1990). Packed towers are the most effective and widely used, packed with material that has a high surface/volume ratio and rises to more than 30 meters in height. Contaminated groundwater is fed through the top of the column while air is blown in through the bottom. Packed towers can remove 90 to 99% of volatile contaminants. In diffused aeration, air is blown through diffuser pipes in the bottom of a tank or pond through which the groundwater passes. Removal rates for diffusers are 70 to 90% for many volatile chemicals. In spray aeration, groundwater is dispersed through nozzles over a pond or basin to strip volatiles into the ambient air. The disadvantage of this system is that it is difficult to contain and treat the contaminated airstream if required by local conditions or regulations. Spray aeration can, however, be used to recharge the aquifer if operated over the recharge area. Tray aeration is sometimes used for pretreatment of groundwater because it is not as efficient as other air stripping methods. In tray aeration, groundwater is fed through the top of a column filled with slat trays while air is blown through the bottom of the column.

Factors affecting the design of an air stripping system include contaminant vapor pressure, water solubility, and concentration; groundwater temperature; metal precipitates; solids; cleanup levels; and emission requirements (Reidy et al. 1990). Air strippers are effective for contaminants that have high vapor pressures and low water solubilities. They are most effective with concentrations above 100 mg/L and become much less effective when influent concentrations drop to less than 1 mg/L. Vapor pressure increases with temperature, so groundwater temperature is an important factor. High concentrations of iron and manganese in the groundwater can precipitate on the packing material and retard contaminant removal. Similarly, suspended solids are a problem with packed towers because they can cause plugging or short-circuiting. Cleanup levels will determine what type and size of air stripper is required. The most stringent cleanup levels will require packed towers. Treatment of the airstream leaving the stripper may be required, depending on regulatory and local conditions. Control devices include demisters, fume or vapor incinerators, carbon columns, and vapor condensers.

Air strippers are capable of treating groundwaters to low-μg/L (microgram per liter) levels. If designed properly, air strippers can remove petroleum-type contaminants to less than 1 μg/L, meeting most regulatory criteria (D. Russell 1992). At Wurtsmith Air Force Base in Michigan, TCE concentrations were reduced from 6000 μg/L to less than 1.5 μg/L (H. Russell 1992). At another site reported by H. Russell, total chlorinated hydrocarbons were reduced from 120,000 μg/L to less than a detection limit of 1 μg/L. At a TCE contamination site in Des Moines, optimum stripper performance was 98% for influent concentrations averaging 900 μg/L (EPA 1990a). Removals of *trans*-1,2-dichloroethene were more variable than for TCE and ranged from 85 to 96%. Vinyl chloride influent concentrations ranged from 1 to 38 μg/L and were nondetected in the effluent stream. Air stripping at a municipal well field site demonstrated removals below 1 μg/L with influent concentrations of less than 3 μg/L for 1,1-dichloroethene, 2 μg/L for 1,1-dichloroethane, 15 μg/L for 1,1,1-trichloroethane, and less than 12 μg/L for TCE (EPA 1990a).

Pretreatment of the groundwater prior to air stripping may be required to equalize flows and concentrations; remove high concentrations of suspended solids, remove immiscible liquids such as NAPLs; remove iron, manganese, or water hardness; remove dissolved metals; or adjust the pH (EPA 1995a).

Carbon Adsorption. Carbon adsorption is widely used in groundwater remediation where very high quality effluent is desired, for example, at or beyond drinking water standards. Carbon adsorption is used on groundwaters that only have low levels of contaminants, making pretreatment of some groundwaters necessary. Carbon is often used to treat the off-gas, or contaminated airstream, from air strippers.

Contaminated groundwater is fed through the unit, where contaminants are adsorbed onto the surface of the carbon. When the adsorptive capacity of the carbon is exceeded, the spent carbon is either regenerated or disposed. GAC, and not PAC, is commonly used in groundwater treatment. A more detailed discussion of general carbon adsorption technology is discussed in the section "Physical–Chemical–Thermal In Situ Treatment."

Factors affecting the design of carbon adsorption units are contaminant or total organic concentration, contaminant molecular weight and water solubility, metal precipitates, and solids (Reidy et al. 1990). Groundwater with high contaminant levels will quickly overload a carbon unit and should be pretreated. Total organic carbon concentrations of less than 1 mg/L are best for carbon units; those greater than 5 mg/L are considered poor candidates for carbon adsorption. Higher-molecular-weight compounds and those with low water solubilities are easily adsorbed on carbon. Dissolved concentrations of iron and manganese greater than 5 mg/L can interfere with adsorption when they precipitate on the carbon surface; concentrations below 0.2 mg/L are desirable. Suspended solids concentrations in the feed to a carbon unit are usually kept below 5 mg/L in order not to plug or short-circuit the unit; concentrations above 20 mg/L are not acceptable.

Other Technologies. Chemical oxidation is often used in groundwater treatment as a pretreatment step: for example, increasing biodegradability by partial oxidation of refractory organics or oxidizing arsenite to arsenate. Chemical oxidants commonly used are hydrogen peroxide, ozone, and chlorine. Ultraviolet radiation may be used in combination with hydrogen peroxide and/or ozone.

Ion exchange is most often used to remove metals from groundwater. It can be used to remove sulfates, nitrates, and radionuclides (EPA 1995a). Pretreatment of the groundwater may be required: for example, 10-μm filtration to remove solids; carbon adsorption to remove organics that foul strong base resins; dechlorination to neutralize chlorine; and aeration, precipitation, or filtration to remove iron and manganese.

Chemical precipitation may be used to remove metals such as arsenic, cadmium, chromium, lead, mercury, selenium, and silver from groundwater.

In photocatalytic oxidation, platinized titanium dioxide is used as a catalyst and ultraviolet radiation (UV) is supplied by UV lamps or the sun to produce hydroxyl radicals that oxidize organic compounds in contaminated groundwater. Pretreatment of the groundwater may be necessary to remove solids that interfere with light transmission or clog the reactor bed and to remove ionic species that can foul the catalyst.

Ex Situ Soil Nonbiological Treatment

Soil Leaching. Soil leaching or acid extraction uses acid to solubilize metals for removal from soils, a technique akin to that in the mining industry. After extraction with an acid such as hydrochloric, sulfuric, or nitric, the soil is separated from the acid, rinsed with water to remove excess acid and metals, dewatered, and neutralized. The extracted metals are separated from the solution during acid regeneration and the acid solution is recycled back to the process. The extracted metals can be precipitated and recovered.

Soil Washing. Soil washing is as simple a process as the name implies. Soil is excavated, physically broken up, washed with a water-based solution, and separated into size fractions (sand/gravel and silt/clay). The goal of soil washing is to transfer pollutants to the wash water and to isolate the finer soil fraction that preferentially adsorbs organics and metals. Various surfactants, extractants, or detergents are added to the wash water to enhance removal of organic contaminants from the soil particles. Soil fines and contaminants are removed from the wash water, which is then recycled back to the process.

The finer soil fraction contains adsorbed organics, small metallic particles, and bound ionic metals. This fraction may be treated further to remove the contaminants, or it may be incinerated or landfilled. The "clean" coarse fraction may contain some residual metallic fragments. With metal contamination, both the fine and coarse soil fractions may be leached with an acid solution to remove the metals (Benker 1995).

Solvent Extraction. Organic contaminants are removed from soil by extraction with solvents such as hexane, triethylamine, acetone, liquefied propane, or even supercritical carbon dioxide. After extraction, the solvent is separated from the soil and treated by processes such as microfiltration and distillation before recycling back to the process. The soil may be further treated by vapor extraction to remove most of the residual solvent, followed by biological treatment. Organics that are amenable to solvent extraction from soils include pesticides, PCBs, petroleum hydrocarbons, chlorinated hydrocarbons, polynuclear aromatic hydrocarbons, polychlorinated dibenzo-*p*-dioxins, and polychlorinated dibenzofurans.

Thermal Desorption. In this process, heat is applied either directly or indirectly to contaminated soil to transfer volatile compounds to an air (or gas) stream. Heat is controlled to prevent combustion or incineration. The volatilized compounds are usually removed by carbon adsorption, destroyed by thermal oxidation, or condensed as recovered materials.

Several innovative thermal desorption systems have been tested or used in field applications for the treatment of volatile and semivolatile organics, PCBs, organometallic complexes, and total petroleum hydrocarbons (EPA 1994b). Soils are heated to temperatures ranging from 200 to 600°C and volatilized organics and metals are removed in a gas stream which is then treated. The Eco Logic process uses a hydrogen-rich carrier gas to transport the volatilized contaminants to a chemical reduction reactor. Prior to the reactor, the gas stream passes through a molten metal bath to remove a portion of the volatilized metals.

REFERENCES

American Petroleum Institute (1995). *In Situ Air Sparging: Evaluation of Petroleum Industry Sites and Considerations for Applicability, Design, and Operation,* Publication 841-46090, API, Washington, DC.

American Society for Testing and Materials (1991). *Annual Book of ASTM Standards,* Vol. 4.08, "Soil and Rock; Dimension Stone; Geosynthetics," ASTM, Philadelphia.

Baker, R. S. (1996). "Bioslurping LNAPL Contamination," *Pollution Engineering,* vol. 28, no. 3.

Benker, K. W. (1995). "Removing Metals from Soil," *Civil Engineering,* vol. 65, no. 10.

Bouwer, E. J., and G. D. Cobb (1987). *In Situ Groundwater Treatment Technology Using Biodegradation,* Report AMXTH-TE-CR-88023, U.S. Army Toxic and Hazardous Materials Agency, Aberdeen Proving Groudn, MD.

Brubaker, G. R. (1995). "The Boom in In Situ Bioremediation," *Civil Engineering,* vol. 65, no. 10.

Chambers, C. D., et al. (1991). *In Situ Treatment of Hazardous Waste-Contaminated Soils,* 2nd ed., Noyes Data Corporation, Park Ridge, NJ.

Clifton, C. E. (1967). *Introduction to Bacterial Physiology,* McGraw-Hill, New York.

Costa, M. J. (1995). "Case Study: Vacuum Extraction Harnessed for Emergency Cleanup," *Environmental Protection,* vol. 6, no. 4, Waco, TX.

Department of Energy (1990a). *Environmental Technology Program: Annual Report,* UCRL-LR-105199, DE91 009070, Lawrence Livermore National Laboratory, Livermore, CA.

——— (1990b). *Surface Sampling of the 300-FF-1 Operable Unit, Hanford Site, Southeastern Washington: A Case Study,* EMO-SA-5501, Richland, WA.

——— (1990c). *Uncontrolled Hazardous Waste Site Cleanup Programs in the U.S.: An Overview,* DE90 002215, prepared by S.-Y. Chiu and S. Y. Tsai, Argonne National Laboratory, Argonne, IL.

——— (1991). *Preliminary Field Demonstration of a Fiber-Optic TCE Sensor,* DE91 008453, prepared by S. M. Angel et al., Lawrence Livermore National Laboratory, Livermore, CA.

Eckenfelder, W. W., Jr. (1966). *Industrial Water Pollution Control,* McGraw-Hill, New York.

Environmental Protection Agency (1980). *Procedures Manual for Groundwater Monitoring at Solid Waste Disposal Facilities,* SW-611, Office of Solid Waste and Emergency Response, Washington, DC.

——— (1985). *A Practical Guide for Groundwater Sampling,* EPA/600/2-85/104, Environmental Research Laboratory, Ada, OK.

——— (1986a). *RCRA Groundwater Monitoring Technical Enforcement Guidance Document,* OSWER 9950.1, Office of Solid Waste and Emergency Response, U.S. EPA, Washington, DC.

——— (1986b). *Test Methods for Evaluating Solid Waste: Physical/Chemical Methods,* 3rd ed., SW-846, Office of Solid Waste and Emergency Response, U.S. EPA, Washington, DC.

——— (1988). *Development of an Environmental Monitoring Technique Using Synchronous Excitation (SE) Fluorescence Spectroscopy,* EPA/600/D-90/060, prepared by D. Stainken and U. Frank for Risk Reduction Engineering Laboratory, Office of Research and Development, U.S. EPA, Cincinnati, OH.

——— (1989a). *Statistical Analysis of Ground-Water Monitoring Data at RCRA Facilities: Interim Final Guidance,* Office of Solid Waste Management Division, U.S. EPA, Washington, DC.

——— (1989b). *Soil-Gas and Geophysical Techniques for Detection of Subsurface Organic Contamination,* ESL-TR-87-67, Las Vegas, NV.

——— (1990a). *Innovative Treatment Technologies for Application to Superfund Sites,* Washington, DC.

——— (1990b). *Enhanced Bioremediation Utilizing Hydrogen Peroxide as a Supplement Source of Oxygen,* EPA/600/2-90/006, Washington, DC.

——— (1990c). *Handbook of Suggested Practices for the Design and Installation of Ground-Water Monitoring Wells,* EPA/600/4-89/034, Environmental Monitoring Systems Laboratory, Las Vegas, NV.

——— (1991). *Soil Sampling and Analysis for Volatile Organic Compounds,* EPA/540/4-911001, Las Vegas, NV.

——— (1993). "EPA Signs License Agreement for TCE-Degrading Microorganisms," EPA/540/N-93/001, no.8, in *Bioremediation in the Field,* Center for Environmental Research Information, Cincinnati, OH.

——— (1994a). *Project Summary: Field Investigation of Effectiveness of Soil Vapor Extraction Technology,* EPA/600/SR-94/142, Cincinnati, OH.

——— (1994b). *SITE Demonstration Bulletin, Thermal Desorption System, Clean Berkshires, Inc.,* EPA/540/MR-94/507, Cincinnati, OH.

——— (1995a). *Manual: Ground Water and Leachate Treatment Systems,* EPA/626/R-94/005, Washington, DC.

——— (1995b). *Project Summary: Assessing UST Corrective Action Technologies: Lessons Learned About In Situ Air Sparging at the Denison Avenue Site, Cleveland, Ohio,* EPA/600/SR-95/040, Cincinnati, OH.

——— (1995c). *SITE Technology Capsule: IITRI Radio Frequency Heating Technology,* EPA/540/R-94/527a, U.S. EPA, Cincinnati, OH.

Environmental Protection Agency, Solar Universal Technologies, Inc., NSI Technology Services Corp, Traverse Group, Inc., and 9th Coast Guard District (1991). "Nitrate for Biorestoration of an Aquifer Contaminated with Jet Fuel, Report for June 1988–Sept. 1990," Washington, DC.

Elizardo, K. (1993). "Biotreatment Techniques Get Chemical Help," *Pollution Engineering,* vol. 25, no. 11.

Fagan, M. R. (1994). "Peroxygens Enhance Biological Treatment," *Environmental Protection,* vol. 5, no. 9, Waco, TX.

Hart, D. P., and R. Couvillion (1986). *Earth-Coupled Heat Transfer,* National Water Well Association, Dublin, OH.

Hasbach, A. (1993). "Moving Beyond Pump-and-Treat," *Pollution Engineering,* vol. 25, no. 6.

Hazen, Terry, (1995). "Savaannah River Site," *Environmental Protection,* Vol. 6, No. 4, Stevens Publishing, Waco, TX.

Kemblowski, M. W., et al. (1987). "Fate and Transport of Residual Hydrocarbon in Groundwater: A Case Study," in *Petroleum Hydrocarbons and Organic Chemicals in Groundwater: Prevention, Detection, and Restoration; Conference and Exposition,* National Water Well Association and American Petroleum Institute, November 17–19.

King, R. B., et al. (1992). *Practical Environmental Bioremediation,* Lewis Publishers, Boca Raton, FL.

Matso, K. (1995). "Mother Nature's Pump and Treat," *Civil Engineering,* vol. 65, no. 10. ASCE,.

Miller, R. N. (1990). "A Field-Scale Investigation of Enhanced Petroleum Hydrocarbon Biodegradation in the Vadose Zone Combining Soil Venting as an Oxygen Source with Moisture and Nutrient Addition," doctoral dissertation submitted to the Civil and Environmental Engineering Department, Utah State University, Logan, UT.

Morgan, P., and R. J. Watkinson (1992). "Factors Limiting the Supply and Efficiency of Nutrient and Oxygen Supplements for the In Situ Biotreatment of Contaminated Soil and Groundwater," *Water Research,*, vol. 26.

National Research Council (1994). *Alternatives for Groundwater Cleanup,* National Academy Press, Washington, DC.

Nelson, C. H. (1995). "Using Ozone to Speed up Air Sparging," *Environmental Protection,* vol. 6, no. 4, Waco, TX.

Raymond, R. L., et al. (1976). *Field Report, Field Application of Subsurface Biodegradation of Gasoline in a Sand Formation,* Project 307-77, Committee on Environmental Affairs, American Petroleum Institute, Washington, DC..

Reidy, P. J., et al. (1990). *Assessing UST Corrective Action Technologies: Early Screening of Cleanup Technologies for the Saturated Zone,* EPA/600/2-90/027, U.S. EPA, Washington, DC.

Russell, D. L. (1996). "Using Horizontal Wells as a Remediation Tool," *Environmental Protection,* vol. 7, no. 1.

Russell, H. H., et al. (1992). *TCE Removal from Contaminated Soil and Groundwater,* Groundwater Issue, EPA/540/S-92/002, U.S. EPA, Washington, DC.

Semprini, L., et al. (1987). *A Field Evaluation of In Situ Biodegradation for Aquifer Restoration,* EPA/600/2-87/096, U.S. EPA, Washington, DC.

Sewell, G. W., et al. (1990). *Anaerobic In Situ Treatment of Chlorinated Ethenes,* EPA/600/D-90/204, U.S. EPA, Washington, DC.

Sims, J. L., et al. (1992). *In Situ Bioremediation of Contaminated Groundwater,* EPA/540/S-92/003, U.S. EPA, Washington, DC.

U.S. Army Corps of Engineers (1991). *A Membrane for In Situ Optical Detection of Organic Nitro Compounds Based on Fluorescence Quenching,* AD-A244 261, prepared by R. W. Seitz, C. Jian, and D. C. Sundberg for U.S. Army Toxic and Hazardous Materials Agency, Aberdeen Proving Ground, MD, January.

——— (1989). *Influence of Well Casing Composition on Trace Metals in Groundwater,* AD-A208 109, prepared by A. D. Hewitt, U.S. Army Cold Regions Research and Engineering Laboratory, Hanover, NH, for U.S. Army Toxic and Hazardous Materials Agency, Aberdeen Proving Ground, MD, April.

Ware, P. J. (1993). "Supplemental Air to Reduce Remediation," *National Environmental Journal,* vol. 3, no. 4.

Water Environment Federation (1994). "Ironing Out Groundwater Contamination," *Water Environment & Technology,* vol. 6, no. 6.

BIBLIOGRAPHY

Adomait, M., and B. Whiffin (1991). "In Situ Volatilization Technologies: R, D & D Scoping Study," *Proceedings of the First Annual Soil and Groundwater Remediation R, D & D Symposium,* Ottawa, Ontario, Canada, January.

Aller, L., et al. (1989). *Handbook of Suggested Practices for the Design and Installation of Ground-Water Monitoring Wells,* National Water Well Association, Dublin, OH.

Aral, M. M. (1990). *Groundwater Modeling in Multilayer Aquifers: Steady Flow,* Lewis Publishers, Boca Raton, FL.

American Society for Testing and Materials, (1988). *Ground-Water Contamination: Field Methods,* Special Technical Publication 963, Collins and Johnson, eds., ASTM, Philadelphia.

——— (1990). *Groundwater and Vadose Zone Monitoring,* Special Technical Publication 1053, D. M. Nielsen and A. I. Johnson, eds., ASTM, Philadelphia.

——— (1991). *Monitoring Water in the 1990's: Meeting New Challenges,* Special Technical Publication 1102, Hall and Glysson, eds., ASTM, Philadelphia.

Canter, L. W., et al. (1987). *Groundwater Quality Protection, Lewis Publishers, Boca Raton, FL.*

Chaudhry, G. R., ed. (1994). *Biological Degradation and Bioremediation of Toxic Chemicals,* Dioscorides Press, Portland, OR.

Department of Energy (1988). *Recommendations for Holding Times of Environmental Samples,* DE88 011831, prepared by M. P. Maskarinec, L. H. Johnson, and S. K. Holladay, Oak Ridge National Laboratory, U.S. DOE, Oak Ridge, TN.

——— (1990). *Performance Evaluation of a Groundwater and Soil Gas Remedial Action,* DE90 017659, prepared by Mary C. Hansen and Suzanne L. Hartnett, Argonne National Laboratory, U.S. DOE, Argonne, IL.

——— (1991). *Bioremediation of Hanford Groundwater,* DE92 000799, prepared by T. M. Brouns et al., Pacific Northwest Laboratory, U.S. DOE, Richland, WA.

——— (1991). *Evaluation of a Multiport Groundwater Monitoring System,* DE91 011073, prepared by T. J. Gilmore et al., Pacific Northwest Laboratory for U.S. DOE, Richland, WA.

——— (1991). *Statistical Approach on RCRA Groundwater Monitoring Projects at the Hanford Site,* DE91 014682, prepared by C. J. Chou for the U.S. DOE, Washington, DC.

Department of Energy–Richland Operations Office (1991). *Annual Report for RCRA Groundwater Monitoring Projects at Hanford Facilities for 1990,* DOE/RL-91-03, DOE–RL, Richland, WA.

Environmental Protection Agency (1983). *Characterization of Hazardous Waste Sites: A Methods Manual,* Vol. 2, "Available Sampling Methods," EPA/600/4-83/040, U.S. EPA, Washington, DC.

——— (1985). *Modeling Remedial Actions at Uncontrolled Hazardous Waste Sites,* EPA/540/2-85/001, U.S. EPA, Cincinnati, OH.

——— (1985). *Remedial Action at Waste Disposal Sites (Revised),* EPA/625/6-85/006, Hazardous Waste Engineering Research Laboratory, Cincinnati, OH, and Office of Emergency and Remedial Response, U.S. EPA, Washington, DC.

——— (1986). *Testing and Evaluation of Permeable Materials for Removing Pollutants from Leachates at Remedial Action Sites,* EPA/600/2-86/074, U.S. EPA, Cincinnati, OH.

——— (1987). *A Handbook on Treatment of Hazardous Waste Leachate,* EPA/600/8-87/006, U.S. EPA, Washington, DC.

——— (1988). *Comparison of Water Samples from PTFE, PVC, and SS Monitoring Wells,* EPA/600/X-88/091, prepared by M. J. Barcelona, G. K. George, and M. R. Schock, Environmental Monitoring Systems Laboratory, U.S. EPA, Las Vegas, NV.

——— (1988). *Ground-Water Modeling: An Overview and Status Report,* EPA/600/2-89/028, U.S. EPA, Cincinnati, OH.

——— (1988). *Guidance for Conducting Remedial Investigation and Feasibility Studies Under CERCLA,* EPA/540/G-89/004 and OSWER-9355.3-01, Office of Emergency and Remedial Response, U.S. EPA, Washington, DC.

——— (1988). *Guidance on Remedial Actions for Contaminated Groundwater at Superfund Sites,* EPA/540/G- 88/003, U.S. EPA, Washington, DC.

——— (1988). *Treatment of Hazardous Landfill Leachates and Contaminated Groundwater,* EPA/600/2-88/064, prepared by R. C. Ahlert and D. S. Kosson, Rutgers University, for Risk Reduction Engineering Laboratory, U.S. EPA, Cincinnati, OH.

——— (1989). *Evaluation of Ground-Water Extraction Remedies,* Vol. 1, "Summary Report," EPA/540/2-89/054, U.S. EPA, Washington, DC.

——— (1989). *Guide on Remedial Actions for Contaminated Groundwater,* EPA/9283-1-02FS, U.S. EPA, Washington, DC.

——— (1989). *Performance of Pump-and-Treat Remediations,* EPA/540/4-89/005, prepared by J. F. Keely, U.S. EPA, Cincinnati, OH.

——— (1989). *Transport Processes Involving Organic Chemicals,* EPA/600/D-89/161, Robert S. Kerr Environmental Research Laboratory, U.S. EPA, Ada, OK.

——— (1989c). *Transport and Fate of Contaminants in the Subsurface,* EPA/625/4-89/019, Center for Environmental Research Information and Robert S. Kerr Environmental Research Laboratory, U.S. EPA, Cincinnati, OH.

——— (1990). *Basics of Pump-and-Treat Ground-water Remediation Technology,* EPA/600/8-90/003, U.S. EPA, Cincinnati, OH.

——— (1990). *Cleanup of Underground Storage Tank Releases Using Pump and Treat Methods,* EPA/101/F-90/037, prepared by M. A. Susavidge, Drexel University, Philadelphia, for Office of Cooperative Environmental Management, U.S. EPA, Washington, DC.

——— (1990). *Emerging Technologies: Bio-recovery Systems Removal and Recovery of Metals Ion from Groundwater,* EPA/540/5-90/005a, Risk Reduction Engineering Laboratory, Office of Research and Development, U.S. EPA, Cincinnati, OH.

——— (1990). *A Guide to Pump and Treat Groundwater Remediation Technology,* EPA/540/2-90/018, Office of Solid Waste and Emergency Response, U.S. EPA, Washington, DC.

——— (1990). *Innovative Operational Treatment Technologies for Application to Superfund Site: Nine Case Studies,* U.S. Office of Solid Waste and Emergency Response, U.S. EPA, Washington, DC.

——— (1991). *Compendium of ERT Groundwater Sampling Procedures,* OSWER Directive 9360.4-06, Environmental Response Team, Office of Emergency and Remedial Response, U.S. EPA, Washington, DC.

——— (1991). *Innovative Treatment Technologies: Semi-annual Status Report,* Office of Solid Waste and Emergency Response Technology Innovation Office, U.S. EPA, Washington, DC.

——— (1991). *In Situ Steam Extraction Treatment,* EPA/540/2-91/005, prepared by Science Applications International Corporation, Cincinnati, OH, for U.S. EPA, Washington, DC.

——— (1991). *Soil Sampling and Analysis for Volatile Organic Compounds,* EPA/540/4-91/001, Las Vegas, NV.

——— (1992). *Silicate Technology Corporation's Solidification/Stabilization Technology for Organic and Inorganic Contaminants in Soils: Applications Analysis Report,* EPA/540/AR-92/010, U.S. EPA, Washington, DC.

——— (1993). *Membrane Treatment of Wood Preserving Site Groundwater by SBP Technologies, Inc., Applications Analysis Report,* EPA/540/AR-92/014, U.S. EPA, Washington, DC.

——— (1994). *Eco Logic International Gas-Phase Chemical Reduction Process, The Thermal Desorption Unit: Applications Analysis Report,* EPA/540/AR-94/504, U.S. EPA, Washington, DC.

——— (1995). *Manual: Groundwater and Leachate Treatment Systems,* EPA/625/R-94/005, U.S. EPA, Washington, DC.

——— (1995). *Project Summary: Removal of PCBs from Contaminated Soil Using the CF Systems Solvent Extraction Process: A Treatability Study,* EPA/540/SR-95/505, U.S. EPA, Cincinnati, OH.

Gas Research Institute (1990). *Laboratory and Pilot-Scale Evaluations of Physical/ Chemical Treatment Technologies for MGP Site Groundwaters,* prepared by Remediation Technologies, Inc., for GRI, Chicago.

Illinois State Water Survey (1983). *A Guide to the Selection of Materials for Monitoring Well Construction and Ground-Water Sampling,* SWS Contract Report 327, prepared by M. J. Barcelona, J. P. Gibb, and R. Miller, ISWS, Champaign, IL.

——— (1985). *Practical Guide for Ground-Water Sampling,* SWS Contract Report 374, prepared by M. J. Barcelona et al., ISWS, Champaign, IL.

——— (1988). *In Situ Aquifer Reclamation by Chemical Means: A Feasibility Study,* HWRIC-RR/028, prepared by G. R. Peyton, M. H. Lefaivre, and M. A. Smith, Hazardous Waste Research and Information Center, ISWS, Champaign, IL.

Matthess, G. (1982). *The Properties of Groundwater,* Wiley-Interscience, Toronto, Ontario, Canada.

Miller, M. E. (1995). "Bioremediation on a Big Scale," *Environmental Protection,* vol. 6, no. 7.

National Research Council (1993). *In Situ Bioremediation: When Does It Work?* National Academy Press, Washington, DC.

Nielsen, D. M. (1991). *Practical Handbook of Ground-Water Monitoring,* Lewis Publishers, Boca Raton, FL.

O'Brien & Gere Engineers, Inc. (1988). *Hazardous Waste Site Remediation,* Van Nostrand Reinhold, New York.

Paff, S. W., B. E. Bosilovich, and N. J. Kardos (1994). *Emerging Technology Summary: Acid Extraction Treatment System for Treatment of Metal Contaminated Soils,* EPA/540/SR-94/513, U.S. EPA, Cincinnati, OH.

U.S. Army (1987). *In-Situ Groundwater Treatment Technology Using Biodegradation,* AD-A244 079, prepared by Edward J. Bouwer and Gordon D. Cobb for U.S. Army Toxic and Hazardous Materials Agency, U.S. Army, Aberdeen Proving Ground, MD.

——— (1988). *Monitoring Well Installation and Groundwater Sampling and Analysis Plan,* prepared by Donahue and Associates, Inc., for U.S. Army Training Reserve–84th Division, Milwaukee, WI.

——— (1989). *Single Fiber Measurements for Remote Optical Detection of TNT,* CETHA-TE-CR-89102, U.S. Army Toxic and Hazardous Materials Agency, U.S. Army, Washington, DC.

Van Der Leeden, F. (1992). *Geraghty & Miller's Groundwater Bibliography,* 5th ed., Lewis Publishers, Boca Raton, FL.

Walton, W. C. (1989). *Numerical Groundwater Modeling: Flow and Contamination Migration,* Lewis Publishers, Boca Raton, FL.

14

ASSESSING AND MANAGING WATER QUALITY

Water quality is protected by setting standards for water bodies and limiting the types and amounts of pollutants in wastewaters that are discharged. Wastewater discharge limits are set to ensure that water quality is protected and are based on either instream water quality standards or levels that treatment technology is capable of achieving, whichever criterion is more restrictive. Technology-based limits have been established by the EPA for many industries in the form of national effluent limitations and standards. In this chapter we discuss water quality standards, effluent limitations and standards, stormwater permits, and pretreatment programs for publicly owned treatment works (POTWs).

WATER QUALITY STANDARDS

Water quality standards are designed to protect water resources and the animals and people that contact or use these resources. The federal water quality standards program is defined by regulatory requirements at 40 CFR Part 131. Standards may be established on a state or federal level. State standards, although customized by the state, must meet the general requirements of Part 131.

For states that have not established adequate water quality standards, EPA has set "national" standards for 126 priority toxic pollutants specific to these states under the National Toxics Rule, which was promulgated in 1992 (57 FR 60910). Included in the list are heavy metals such as arsenic, mercury, and lead; volatile organics such as benzene, toluene, methylene chloride, and vinyl chloride; acid and base/neutral extractables such as naphthalene, benzidine, nitrobenzene, and dimethyl phthalate; polychlorinated biphenyls (PCBs) and pesticides such as PCB-1242, chlordane, 4,4′-DDT, and aldrin; and miscellaneous pollutants.

The federal requirements at 40 CFR Part 131 are divided into four subparts. Subpart A covers general provisions such as scope, purpose, definitions, minimum requirements for state program submittals, and special considerations of Indian tribes. Subpart B contains the requirements for state water quality standards, including designated uses of water resources, water quality criteria, and an antidegradation policy. Subpart C sets out the requirements for reviewing and revising state water quality standards every three years. Subpart D contains the national standards established for states that the EPA believes do not have a standards program that meets the requirements of the Clean Water Act (CWA).

Designated uses of a stream, river, lake, bay, or other water resource determine the water quality standards that must be developed to protect those uses. Designated uses may include public water supply; protection and propagation of fish, shellfish, and wildlife; recreation; agriculture; industry; and navigation. Other less commonly considered uses may include aquifer protection, groundwater recharge, coral reef preservation, hydroelectric power, and marinas.

Water quality standards contain both narrative and numeric criteria. *Narrative criteria* are statements such as the following, suggested by the EPA in its *Water Quality Standards Handbook* (EPA 1994b):

> All waters, including those within mixing zones, shall be free from substances attributable to wastewater discharges or other pollutant sources that:
>
> (1) Settle to form objectionable deposits;
> (2) Float as debris, scum, oil, or other matter forming nuisances;
> (3) Produce objectionable color, odor, taste or turbidity;
> (4) Cause injury to, or are toxic to, or produce adverse physiological responses in humans, animals or plants; or
> (5) Produce undesirable or nuisance aquatic life (54 FR 28627, July 6, 1987).

Numeric water quality criteria are specific limits on pollutants such as metals, and toxic organics or instream parameters such as dissolved oxygen, pH, chlorides, and sulfates. Numeric criteria are usually developed as aquatic life and human health criteria, depending on the type of water body and designated use. Aquatic life criteria are developed separately for fresh- and saltwater systems and for short-term (acute) and long-term (chronic) exposure conditions. Human health criteria are based on water, fish, and shellfish consumption in freshwater systems and fish and shellfish consumption in saltwater systems; they may also be based on contact and noncontact recreation.

Although not nearly as well developed as aquatic and human health criteria, other types of water quality criteria may include sediment criteria, biological criteria, wildlife criteria, and numeric criteria for wetlands. Sediment and wildlife criteria will be EPA's next areas of focus.

Antidegradation

Federal regulations outline three levels or tiers of antidegradation of water quality as discussed in EPA's *Water Quality Standards Handbook* (EPA 1994b), which are outlined in Table 14.1.

Aquatic Life Criteria

Aquatic life criteria are developed for acute and chronic exposure of aquatic organisms in both fresh- and salt-water (marine) systems. Criteria are derived from toxicity tests of single chemicals on a variety of aquatic plants and animals. Toxicity tests for acute criteria are usually 48- to 96- hour tests that measure lethality or immobilization of the organism. Toxicity tests for chronic criteria are run for much longer periods, often greater than 28 days, and measure effects on survival, growth, or reproduction. Toxic concentrations are referred to as the *criterion maximum concentration* (CMC) for acute exposure and the *criterion continuous concentration* (CCC) for chronic exposure.

EPA recommends limiting exposure to the acute CMC to a 1-hour period (EPA 1994b), although in practice, regulators use 1-day periods for setting waste load allocations and permit limits (EPA 1991). EPA recommends lim-

TABLE 14.1 Antidegradation of Water Quality

Tier	Description of Protection	Comments
I	Minimum level of protection; protects existing uses; water quality in a water body can be lowered only if existing uses will still be fully protected and existing water quality downstream water quality standards are not exceeded.	This tier is designated when a better use of the water body (which would require higher water quality) cannot be achieved.
II	Protects high-quality waters from degradation; water quality exceeds that necessary to protect aquatic life, wildlife, and recreational uses (termed *fishable, swimmable waters*); water quality may be lowered only after an antidegradation review.	Antidegradation review must show that lowering water quality is necessary for important economic and social development; point sources must meet highest pollutant control requirements.
III	Protects highest-quality waters, which are designated outstanding natural resource waters having exceptionally high water quality or ecological value; prohibits any degradation of water quality.	No new or increased discharges allowed; some temporary degradation may be allowed by states.

iting exposure to the chronic CCC to a 4-day period (EPA 1994b), which can then be statistically translated into a monthly average permit limit (EPA 1991). Many states, however, establish chronic criteria as 7-day averages.

Criteria for certain chemicals or types of chemicals may include other water quality parameters, such as pH (ammonia, pentachlorophenol) and hardness (metals). Metals criteria are often given as total recoverable metals, which represent both the particulate and dissolved fractions. Because the dissolved form is believed to represent the form of the metal that is most available to the organism biologically (bioavailable), such as through adsorption on the gill surface, EPA recommends that compliance with water quality standards be based on the dissolved fraction (EPA 1994b).

Human Health Criteria

Human health criteria are designed to protect against chronic effects from long-term exposure. For carcinogenic chemicals, the exposure period is a person's lifetime, assumed to be equal to 70 years. Criteria are usually based on consumption of water and contaminated fish or shellfish, depending on the type and designated uses of the water source. For example, estuarine or marine waters are not used for drinking water, so the human health criteria for salt water would be based on food consumption alone. Consumption of contaminated organisms is a particular concern where a chemical bioaccumulates or concentrates in the organism, as does PCBs in the lipid or fatty tissues of fish.

Sediment Criteria

EPA is currently developing sediment quality criteria, focusing initially on nonionic organic chemicals and metals for aquatic life protection. EPA is considering various applications of sediment criteria: assessing risks of contaminated sediments, site monitoring, ensuring control of contaminants in discharges, maintaining uncontaminated sediments, early warning of potential problems, establishing permit limits for point source discharges and target levels for nonpoint sources, and aiding remediation activities (EPA 1994b). EPA's contaminated sediment management strategy is designed to:

- Prevent further sediment contamination that may cause unacceptable ecological or human health risks.
- When practical, clean up existing sediment contamination that adversely affects the nation's water bodies or their uses, or that causes other significant effects on human health or the environment.
- Ensure that sediment dredging and dredged material disposal continue to be managed in an environmentally sound manner.

- Develop and consistently apply methodologies for analyzing contaminated sediments (EPA 1994a).

Criteria for nonionic organic chemicals will probably be based on the *equilibrium partitioning* (EqP) *approach,* which assumes that a chemical partitions between the organic carbon in the sediment and the pore water in the interstices of the sediment and is at equilibrium. Criteria for metals will probably be based on acid volatile sulfides (AVS) and other binding factors.

Mixing Zones

A mixing zone is an area where a wastewater discharge is initially allowed to mix with a receiving water without having to meet a water quality standard. The idea behind a mixing zone is that a small area around an outfall where a standard may be exceeded is not harmful to the overall receiving water. A mixing zone must be carefully delineated so that:

- It does not impair the integrity of the water body as a whole.
- There is no lethality to organisms passing through the mixing zone.
- There are no significant health risks, considering probable pathways of exposure (EPA 1994b).

The type of mixing zone relates to the applicable water quality criteria (acute aquatic, chronic aquatic, or human health). The immediate area around or downstream of the discharge is the zone of initial dilution (ZID). Acute aquatic criteria must be met at the edge of the ZID, while within the ZID the criteria may be exceeded. Beyond the ZID is the secondary mixing zone. Chronic aquatic criteria may be exceeded within this zone, and compliance is determined at the edge of the mixing zone. A mixing zone may also be defined for human health criteria; the edge of this zone is usually considered to be the point of complete (100%) mixing with the receiving water.

Federal regulations allow states to incorporate mixing zones into their water quality standards. Although many do, states are not obligated to use mixing zones. Without mixing zones, water quality criteria apply directly to the wastewater at the point of discharge (end-of-pipe). Because water quality criteria for many pollutants are quite restrictive, lack of a mixing zone can result in discharge limits that are very difficult and expensive to meet.

Mixing zones may be defined by specific size areas around or distances from the discharge point translated into dilution factors. For example, the mixing zone may be one-fourth of the stream width for an acute aquatic criterion, representing a dilution factor of 25%. In this case, the effluent would have to meet the acute aquatic criterion at the edge of the ZID while the permit limit would be four times the acute criterion (the acute criterion divided by 0.25). Where multiple criteria apply to a discharge (acute aquatic, chronic aquatic, human health), the most stringent criterion determines the permit limit.

Whole Effluent Toxicity

Whole effluent toxicity (WET) testing directly measures the toxicity of an effluent on aquatic organisms. An organism is exposed to wastewater effluent in diluted or undiluted form, depending on the type of WET test, and evaluated for growth, survival, fecundity, or other characteristics. The WET test attempts to evaluate toxicity by a more holistic approach because it incorporates all the effluent's characteristics; hence, the term *whole effluent*. In this way, it differs from numeric criteria, which are based on studies using clean water samples spiked with a single toxic chemical. WET testing is one way of ensuring that narrative criteria to protect water quality are being met.

Wastewater discharge permits include WET testing requirements. Typically, two species must be tested, one a feeder type such as the crustacean, *Ceriodaphnia dubia,* the water flea, and a one fish such as *Pimephales promelas,* the fathead minnow. Both of these are freshwater species and are used only to test effluents that discharge into fresh water. Marine or estuarine species are used for effluent discharging into marine or estuarine environments. Saltwater species often used for WET testing are *Mysidopsis bahia,* the mysid shrimp, *Menidia beryllina,* the Atlantic silverside, and *Cyprinodon variegatus,* the sheepshead minnow.

States may set WET limits in wastewater permits which are analogous to numeric permit limits for a specific chemical. Other permits may only include requirements for conducting the WET test and reporting the results. Test frequency varies and usually depends on how often the test results demonstrate no toxicity. For example, a discharger may have to test monthly at the beginning of the permit and be allowed to reduce testing to a quarterly or semiannual basis if no toxicity is shown. If repeated tests show toxicity, the discharger will be required to investigate the cause and implement changes to eliminate the toxicity by means of a toxicity reduction evaluation (TRE).

Great Lakes Water Quality Guidance

In 1995, EPA published regulations at 40 CFR Part 132 outlining its guidance for water quality in the Great Lakes system. These regulations identify minimum water quality standards, antidegradation policies, and implementation procedures to protect human health, aquatic life, and wildlife in the Great Lakes system. The Great Lakes system is defined as all the streams, rivers, lakes, and other bodies of water within the drainage basin of the Great Lakes. States that are subject to the regulations are Illinois, Indiana, Michigan, Minnesota, New York, Ohio, Pennsylvania, and Wisconsin; Indian tribes within the drainage basin of the Great Lakes are also subject to the regulations. These states and tribes were required to submit criteria, methodologies, policies, and procedures for water quality standards to EPA by September 23, 1996. The majority of the Part 132 regulations is contained in six appendices outlining the Great Lakes antidegradation policies, implementation procedures, and methodologies for developing criteria for aquatic life, human health, wildlife, and bioaccumulation factors. Wildlife criteria focus on bioaccumulative chemicals of concern (BCCs). A wildlife BCC is one that is persistent in the environment and tends to accumulate in the aquatic food chain which may cause adverse effects on the reproduction, development, and population survival of birds and mammals.

The Great Lakes water quality program is very similar to, while quite different from the national water quality program. The Great Lakes program includes three antidegradation categories, similar to tiers I to III of the national water quality program, covering (1) waters not attaining designated uses, (2) high-quality waters, and (3) outstanding national or international resource waters. The Great Lakes guidelines also provide for mixing zones, however, mixing zones for BCCs must be eliminated by the year 2005, although exceptions will be allowed. The Great Lakes rule includes minimum water quality criteria for 29 pollutants. When developing new or additional numbers for their own programs, the states are required to use either tier I or tier II methodologies. Tier I is used to calculate numeric *criteria.* When there are not enough data to satisfy tier I requirements, tier II is used to calculate numeric *values.* Because tier II methodologies depend on fewer data, they incorporate a high degree of conservatism and generally produce numbers more stringent than those produced in tier I.

EPA guidance for the Great Lakes water quality program incorporates bioaccumuulation factors (BAFs) in human health and wildlife criteria (EPA 1995). BAFs reflect the net accumulation of a substance by an aquatic organism from water, sediment, and food. The accumulation of toxic chemicals in aquatic organisms can increase to levels that are harmful to the wildlife and humans that eat them. A BAF represents the ratio of a substance's concentration in the tissue of aquatic organisms to its concentration in the ambient water in situations where both the organism and its food are exposed and the ratio does not change substantially over time. BAFs can be determined by several different methods. EPA recommends four methods, in the following order of preference:

1. A BAF measured in the field, in fish collected at or near the top of the food chain.
2. A BAF derived from a biota-sediment accumulation factor (BSAF). A BSAF relates the concentration of a substance in the tissue of an aquatic organism (normalized by lipid content) to the concentration of the substance in the sediment (normalized by organic carbon content).
3. A BAF calculated by multiplying a bioconcentration factor (BCF) measured in the laboratory, preferably on an indigenous fish species, by a food chain multiplier (FCM) factor. A BCF represents the net accu-

mulation of a substance by an aquatic organism from the ambient water only through gill membranes or other external body surfaces. An FCM is a multiplying factor that accounts for the biomagnification of a chemical through trophic levels in the food chain.

4. A BAF calculated by multiplying a BCF, calculated from the octanol–water partition coefficient, by an FCM.

BAFs and the methodologies for deriving BAFs are not limited to the Great Lakes. The EPA will be incorporating them into wildlife and human health criteria for other bodies of water.

EFFLUENT DISCHARGE LIMITATIONS AND STANDARDS

EPA has developed national limitations and standards for wastewater from 51 specific industrial categories. These categories cover a diverse range of industries, including food processing, metal manufacturing, electrical components, inorganic and organic chemicals, plastics, and mining. These effluent guidelines apply to both direct and indirect dischargers. A *direct discharge* is one that goes into a receiving body of water such as a creek, stream, river, lake, bay, or ocean and must have a permit under the National Permit Discharge Elimination System (NPDES) or a state equivalent NPDES permit. An *indirect discharge* is one that goes to a publicly owned treatment works (POTW) such as a municipal wastewater plant.

There are several classes of effluent guidelines. There are separate effluent guidelines for direct and indirect dischargers, as well as different effluent guidelines for existing and new dischargers. There are effluent guidelines that cover common pollutants such as BOD, TSS, and pH and others that cover toxic pollutants. Table 14.2 is a generalized description of these different types of guidelines and limitations for wastewater discharges.

Effluent limitations guidelines and standards are revised periodically. The general trend is toward more stringent limits for all types of pollutants (conventional, nonconventional, toxic) and the development of new limits for toxic pollutants. Effluent permit limits are set as either concentration limits or mass loads based on manufacturing production levels or wastewater flow.

STORMWATER STANDARDS

Stormwater associated with industrial activity and certain size municipal systems must have a discharge permit according to federal regulations at 40 CFR §122.26. Industrial activity is defined very specifically in these regulations. Municipal stormwater discharges that must be permitted are those classified as medium to large. Generally speaking, large systems are those with a population of 250,000 or more and medium systems are those with a population

TABLE 14.2 General Guidelines and Limitations for Wastewater Discharges

Guidelines for Direct Dischargers

- *BPT:* effluent limits for existing dischargers that are based on best practicable technology for wastewater treatment in a particular industrial category. These limits apply to conventional pollutants which are defined at 40 CFR §401.16 as BOD, TSS, fecal coliform bacteria, pH, and oil and grease and nonconventional pollutants such as COD and ammonia.
- *BCT:* effluent limits for existing dischargers that are based on the best conventional technology for conventional pollutants (usually BOD and TSS). BCT limits may be the same or more stringent than BPT limits. They are more stringent if the cost of providing a higher level of treatment is *reasonable,* which is decided by a two-part cost/benefit test.
- *BAT:* effluent limits for existing dischargers that are based on the best available technology. These limits apply to toxic pollutants and nonconventional pollutants.
- *NSPS:* performance standards (effluent limits) for new direct sources (discharges). New sources are specifically defined at 40 CFR §122.2 for direct dischargers and at 40 CFR §403.3 for indirect dischargers. In general, they are wastewater discharges that are generated by new construction or major modifications of facilities after effluent limits are proposed for a particular industrial category. NSPS are usually more stringent than BPT, BCT, and BAT limits because new facilities have the opportunity to install more efficient pollution control and treatment technology. NSPS apply to all types of pollutants: conventional, nonconventional, and toxic.

Guidelines for Indirect Dischargers

- *PSES:* pretreatment standards for existing sources from indirect dischargers. These standards cover all types of pollutants: conventional, nonconventional, and toxic. PSES are not usually developed for the conventional pollutants BOD and TSS because they are assumed to be readily treated at POTWs and adequately controlled by local POTW limits or restrictions. PSES is analogous to BPT and BAT limits for direct dischargers.
- *PSNS:* pretreatment standards for new sources from indirect dischargers. These standards cover all types of pollutants: conventional, nonconventional, and toxic. Like PSES, PSNS are not usually developed for the conventional pollutants BOD and TSS because they are assumed to be readily treated at POTWs and adequately controlled by local POTW limits or restrictions. PSNS is analogous to NSPS for direct dischargers.

Other Types of Effluent Limits

- *BMP:* best management practices. These are procedures that usually address maintenance and good housekeeping to minimize spills and pollutant concentrations in wastewaters.
- *BPJ:* effluent limits based on professional judgment. BPJ is used when limits cannot be set entirely on categorical effluent guidelines. For example, BPJ is used to set limits for wastewaters that are a mixture of categorical streams and utility wastewaters from cooling towers, boilers, and deminieralizer units. BPJ is also used to set limits for pollutants not covered by effluent guidelines.

between 100,00 and 250,000. Facilities that discharge stormwater from industrial activity into stormwater systems of medium or large municipalities must notify the operator of the municipal system and make their stormwater pollution prevention plans available to the operator upon request.

Various types of stormwater permits are issued by EPA. An individual permit can be issued that is tailored to a specific facility. The baseline general permit, as the name implies, sets forth general requirements that EPA considers fundamental to proper stormwater management. Because there are so many stormwater discharges that must be permitted, to promote its use, EPA has made the baseline general permit easy to obtain. The requirements set forth by the baseline general permit are not trivial, however, as discussed later in this section. Multisector permits are similar to the baseline general permit except that they have been tailored to specific industrial and facility sectors. Table 14.3 lists the 29 sectors for which EPA has developed multisector permits.

Industrial Activity

EPA's definition of industrial activity with respect to stormwater discharges is lengthy and subject to interpretation to a certain degree. In defining industrial activity, EPA focuses on stormwater that can become contaminated with chemicals, oils, and dirt at industrial facilities and construction sites. Areas that are likely to generate stormwater from industrial activity include plant yards, access roads and rail lines, product and chemical storage areas, waste management areas, material handling equipment areas, loading/unloading and shipping/receiving areas, tank storage areas, and large construction sites. There are 11 types of facilities and activities that are defined as industrial activity in 40 CFR§122.26(b)(14). These are:

1. Facilities subject to stormwater effluent limitations guidelines, new source performance standards, or toxic pollutant effluent standards under 40 CFR subchapter N. Certain standard industrial classifications (SIC) with toxic pollutant standards are exempted (refer to the SICs at the end of this list).
2. Facilities under SIC 24 (except 2434), 26 (Except 265 and 267), 28 (except 283), 29, 311, 32, (except 323), 33, 3441, 373.
3. Facilities in the mineral industry under SIC 10 through 14, including active or inactive mining operations and oil and gas exploration, production, processing, or treatment operations, or transmission facilities that discharge stormwater contaminated by contact with or that has come into contact with any overburden, raw material, intermediate products, finished products, by-products or waste products located on the site of such operations. Certain mining operations that have been

TABLE 14.3 Industrial and Other Facilities Sectors with Multisector Stormwater Permits

- Asphalt paving and roofing materials manufacturing and lubricant manufacturing
- Automobile salvage yards
- Chemical and allied products manufacturing
- Coal mines and coal mining related facilities
- Electronic and electrical equipment and components, photographic and optical goods manufacturing
- Fabricated metal products
- Food and kindred products
- Glass, clay, cement, concrete, and gypsum product manufacturing
- Hazardous waste treatment, storage, or disposal facilities
- Landfills and land application sites
- Leather tanning and finishing
- Metal mining (ore mining and dressing)
- Mineral mining and processing
- Oil and gas extraction
- Paper and allied products manufacturing
- Primary metals
- Printing and publishing facilities
- Publicly owned treatment works
- Rubber, miscellaneous plastic products, and miscellaneous manufacturing
- Scrap and waste recycling facilities
- Ship and boat building or repairing yards
- Steam electric power generating facilities, including coal handling areas
- Textile mills, apparel, and other fabric product manufacturing
- Timber products
- Transportation equipment, industrial, or commercial machinery manufacturing
- Vehicle maintenance areas and or/equipment cleaning operations at water transportation facilities
- Vehicle maintenance areas, equipment cleaning areas, or deicing areas located at air transportation facilities
- Vehicle maintenance or equipment cleaning areas at motor freight transportation facilities, passenger transportation facilities, petroleum bulk oil stations and terminals, rail transportation facilities, and the U.S. Postal Service transportation facilities
- Wood and metal furniture and fixture manufacturing

Source: Information from 60 FR 50808-9, September 29, 1995.

reclaimed or have been released reclamation requirements are not considered industrial activity.

4. Hazardous waste treatment, storage, or disposal facilities, including those that are operating under interim status or a permit under Subtitle C of RCRA.

5. Landfills, land application sites, and open dumps that receive or have received any industrial wastes form industrial activities, including those that are subject to regulation under Subtitle D of RCRA.
6. Recycling facilities, including metal scrap yards, battery reclaimers, salvage yards, and automobile junkyards, including but limited to those classified as SCI 5015 and 5093.
7. Steam electric power generating facilities, including coal handling sites.
8. Transportation facilities classified as SIC 40, 41, 42 (except 4221-25), 43, 44, 45, and 5171 which have vehicle maintenance shops, equipment cleaning operations, or airport deicing operations.
9. Domestic sewage treatment plants or any other sewage sludge or wastewater treatment device or system, used in the storage, treatment, recycling, and reclamation of municipal or domestic sewage, including land dedicated to the disposal of sewage sludge, that are located within the confines of the facility, with a design flow of 1.0 MGD or more, or required to have an approved pretreatment program under 40 CFR Part 403. Facilities that are not included in this category are farmlands, domestic gardens, or lands used for sludge management where sludge is beneficially reused and which are not physically located in the confines of the facility, or areas that are in compliance with section 405 of the CWA (disposal or use of sewage sludge).
10. Construction activities, including clearing, grading, and excavation, extending over 5 or more acres. Areas that are less than 5 acres are not considered industrial activity unless they are part of a larger common plan of development or sale.
11. Facilities under SIC 20, 21, 22, 23, 2434, 25, 265, 267, 27, 283, 285, 30, 31 (except 311), 323, 34 (except 3441), 35, 36, 37, (except 373), 38, 39, and 4221-25, and which are not otherwise included in any of the industrial categories above. Stormwater from these facilities (SIC 20, etc.) is not considered to be from industrial activity, however, if the stormwater has not come in contact with industrial materials or activities. In contrast, stormwater from the other categories of industrial activity are assumed to be associated with industrial activity regardless of actual exposure.

Baseline General Permit

Baseline general permits from EPA are available only to facilities in those states and territories that do not have authorized NPDES programs, although states with authorized NPDES programs may develop their own type of general permit. There are two types of EPA baseline general permits: one for construction sites and one for other types of industrial activity. Either type of permit may be obtained by submitting a notice of intent (NOI). In a similar fashion, facilities may terminate their general permits by submitting a notice of termination (NOT).

An important element of general permits is the stormwater pollution prevention plan. These plans must be prepared prior or submitting the NOI and updated as conditions of the site. Stormwater pollution prevention plans for construction activities must include a description of the site, stabilization and other practices to prevent erosion and sediment runoff, management of stormwater, waste disposal, other state or local requirements, maintenance of sediment control measures, inspections, and a list of any nonstormwater discharges that are combined with the stormwater.

Stormwater pollution prevention plans for industrial activities are more extensive than those for construction sites. A pollution prevention team must be created and identified in the plan. The plan must also identify potential pollutant sources stormwater management controls and a comprehensive site compliance evaluation. Basic and other elements that might be included in the stormwater pollution prevention plan are presented in Table 14.4. There are additional requirements for stormwater that is discharged through municipal systems serving a population of 100,000 or more as well as for those facilities that are subject to EPCRA Section 313 requirements and for those that have salt storage piles.

Facilities other than construction sites must sample and analyze their stormwater discharges either annually or semiannually, depending on the type of facility. Facilities that must monitor semiannually must also submit their results once a year to EPA; facilities with annual monitor requirements need only retain their data on site. Monitoring and reporting frequencies depend on the type of facility; however, common pollutants that are monitored include oil and grease, BOD, COD, TSS, and pH.

Multisector Permits

Multisector stormwater permits are similar to baseline general permits except that EPA has specifically tailored them to certain types of industries and facilities (see Table 14.3). Because of time constraints, EPA issued the baseline general permit before multisector permits. Multisector permits are available from EPA for 29 sectors; however, like the baseline general permit, they are available only in those states or areas that do not have authorization from EPA to administer the NPDES program. EPA encourages states that do have NPDES authorization to use the multisector permit as a model for their own stormwater permits. Like the baseline general permit, multisector permits are obtained by submitting an NOI and require a stormwater pollution prevention plan.

Multisector permits have fewer requirements than the baseline general permit, because EPA had the time to tailor the general requirements to each sector. The rubber manufacturing sector is an exception in that its multisector permit requires BMPs, and monitoring for requirements authorization for mul-

TABLE 14.4 Elements That Might Be Included in a Stormwater Pollution Prevention Plan

Pollution Prevention Team

- Member's name
- Member's department (i.e., facilities services, environmental engineering, security, fire protection)
- Member's office and home phone numbers
- Member's pager numbers
- Roles of each team member
- Goals of the team

Description of Potential Pollutant Sources

- Site map showing areas such as bulk storage tanks, chemical and waste storage areas, load/unload areas, wastewater treatment area, central utilities plant, fueling stations, vehicle and equipment maintenance and/or cleaning airtight manufacturing areas
- Materials identification/inventory
- Identification of activities that can generate excessive dust or particulate
- Identification of outdoor activities that can be a potential source of pollution
- Identification of outfalls and drainage areas
- Certification that all stormwater outfalls have been tested or evaluated for the presence of nonstormwater discharges and that none are occurring
- Identification of structural stormwater controls
- Map of sewer system
- Identification of surface waters near the site
- History of past leaks and spills
- Description of any dye testing or other methods for detecting improper connections
- Written assessment of potential pollutant sources

Identification of EPA-Authorized Nonstormwater Discharges (Under General Permits)

- Discharges from firefighting activities
- Fire hydrant flushing
- Potable water sources, including water-line flushing
- Irrigation drainage
- Uncontaminated groundwater
- Foundation or footing drains where flows are not contaminated with process materials
- Routine exterior building washdown that does not use detergents or other compounds
- Pavement wash waters where spills or leaks of toxic or hazardous materials have not occurred and where detergents are not used
- Air conditioning condensate

TABLE 14.4 (*Continued*)

Data
• Stormwater discharge sampling data
• Other relevant test data, such as groundwater data when groundwater is routinely or periodically discharged to storm
Best Management Practices (BMP) and Measures and Controls
• Housekeeping procedures and practices
• Preventive maintenance program and procedures
• Visual inspection program and procedures
• Spill prevention program and procedures
• Sediment and erosion control measures
• Runoff management
• Employee training programming
• Procedures for record keeping and reporting
Implementation Documentation
• Written stormwater pollution prevention plan
• Training records
• Records of comprehensive site evaluation
• Inspection records
• Spill and release records

Source: Information from EPA (1993).

tisector permits are based on the potential for pollutants to contaminate stormwater at a particular facility. For facilities with low potential, only visual monitoring of stormwater discharges is required. Other facilities must sample and analyze their discharges. EPA has set numeric limits for only a few such facilities; several states have added other numeric limits for facilities in their own domain. Monitoring for pollutants that do not have numeric limits must be done quarterly in the second year of the five-year permit. If the average value exceeds the benchmark value for that pollutant, monitoring must be repeated in the fourth year to see if pollution controls have improved stormwater quality.

PRETREATMENT REGULATIONS

Regulations implementing the federal pretreatment program were first published in 1978 when EPA issued its national pretreatment strategy. Since then, the pretreatment regulations have been amended several times. The federal pretreatment regulations at 40 CFR 403 require publicly owned treatment works (POTWs) to control the discharge of nondomestic wastewaters. POTWs are typically wastewater treatment plants operated by a city, town, or other municipality. They may, however, also include facilities that are specifically designed to treat industrial wastewaters if they have been so structured to meet the legal definition of a POTW. Nondomestic wastewaters, although not strictly defined in the regulations, are wastewaters that are not normally thought of as domestic, that is, originating from residential dwellings.

The term *industrial discharger,* which is used in the regulations to describe dischargers of nondomestic wastewater, is somewhat of a misnomer. Although industrial discharges from manufacturing processes are clearly nondomestic, there are numerous other sources of nondomestic wastewater from commercial trade and institutional establishments. For example, wastewaters from restaurants are generally considered nondomestic because they contain more organics, oils, and greases, than household wastes.

Nondomestic dischargers are also sometimes referred to as indirect dischargers. An *indirect discharge* is any nondomestic discharge to a POTW that is regulated under §§307(b), (c), or (d) of the Clean Water Act (sections that provide for the establishment of national pretreatment standards).

POTWs control nondomestic discharges through highly structured pretreatment programs required by federal regulations. However, only certain POTWs are required to establish pretreatment programs. POTWs that must develop pretreatment programs are those with a total flow (domestic and nondomestic) greater than 5 million gallons per day (MGD). If multiple POTWs are controlled by the same authority and their combined flow is greater than 5 MGD, they too must develop pretreatment programs. Additionally, the nondomestic wastewater must contain discharges that are either subject to federal pretreatment standards or contain pollutants that will cause problems for the POTW. In reality, almost any city with a wastewater flow greater than 5 MGD will be large enough to have some industrial development that will require the city to establish a pretreatment program. If a POTW has a total flow of 5 MGD or less, it may still be required to establish a pretreatment program if there are treatment problems, wastewater permit violations, contamination of the treatment sludge, or discharges that warrant stricter control.

POTWs must submit their pretreatment programs to EPA for approval. To be approved, pretreatment programs must contain several required elements. The POTW must demonstrate that it has the proper legal authority to carry out its pretreatment regulations, including issuing pretreatment permits or equivalent means of discharge control and assessing civil and criminal penalties for noncompliance. The POTW must inspect and sample the discharges of its users, track their compliance status and publish the names of those with significant noncompliances, and develop discharge limits based on the characteristics of the treatment facility (local limits).

The POTW must enforce its pretreatment program and is routinely monitored in this regard by either the EPA or the state. The state performs the monitoring function in lieu of EPA if the state has received authorization for the pretreatment program from EPA. The federal pretreatment regulations set out the requirements that states must satisfy in order to receive authorization.

Purpose of Pretreatment Programs

The primary purpose of a pretreatment program is to prevent the discharge of pollutants that pass through inadequately treated or that interfere with the operation of the treatment facility. *Pass-through* occurs when the treatment facility cannot remove or degrade a compound sufficiently to meet its discharge permit limits. *Interference* is broadly defined in the regulations to include damage to or obstruction of the sewer system, conditions in the sewer that endanger workers or can cause explosions, negative impacts on process units such as excessive oil loadings or inhibitory effects on the biological units, and degradation of the biological sludge (biosolids) quality that prevents its use for land application or composting.

An additional goal of a pretreatment program is to increase opportunities to reuse wastewaters and biosolids. The fewer pollutants that these materials contain, the more opportunities there are for reuse. Thus, preventing pass-through and interference of pollutants is critical to meeting this goal.

Elements of a Pretreatment Program

POTW Requirements. POTWs must demonstrate that they have the legal authority to carry out their pretreatment program. They must have the legal authority to enforce compliance by their users, to deny or set conditions on discharges, and to issue pretreatment permits to users or control their discharges in a similar fashion. In addition, they must have the legal authority to require users to establish compliance schedules and to submit notices and discharge monitoring reports and to inspect monitor users.

Pretreatment programs must include certain procedures. Procedures must be established to identify all users subject to the pretreatment program and to characterize their discharges, to notify these users of their requirements, to review discharge monitoring reports and notices sent in by users, to sample and analyze user discharges, to investigate noncompliances, and to publish the names of users in significant noncompliance at least annually in the largest daily local newspaper. There are special requirements regarding what are called *significant industrial users* (SIUs). Stated simply, SIUs are those users that have high-enough flows or pollutant loadings that might affect the treatment facility. Of special distinction are SIUs that are categorical users, that is, have federal pretreatment standards established for their particular industrial category. The POTW must prepare a list of SIUs, notify SIUs of their SIU status, inspect SIU facilities and sample their discharges annually, and evaluate at least every two years whether these SIUs need to develop slug control plans to control spills and accidental discharges.

POTWs must show that they have adequate resources and personnel to implement their pretreatment program. They must also prepare an enforcement response plan that details how they will investigate noncompliances, what enforcement actions will be taken, and how these actions will be linked to the severity of the noncompliance (notice of violation, administrative order, suspension of service, termination of permit).

Each year, POTWs must submit to the approval authority a status report on the pretreatment program. At a minimum, the POTW must include a current list of users and the discharge standards that are applicable to each (categorical, local limits, other); the compliance status of each user over the last year; a summary of compliance, enforcement, and inspection activities; and any other information relevant to the pretreatment program.

Of particular importance to the POTW is the development of local limits. *Local limits* are discharge limits for users that are based on the treatment capacity of the POTW facility and the disposal method for biosolids. Local limits and how they may be calculated are discussed later in more detail in the chapter. EPA has published a number of guidance manuals to help POTWs develop their pretreatment programs. These manuals are listed in the bibliography at the end of the chapter.

Discharger Requirements. Dischargers are subject to wastewater limitations and requirements for monitoring, notification, and reporting. Of the three types of limitations to which dischargers are subject, the first type are discharge prohibitions, which include both numeric and narrative restrictions. There is a general prohibition against any discharge that causes pass-through or interference. In addition, there are several specific prohibitions. It is prohibited to discharge a wastewater that can create a fire or explosion hazard; the flash point must always be 140°F (60°C) or greater. Corrosive wastewaters are prohibited, including any wastewater with a pH below 5. High-temperature wastewaters are prohibited if they will inhibit the treatment facility's biological processes or raise the temperature at the headworks above 104°F (40°C). Also prohibited are wastewaters that can obstruct flow in the sewer, contain petroleum or mineral oils that will cause interference, or generate emissions that endanger the health or safety of workers. The last specific prohibition is against trucked- or hauled-in wastewaters, unless the POTW has approved the discharge at a designated point into the treatment facility.

The second type of limitations are the national categorical pretreatment standards for specific industrial categories. These limitations are detailed in subchapter N of 40 CFR Chapter I and are included in the pretreatment regulations only by reference. Users subject to this type of limitation are called *categorical users* and, by definition, are considered SIUs.

Categorical limitations for a particular categorical user may be modified by the combined wastestream formula, removal credits, intake credits, or a fundamentally different factors variance. The combined wastestream formula is used where a categorical wastewater is combined with a noncategorical wastewater. The discharge limit is calculated by flow-weighting the concentration or mass of pollutant in each waste stream; because the noncategorical waste streams usually have lower levels of the pollutant, the combined discharge limit is typically lower than the categorical limit. A POTW can increase a categorical limit by granting a removal credit; the credit is based on amount of the pollutant the POTW can remove (degrade or settle out) within its own treatment facility. A categorical limit can also be increased by an intake credit when the pollutant is found in the user's water supply; the limit can be increased up to the amount of the intake water. A variance granted from a categorical limit because a user is fundamentally different from facilities used to develop the categorical standard may be requested by the user or other interested party; the variance can either lower or raise the categorical limit.

Local limits are the third type of limitation. As the term implies, local limits are specific to each POTW and are based on the POTW's treatment capabilities and regulations that affect the POTW's operation. A local limit for a pollutant must satisfy all of the following criteria, as applicable: the POTW's wastewater permit limit, the water quality standard in the receiving water, prevention of treatment process upset or inhibition, sludge (biosolids) disposal limit, and protection of worker safety and health. The lowest concentration that satisfies all of these criteria becomes the local limit for that pollutant.

All SIUs are required to sample and analyze (self-monitor) their discharges. SIUs must submit self-monitoring reports summarizing their results at least once every six months. In addition to these periodic six-month reports, categorical SIUs are required to submit a one-time baseline monitoring report (BMR) and a categorical compliance deadline report. A BMR is required after a federal categorical pretreatment standard becomes effective; its purpose is to provide a baseline description of the discharge flow and pollutants. The compliance deadline report is similar to a BMR but must be submitted 90 days after the date for final compliance with the categorical standard. Self-monitoring and reporting for users who are not SIUs are determined based on the type and size of discharge.

Users are required to provide notification of certain events to the POTW. A user must notifiy the POTW immediately if the user's discharge could cause problems, such as pass-through or interference. A user must notify the POTW prior to making a substantial change in the volume or character of its discharge. If a user unintentionally bypasses any part of its pretreatment facilities, the user must notify the POTW orally within 24 hours, followed by a written report within 5 days. If a user anticipates a bypass, for example, for equipment maintenance, the user should notify the POTW at least 10 days before, if possible. Categorical users must notify the POTW of violations of a categorical limit within 24 hours. A user must also notify the POTW of any discharge that would be a hazardous waste if it were not exempted by the domestic sewage exclusion (DSE) within 180 days after the discharge starts.[1]

[1]Under the DSE, a mixture of domestic sewage and other waste that passes through the sewer to a POTW is not a solid waste under RCRA and, therefore, cannot be defined as a hazardous waste under RCRA. The domestic sewage exclusion is included in RCRA because the federal pretreatment program was thought to adequately regulate the discharge of hazardous nondomestic waste to POTW's, and requiring additional pretreatment under RCRA would be redundant.

Users discharging 15 kilograms per month or less of nonacute hazardous wastes are exempt from this notification.

Development of Local Limits. Local limits on users' discharges are based on the characteristics of the individual POTW. Such characteristics normally include the types of wastewater treatment and sludge management processes, effluent limits in the POTW's wastewater discharge permit, water quality standards for the water body that receives the POTW's discharge, sludge disposal regulations, and worker health and safety standards (EPA 1987).

Each POTW must develop local limits or demonstrate that they are not necessary; however, a POTW is unlikely to be able to exempt all pollutants. At a minimum, EPA expects the POTW to develop local limits for cadmium, chromium, copper, lead, nickel, and zinc because they are common to industrial discharges and have the potential to impact the POTW. In addition to these six metals, EPA recommends that the POTW develop local limits for arsenic, cyanide, mercury, and silver. Beyond these 10 basic pollutants, the POTW uses its own judgment to add any other pollutants to its local limits list.

REFERENCES

Environmental Protection Agency (1987). *Guidance Manual on the Development and Implementation of Local Discharge Limits Under the Pretreatment Program,* EPA/833/B87/202, U.S. EPA, Washington, DC.

——— (1991). *Technical Support Document for Water Quality-Based Toxics Control,* EPA/505/2-90-001, U.S. EPA, Washington, DC.

——— (1993). *Storm Water Management and Technology,* Noyes Data Corporation, Park Ridge, NJ.

——— (1994a). *EPA's Contaminated Sediment Management Strategy,* EPA/823/R-94/001, U.S. EPA, Washington, DC.

——— (1994b). *Water Quality Standards Handbook,* EPA/823/B-94/005a, U.S. EPA, Washington, DC.

——— (1995). *Water Quality Guidance for the Great Lakes System: Supplementary Information Document (SID)* EPA/820/B-95/001, U.S. EPA, Washington, DC.

BIBLIOGRAPHY

Environmental Protection Agency (1985). *Guidance Manual for Implementing the Total Toxic Organics (TTO) Pretreatment Standards,* EPA/440/185/009g, U.S. EPA, Washington, DC.

——— (1985). *Guidance Manual for Preparation and Review of Removal Credit Applications,* EPA/833/B85/200, U.S. EPA, Washington, D.C.

——— (1985). *Guidance Manual for the Use of Production-Based Standards and the Combined Wastestream Formula,* EPA/833/B85/201, U.S. EPA, Washington, DC.

——— (1985). *RCRA Information on Hazardous Wastes for Publicly Owned Treatment Works,* EPA/833/B85/202, U.S. EPA, Washington, DC.

——— (1987). *Guidance Manual for the Identification of Hazardous Wastes Delivered to POTWs by Truck, Rail, or Dedicated Pipe,* U.S. EPA, Washington, DC.

——— (1987). *Guidance Manual for Preventing Interference at POTWs,* EPA/833/B89/201, U.S. EPA, Washington, DC.

——— (1989). *Guidance for Developing Control Authority Enforcement Response Plan,* U.S. EPA, Washington, DC.

——— (1989). *Industrial User Permitting Guidance Manual,* EPA/833/B89/001, U.S. EPA, Washington, DC.

——— (1990). *Guidance Manual for POTWs to Calculate the Economic Benefit of Noncompliance,* U.S. EPA, Washington, DC.

——— (1991). *Control of Slug Discharges to POTW's Guidance Manual,* EPA/21W/4003, U.S. EPA, Washington, DC.

——— (1991). *Storm Water Guidance Manual for the Preparation of NPDES Permit Applications for Storm Water Discharges Associated with Industrial Activity,* Government Institutes, Rockville, MD.

——— (1991). *Supplemental Manual on the Development and Implementation of Local Discharge Limitations Under the Pretreatment Program: Residential and Commercial Toxic Pollutant Loadings and POTW Removal Efficiency Estimation,* EPA/21W/4002, U.S. EPA, Washington, DC.

——— (1992). *Guidance to Protect POTW Workers from Toxic and Reactive Gases and Vapors,* EPA/812/B92/001, U.S. EPA, Washington, DC.

——— (1992). *Model Pretreatment Ordinance,* EPA/833/B92/003, U.S. EPA, Washington, DC.

——— (1992). *Storm Water Management for Industrial Activities: Developing Pollution Prevention Plans and Best Management Practices,* Government Institutes, Rockville, MD.

——— (1993). *Guidelines for Deriving Site-Specific Sediment Quality Criteria for the Protection of Benthic Organisms,* EPA/822/R-93/017, U.S. EPA, Washington, DC.

——— (1993). *Guides to Pollution Prevention, Municipal Pretreatment Programs,* EPA/625/R-93/006, U.S. EPA, Washington, DC.

——— (1993). *Technical Basis for Deriving Sediment Quality Criteria for Nonionic Organic Contaminants for the Protection of Benthic Organisms by Using Equilibrium Partitioning,* EPA/822/R-93/011, U.S. EPA, Washington, DC.

——— (1994). *Industrial User Inspection and Sampling Manual for POTWs,* EPA/831/B94/001, U.S. EPA, Washington, DC.

Federal Register (1990). "40 CFR Parts 122, 123, and 124, National Pollutant Discharge Elimination System Permit Application Regulations for Storm Water Discharges; Final Rule."

——— (1992). "Final NPDES General Permits for Storm Water Discharges Associated with Industrial Activity; Notice."

——— (1992). "Final NPDES General Permits for Storm Water Discharges from Construction Sites; Notice."

——— (1995). "Final National Pollutant Discharge Elimination System Storm Water Multi-sector General Permit for Industrial Activities; Notice."

15

AIR QUALITY ASSESSMENT AND CONTROL

The Clean Air Act (CAA) Amendments of 1990 touch on virtually every aspect of air pollution law. Mobile and stationary sources, large and small businesses, routine and toxic emissions, and consumer products are all regulated under the CAA. In addition to regulating air pollution sources, it is the intent of the act to focus on areas of poor air quality, particularly large cities that have had serious air pollution problems of many years. Air emissions are regulated by EPA through the Clean Air Act and the Resource Conservation and Recovery Act (RCRA) in 40 CFR Parts 0–99 and 40 CFR Part 264, Subparts AA and BB, respectively. In this chapter we focus on air emissions modeling, sampling methods, and air pollution control technologies.

MODELS FOR ASSESSING IMPACTS TO AIR QUALITY FROM INDUSTRIAL SOURCES

Numerous models are used for assessing air quality impacts from industrial sources. These models typically assess contaminant concentrations at the facility property line and other selected receptor points. Features of a typical air model include:

- Polar or Cartesian coordinate systems for receptors
- Rural or urban options, with a choice of three urban options
- Plume rise due to momentum and buoyancy as a function of downwind distance for stack emissions
- Evaluation of building wake effects with an option of using one of two downwash calculation methods
- Procedures for evaluation of stack-tip downwash

- Consideration of the effects of gravitational settling and dry deposition on ambient particulate concentrations and capability to calculate dry deposition
- Consideration of wind profile, including a procedure for calm-wind processing
- Concentration estimates for periods of 1-hour to annual averaging periods
- Adjustment procedures for elevated terrain
- Consideration of time-dependent exponential decay of pollutants
- Consideration of buoyancy-induced dispersion
- A default option to use EPA-established options and parameters
- Capability to treat height of receptor aboveground ("flagpole" receptors)

Typical input parameters for air emissions models are listed in Table 15.1. The EPA sponsors an electronic bulletin board service through the Office of Air Quality Planning and Standards that makes EPA-approved models available to the public.

Specialized models that predict effects of specific source emissions such as gas and liquid jet releases and dense gas releases at grade are documented (Hanna and Drivas 1987). Incidents such as that at Bophal have emphasized the need for facilities to model potential effects from accidental releases of toxic gases and other chemicals in order to perform process risk assessments. In addition, Local Emergency Planning Committees (LEPCs) and State Emergency Response Commissions (SERCs) must consider worst-case accidental releases in emergency plans, and models such as CHARM (TM) and others have been used for these purposes. The topics of hazard assessment and emergency planning are explored further in Chapters 18 and 19.

AIR MONITORING

Industrial Source Monitoring

Unlike source emissions and transport modeling, which predicts concentrations of contaminants at facility boundaries or other receptors, emissions monitoring of industrial sources will assess actual emissions during facility operations. There are basically two types of emissions from an industrial facility: point source emissions and fugitive emissions. For many years, the main focus of air pollution management was on point source emissions, and much attention was given to monitoring these sources. However, when the toxic release inventory (TRI) reports submitted by facilities in the late 1980s showed fugitive emissions to be a substantial portion of releases to the atmosphere, these emissions came under focus.

Point Source Emissions Monitoring. EPA has specified approved monitoring methods for point sources in 40 CFR Part 60, Appendix A. Several of these

TABLE 15.1 Typical Input Required for Air Emissions Models

Meteorological data	Source data
Wind velocity	Physical characteristics
Temperature	Chemical characteristics
Relative humidity	Geometry of source
Turbulence	Release rates
Net radiation flux	Receptor information
Site information	Receptor locations
Topography	Distance between receptors
Equipment operating parameters	
Property boundaries	

methods are described in Table 15.2. Examples of several sampling trains are shown in Figures 15.1 to 15.4.

A series of samples may be taken at the inlet and outlet of abatement equipment to ascertain equipment efficiencies as well as contaminant emissions during specific operational modes such as startup, full production, and emergency shutdown. In addition, unabated stack emissions may require monitoring. The major drawback to stack sampling using EPA-approved methods is that the data gathered are from a noncontinuous, one-time event. Since the sampling is noncontinuous, operational changes or system efficiency problems

TABLE 15.2 Selected EPA-Approved Monitoring Methods

Method	Description/Applications
2	Uses a type S Pitot tube to determine stack velocity and volumetric flow rate
4	Uses sampling train with probe heater, condenser, and vacuum system to determine moisture content of stack gases
5	Uses sampling train with a glass wool filter and impingers to collect particulates for analysis; method can be modified to collect volatile organic compounds
6C and 7E	Allow for use of a gas analyzer to measure concentrations of sulfur dioxide and nitrogen oxides emissions, respectively, from stationary sources
12	Allows for use of inorganic lead sample train to collect samples for analysis by atomic absorption spectrophotometer
17	Allows for in-stack filtration (collection) of particulates
21	Allows for determination of volatile organic compound leaks
25A	Allows for determination of total gaseous organic concentration using a flame ionization analyzer

Source: Information from 40 CFR Part 60, Appendix A.

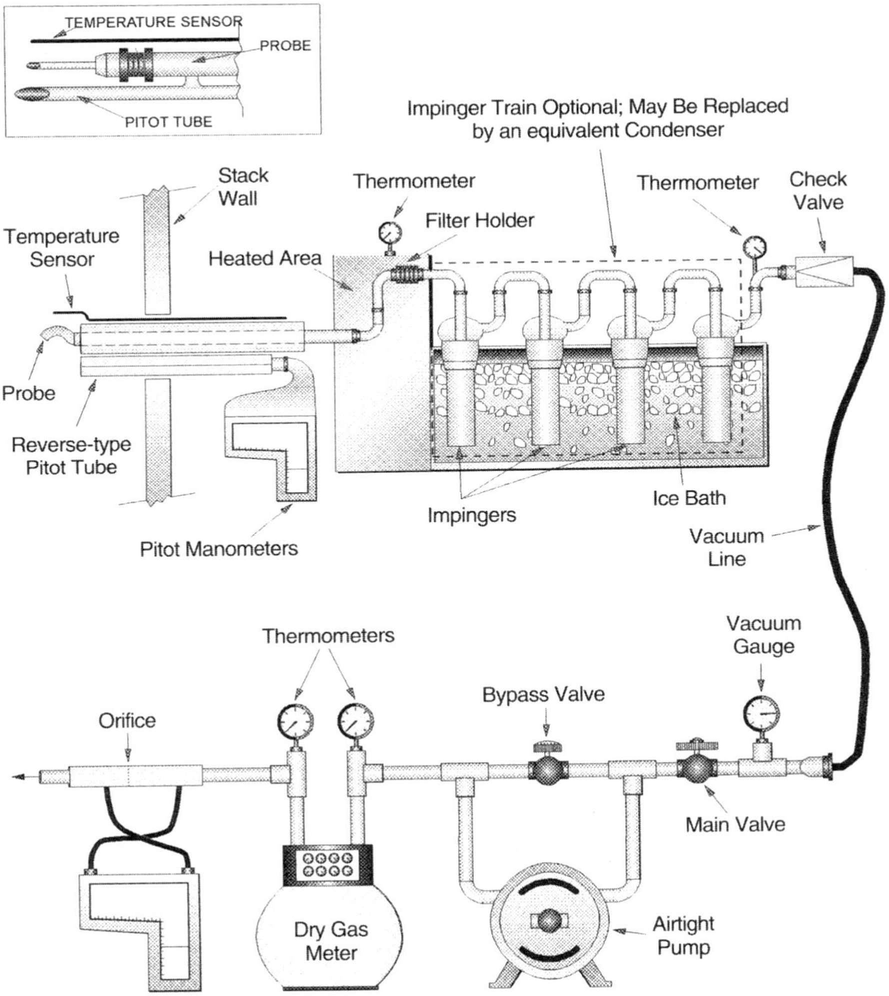

Figure 15.1 Particulate sampling train.

may not be detected until the next stack sampling event. Because of sampling expense, it is not uncommon to sample major stacks only annually or biennially, or as specified in the facility's permit.

On-line monitors often are used to bridge data gaps between stack sampling events. These monitors give real-time emissions data and can be installed with programmable logic control system that will sound an alarm at higher-than-normal emissions. Some on-line monitors, such as flame ionization detectors (FIDs), analyze an airstream for a total concentration, in this case total

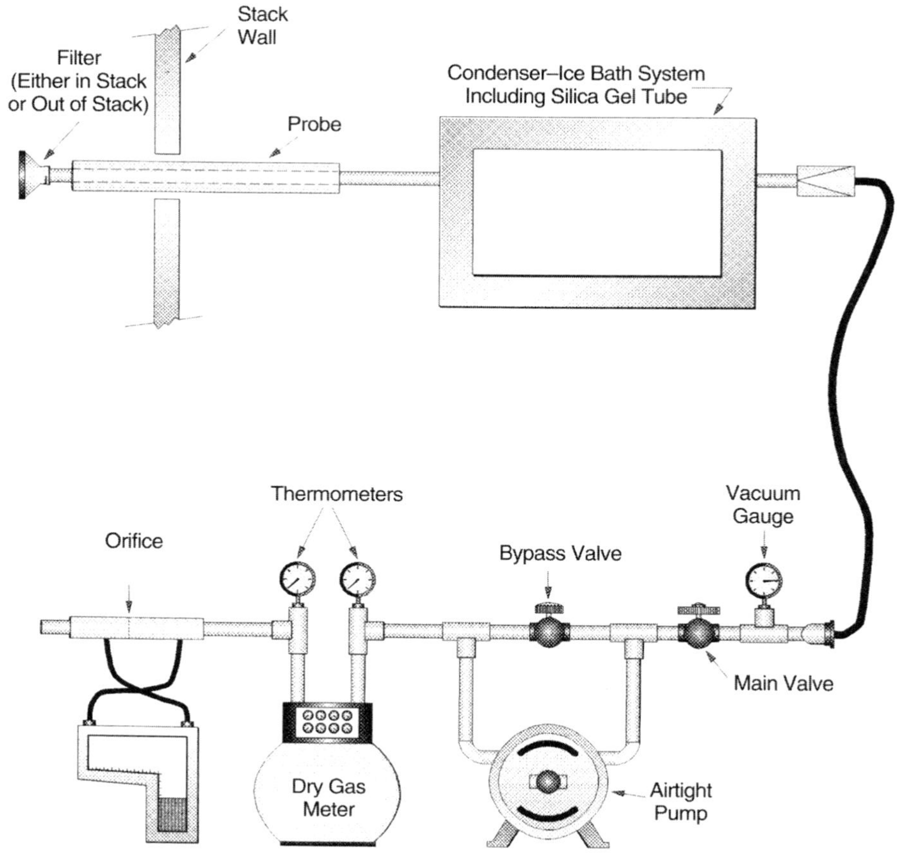

Figure 15.2 Moisture sampling train.

hydrocarbons. Other monitors, such as oxygen or carbon monoxide monitors, can analyze for a particular chemical. Many on-line analyzers described in Chapter 5 can be used for continuous monitoring of stacks.

Fugitive Emissions Monitoring. Fugitive emissions are more difficult to monitor since they are not captured and released through a stack or vent but, instead, are dispersed into the air from small sources spread throughout a manufacturing facility. Industrial sources for fugitives include leaking valves and connections, pumps, closed but leaking vents, door seals, individual cleaning operations that use small bottles of volatile solvents, poorly covered or uncovered process equipment, and poorly enclosed or inadequately vented spraying operations. Sources for fugitive emissions such as valves and connections can be monitored with portable instruments such as an organic vapor

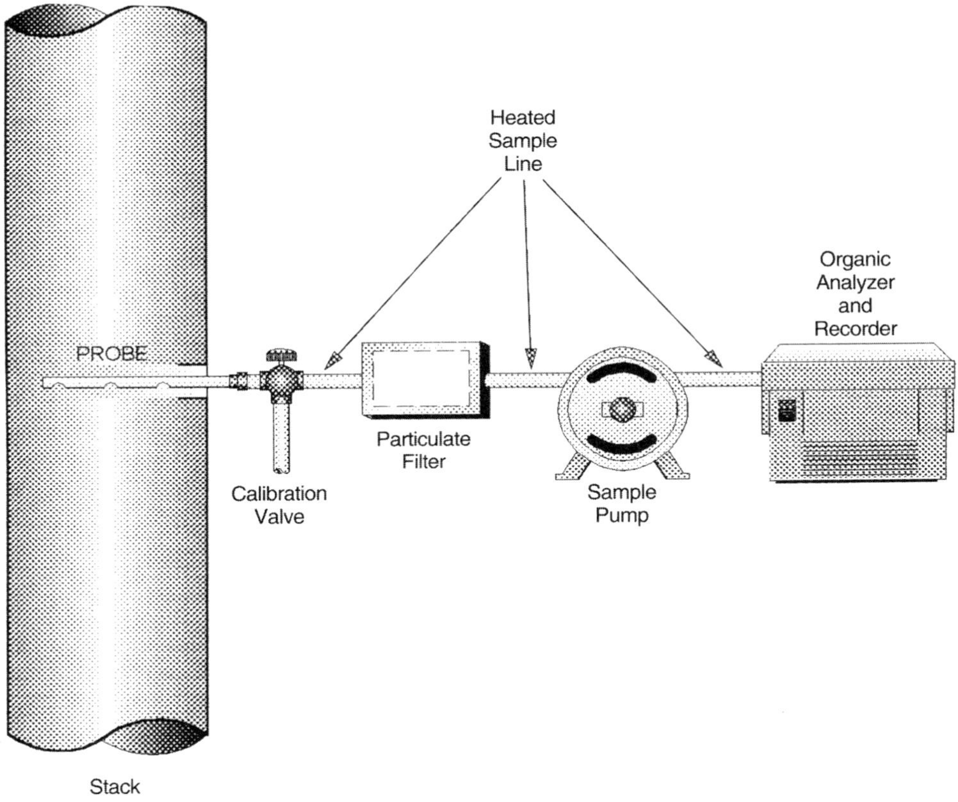

Figure 15.3 Organic concentration measurement system.

analyzer (OVA), a portable FID, or a photoionization detector. EPA Method 21 provides an accepted method for determination of volatile organic compound leaks, particularly from valves, flanges, other connections, and pumps. Fugitives from industrial operations often can be quantified through workplace sampling using a portable infrared spectrometer (IR) or diffusional monitors.

Ambient Air Monitoring

In addition to industrial emission sources, air quality can be affected by other sources, such as mobile combustion sources, power-generating facilities, municipal and other incinerators, and construction activities. Thus ambient air monitoring is performed to obtain data about the quality of the air in an entire region. Key parameters that are used to determine the general state of ambient air quality include carbon dioxide and nitrogen dioxide (greenhouse gases), ozone, carbon monoxide, nitrogen oxides, sulfur compounds, nonmethane hy-

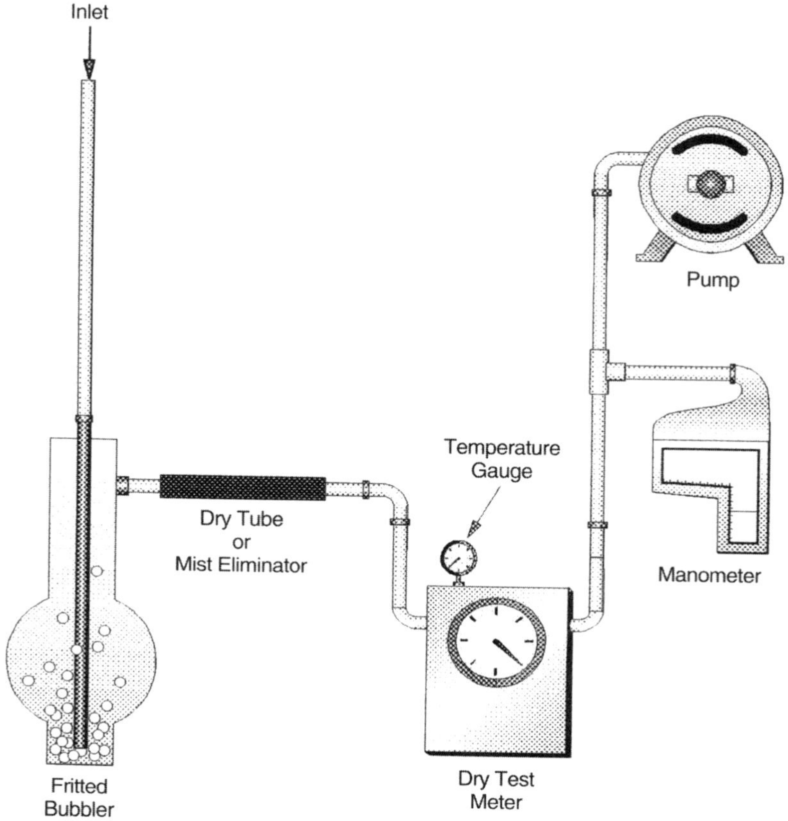

Figure 15.4 Nitrogen dioxide sampling train.

drocarbons, particulate matter, including total suspended particulate matter and particulate matter of less than 10 μm and less than 2.5 μm in aerodynamic diameter (PM_{10} and $PM_{2.5}$, respectively), and detectable concentrations of toxic pollutants, including organics and metals. In addition, in some regions acid rain and the deposition of acidic compounds on soils are key indicators of pollution.

Methods of monitoring for the toxic organic compounds in ambient air are documented by EPA (1988), and selected test methods and applicabilities of these are presented in Table 15.3. In addition, examples of commonly used analytical methods and applications for monitoring ambient air are presented in Table 15.4.

Other analytical methods for analyzing compounds in ambient air are as follows:

- *Ultraphotometric ozone analysis:* used for continuous analysis of ozone

TABLE 15.3 EPA Test Methods and Applicabilities for Sampling Toxic Organic Compounds in Ambient Air

Method TO1

Description: Tenax gas chromatograph (GC) adsorption and gas chromatograph/mass spectrometer (GC/MS) analysis
Type of compounds determined: volatile nonpolar organics with boiling points in the range 80 to 200°C
Examples of specific compounds determined: benzene, carbon tetrachloride, chlorobenzene, chloroform, chloroprene, 1.4-dichlorobenzene, ethylene dichloride, methyl chloroform, nitrobenzene, perchlorethylene, toluene, trichloroethylene, and *o*-, *m*-, and *p*-xylene

Method TO2

Description: carbon molecular sieve adsorption and GC/MS analysis
Type of compounds determined: highly volatile nonpolar organics with boiling points in the range −15 to +120° C
Examples of specific compounds determined: acrylonitrile, allyl chloride, benzene, carbon tetrachloride, chloroform, ethylene dichloride, methyl chloroform, methylene chloride, toluene, trichloroethylene, vinyl chloride, and vinylidene chloride

Method TO3

Description: cryogenic trapping and GC/FID or ECD analysis
Type of compounds determined: volatile nonpolar organics with boiling points in the range −10 to +200°C
Examples of specific compounds determined; acrylonitrile, allyl chloride, benzene, carbontetrachloride, chlorobenzene, chloroform, 1.4-dichlorobenzene, ethylene dichloride, methyl chloroform, methylene chloride, nitrobenzene, perchlorethylene, toluene, trichloroethylene, vinyl chloride, vinylidene chloride, and *o*-, *m*-, and *p*-xylene

Method TO4

Description: high-volume PUF sampling and GC/ECD analysis
Types of compounds determined: organochlorine pesticides and PCBs
Examples of specific compounds determined: 4,4′ -DDE and 4,4′ - DDT

Method TO5

Description: dinitrophenylhydrazine liquid impinger sampling and HPLC/UV analysis
Types of compounds determined: aldehydes and ketones
Examples of specific compounds determined: acetaldehyde, acrolein, benzaldehyde, and formaldehyde

Method TO6

Description: HPLC analysis
Compound determined: phosgene

TABLE 15.3 (***Continued***)

Method TO7

Description: thermosorb/N adsorption
Compound determined: N-nitrosodimethylamine

Method TO8

Description: sodium hydroxide liquid impinger with high-performance liquid chromatography
Compounds determined: cresol and phenol

Method TO9

Description: high-volume PUF sampling with HRGC/HRMS analysis
Compound determined: dioxin

Method TO10

Description: low-volume PUF sampling with GC/EDC analysis
Type of compounds determined: pesticides
Examples of specific compounds determined: Alochlor 1242, 1254, and 1260, captan, chlorothalonil, chlorpyrifos, dichlorovos, dicofol, dieldrin, endrin, endrin aldehyde, folpet, heptachlor, heptachlor epoxide, hexachlorobenzene, α-hexachlorocyclohexane, hexachlorocyclopentadiene, lindane, methoxychlor, mexacarbate, mirex, *trans*-nonachlor, oxychlordane, pentachlorobenzene, pentachlorphenol, *p,p'*-DDE, *p,p'*-DDT, ronnel, 1,2,3,4-tetrachlorobenzene, 1,2,3-trichlorobenzene, 2,4,5-trichlorophenol

Method TO11

Description: adsorbent cartridge followed by HPLC detection
Type of compounds determined: aldehydes
Examples of specific compounds determined: acetaldehyde, acrolein, butyraldehye, crotonaldehyde, 2,5,-dimethylbenzaldehyde, formaldehyde, hexanaldehyde, isovaleraldehyde, propionaldehyde, *o*-toluene, *m*-toluene, *p*-toluene, and valeraldehyde

Method TO12

Description: cyrogenic PDFID
Compounds determined: nonmethane organic compounds

Method TO13

Description: PUF/XAD-2 adsorption with GC and HPLC detection
Type of compounds determined: polynuclear aromatic hydrocarbons
Examples of specific compounds determined: acenaphthene, acenaphthylene, anthracene, benzo[a]anthracene, benzo[a]pyrene, benzo[b]fluoranthene, benzo[e]pyrene, benzo([g.h.i.]perylene, benzo[k]fluoranthene, chrysene, dibenzo[a,h]anthracene, fluoranthene, fluorene, indeno[1,2,3-cd]pyrene, naphthalene, phenanthrene, and pyrene

TABLE 15.3 ***(Continued)***

Method TO14

Description: SUMMA passivated canister sampling with gas chromatography
Types of compounds determined: semivolatile and volatile organic compounds
Examples of specific compounds determined: acenaphthene, acenaphthylene, benzene, benzylchloride, carbon tetrachloride, chlorobenzene, chloroform, 1,2-dibromomethane, 1,2-dichlorobenzene, 1,3-dichlorobenzene. 1,4-dichlorobenzene, 1,1-dichloroethane, 1,2-dichloroethylene, 1,2-dichloropropane, 1,3-dichloropropane, ethyl benzene, ethyl chloride, ethylene dichloride, 4-ethyl toluene, Freon 11, Freon 12, Freon 113, Freon 114, methyl benzene, methyl chloride, methyl chloroform, perchlorethylene, 1,1,2,2-tetrachlorobenzene, toluene, 1,2,3-trichlorobenzene, 1,2,4- trichlorobenzene, 1,1,2-trichlorobenzene, trichloroethylene, 1,2,4-trimethylbenzene, 1,3,5-trimethylbenzene, vinyl benzene, vinyl chloride, vinylidene chloride, and *o*-, *m*-, and *p*-xylene

Source: Information from EPA (1988).

Note: Acronyms used are as follows: ECD, electron capture detection; GC, gas chromatography; GC/MS, gas chromatography/mass spectrometry; GC/FID, gas chromatography/flame ionization detection; HPLC, high-performance liquid chromatography; HRGC/HRMS, high-resolution gas chromatography/high-resolution mass spectrometry; PDIFD, preconcentration and direct flame ionization detection; PUF, polyurethane foam; UV, ultraviolet; XAD-2, type of resin adsorbent.

- *Nondispersive infrared spectrometry:* used for continuous analysis of carbon monoxide
- *Visible absorption spectrometry:* used for analyzing metals such as hexavalent chromium that are collected on a polyvinyl chloride membrane or equivalent filter
- *Inductively coupled plasma–atomic emission spectrometry:* used for analyzing metals that are collected on a mixed cellulose ester filter or equivalent
- *Constant-volume sampling analysis:* used to analyze carbon monoxide, carbon dioxide, and nitrogen oxides, generally from automobile emissions
- *Superfluid extraction/gas chromatography:* used to analyze volatile and semivolatile compounds collected on a sorbent bed
- *Colorimetric analysis:* used to analyze mercaptans collected with a midget bubbler with coarse porosity frit
- *Gravitation measurements:* used to analyze for dusts and other heavy particulates that are collected on filters
- *Optical sizing:* used for quantifying small particulates that are collected on filters

AIR POLLUTION ABATEMENT EQUIPMENT

Air pollution abatement equipment is used for sources that are regulated or that have the potential to emit significant amounts of toxic or other com-

TABLE 15.4 Selected Examples of Analytical Methods Used for Monitoring Ambient Air Quality

Analytical Method	Application
Gas chromatograph/mass spectrometer (GC/MS)	Used for analyzing organic compounds collected in a Tedlar bag, charcoal filter tube, or air canister
Preconcentration direct flame ionization detector (PDFID)	Used for analyzing nonmethane hydrocarbons
Ultraviolet–photometeric ozone analyzer	Used for continuous analysis of ozone
Nondispersive infrared spectrometry	Used for continuous analysis of carbon monoxide
Visible absorption spectrometry (VAS)	Used for analyzing metals such as hexavalent chromium that are collected on a PVC membrane or equivalent filter
Inductively coupled plasma–atomic emission spectrometry	Used for analyzing metals that are collected on a mixed cellulose ester filter or equivalent
Constant-volume sampling (CVS) analyzer	Used to analyze carbon monoxide, carbon dioxide, and nitrogen oxides, generally from automobile emissions
Superfluid extraction/gass chromatograph	Used to analyze volatile and semi-volatile compounds collected on a sorbent bed
Colorimetric analysis	Used to analyze mercaptans collected with a midget bubbler with coarse porosity frit
Gravitation measurements	Used to analyze for dusts and other heavy particulates that are collected on filters
Optical sizing	Used for quantifying small particulates that are collected on filters

Source: Information from ASTM (1992), NIOSH (1984), and EPA (1988).

pounds into the atmosphere. Examples of types of air pollution abatement equipment include gravitational and inertial separation, filtration, gas–solid adsorption, electrostatic precipitators, liquid scrubbers, combustion units, and combination systems (Clayton and Clayton 1991).

Gravitational and Inertial Separation

Gravitational and inertial separators are typically used for particulate removal. Also known as *mechanical collectors,* these abatement units use gravity or inertia to remove relatively large particles (i.e., 50 μm or greater) from suspension in a moving gas stream. Abatement unit types include settling chambers, inertial separators, dynamic separators, and cyclones.

Settling Chambers. This type of air abatement equipment represents the oldest type of air pollution control device. Essentially, a settling chamber uses the forces of gravity to separate large particles such as dust and mist from a gas stream. The device is typically used as a precleaning device before the gas stream is sent to a more energy intensive and sophisticated air pollution control device. An example of this type of emission control is presented in Figure 15.5.

Inertial Separators. Inertial separators are a simple and cost-effective air pollution control device; however, they are not particularly efficient, and consequently are used mainly as precleaning devices. Inertial separators include dry-type collectors that utilize inertia of particles to provide for particulate–gas separation.

Dynamic Separators. Dynamic separators are moving devices. Also termed *rotary centrifugal separators,* these devices use centrifugal force to separate particulates from the airstream. The particulates are concentrated on impellers, and the filter cake is dropped into a hopper.

Cyclones. Cyclones are a simple and economical mechanical collector. There are three major elements of a cyclone: a gas inlet that produces the vortex; an outlet for discharged, cleaned gas; and a discharge for particulates (dust). An example of a cyclone that has an tangential inlet and an axial discharge is presented in Figure 15.6.

Wet Cyclones. Although typically, cyclones are thought of as dry mechanical devices, there are also wet cyclones, which have an inlet for water or some other fluid to be contacted by incoming particles. The fluid rinses the particles from the airstream. Although the cost to operate this type of cyclone is more than that to operate a dry cyclone, the efficiencies are typically greater.

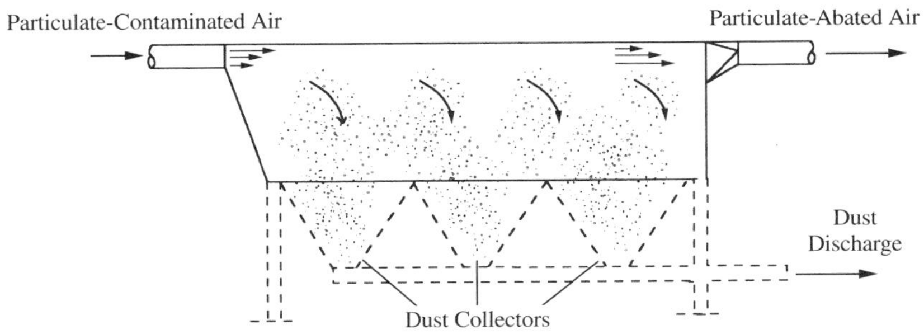

Figure 15.5 Schematic of a settling chamber.

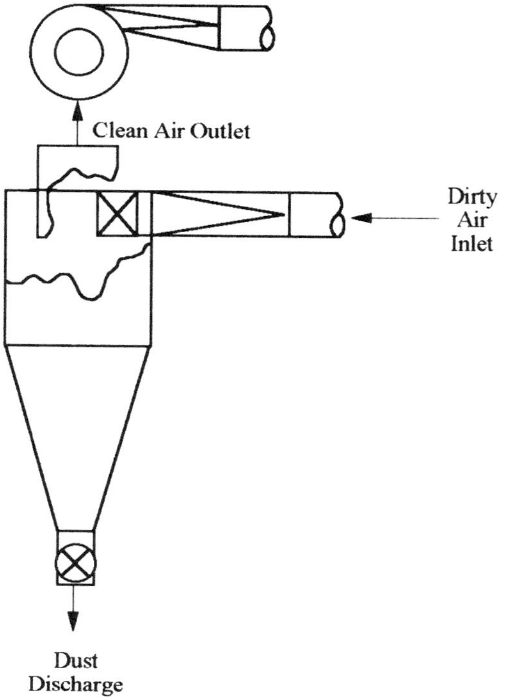

Figure 15.6 Schematic of a cyclone.

Multiple Cyclones. Multiple cyclones consist of several cyclones—usually in parallel, and infrequently, in series. Actually, this configuration is not typical, but might be used in cases where there is heavy dust load and a single cyclone would prove inefficient.

Filtration

Filtration uses a filtering media for removal of particulate matter from gas streams by retention of particles in and around the pores of the filter. Air emissions filtration devices include woven fibrous mats and aggregate beds; paper filters; and fabric filters. Filtration units are typically considered reliable and inexpensive to operate.

Fibrous Mats and Aggregate Beds—General. These types of filtration devices have a high porosity, with predominant forces for filtration being impaction, impingement, and surface attraction. Mats may be made of glass fibers, stainless steel, brass, and aluminum. Efficiencies can be very high, even with low dust loadings in the input airstream.

Designs for aggregate beds are numerous. Sand is often used, although glass fiber beds have been demonstrated. Additionally, many designs are available for self-cleaning the beds or filters. An example of a self-cleaning mat that uses a water spray is shown in Figure 15.7. This type of filter, commonly termed a *wet filter,* is sprayed continuously while the airstream flows through the filter. The particles are dislodged by the water and flow with it into a sump.

Paper Filters. These filters are typically used for applications that require ultrahigh efficiencies, such as clean rooms in semiconductor manufacturing or other equivalent applications in the aerospace industry, hospitals, and data processing centers. Although termed "paper," these filters can be made of glass microfibers, minerals, or other materials. Sizes and shapes vary depending on the application. Diffusion is the common capture mechanism for this type of filter, and efficiencies are typically as high as 99.97%.

Fabric Filters. Fabric filters, also termed *baghouses,* are suitable for removing particles as small as 0.5 μm. The filters are typically made of woven or felted synthetic fabric. The particle-ladened gas passes through the filter and the particles are collected through several mechanisms. These include inter-

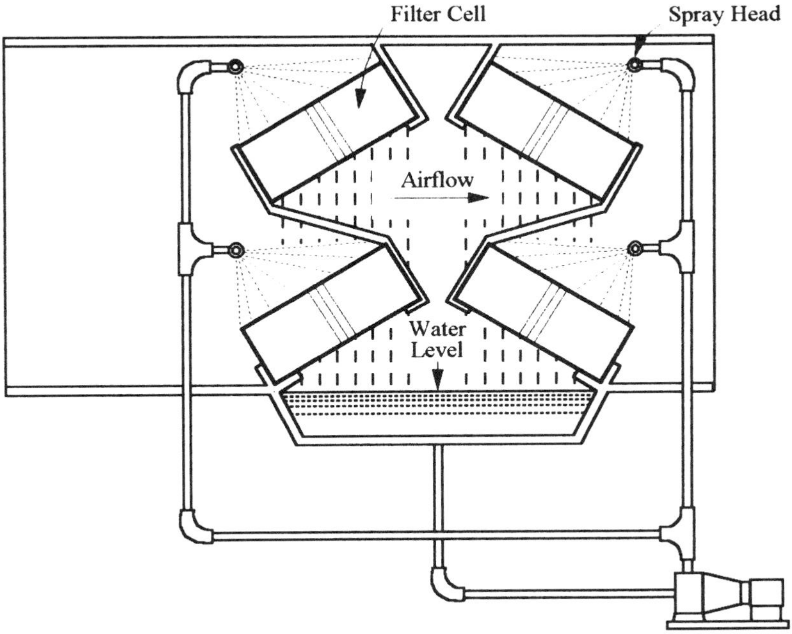

Figure 15.7 Schematic of a filtration unit. (From Clayton and Clayton, 1991.)

ception and impaction of the particle on the fabric filters, diffusion, electrostatic attraction, and gravitational setting within the filter pores. Once a cake is built up, there is a mechanism such as a shaker or air pressure to remove the cake, which falls into a hopper or other containment. An example of this technology is presented in Figure 15.8.

Gas–Solid Adsorption

Adsorption is a diffusion process whereby certain gases are retained selectively on the surface or in the pores of a selected solid. Typical materials used in this method of air pollution abatement include activated carbon, alumina, bauxite, and silica gel; activated carbon is the most popular type because of its low cost and its ability to be regenerated. In addition, the micropore structure of the carbon allows for a large surface area for a given weight of carbon.

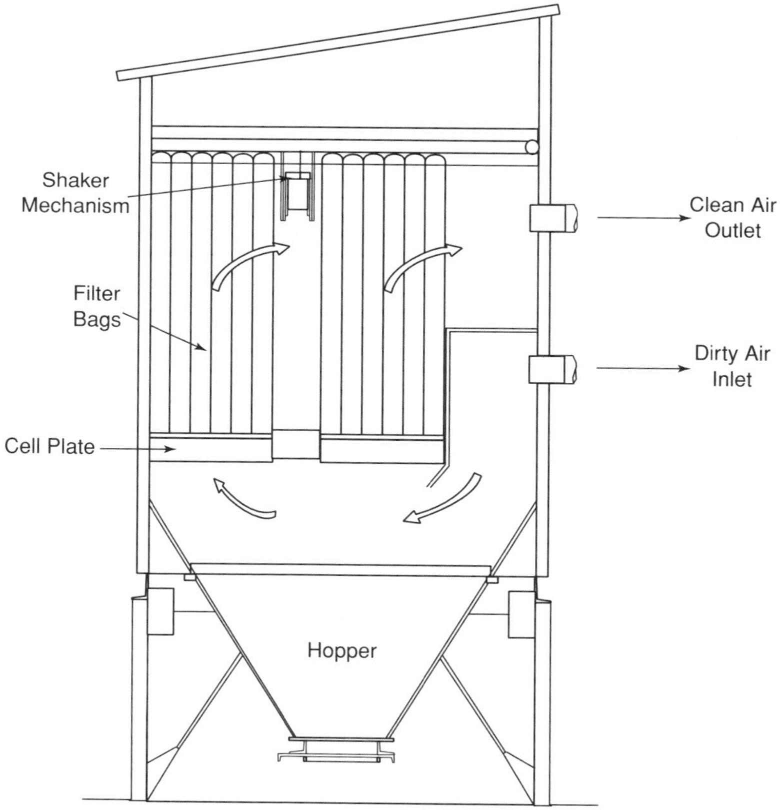

Figure 15.8 Schematic of a fabric filter.

Types of gas–solid adsorption include physical adsorption, polar adsorption, and chemical adsorption.

Physical Adsorption. This process involves a stream of gas passing through the adsorption material, the result being adsorbent molecules in the gas stream being "driven" from the relatively homogeneous gas phase into the pores of the material. At the surface of the adsorbing solid, molecules are attracted and deposited at a rate that depends on concentration, ease of saturation of the material, and residence time. Once the solvent adsorbed on the sorbent material reaches a saturation point, breakthrough occurs.

Polar Adsorption. Polar adsorption is a process whereby molecules of a given adsorbate are attracted to and deposited on the surface of an adsorbing solid by virtue of the polarity of the adsorbate. The medium is typically siliceous adsorbents, which provide for selective removal of gas molecules from an air stream. Some synthetic materials of uniform pore sites have the ability to segregate molecules based on their shape.

Chemical Adsorption. Chemical adsorption takes place when a chemical reaction occurs between the adsorbed molecule and the solid adsorbent, resulting in the formation of a chemical compound. Typically, in this type of adsorption the forces holding the adsorbate to the solid are so strong that the reaction is irreversible, so that on-site regeneration is usually not an option.

Electrostatic Precipitation

During the electrostatic precipitation process, particulates are separated from a gaseous stream using electrostatic forces. As with several other types of air pollution control devices, the particulates are deposited on solid surfaces for subsequent removal. Removal of particles 20 micron (μm) or less (with removals to the submicro level, depending on the device) are common. Airstreams that have particles greater than 20 μm must first pass through a mechanical precleaning device, such as a settling chamber or cyclone, before being sent to the electrostatic precipitation unit. The devices are often operated in sequential stages to afford very high efficiencies.

The mechanisms involved in electrostatic precipitation include:

- Electrically charging entrained particulate matter in the first stage of the unit
- Providing a voltage gradient to propel the charged particulate matter onto a grounded surface, on which it is collected
- Removal of the particulate matter from the collection surface, generally through "rapping" of the surface at selected points

Plate-Type Precipitators. This type of electrostatic precipitator uses a series of parallel, vertical grounded steel plates through which the gas flows between. Spacing, size, and shape of the plates is dependent on application.

Pipe-Type Precipitators. This type of precipitator is typically configured with high-voltage electrodes positioned at the centerline of a grounded pipe. Dust is deposited on the inner walls of the grounded pipe. Sizing of the unit is based on airstream parameters such as particle size and air velocity as well as dust resistivity and total collection area.

Liquid Scrubbing

During liquid scrubbing, a suitable liquid is used to remove contaminants such as particulates, chlorides, sulfur compounds, and other compounds from a gas stream. Removals can be in the submicrometer range. Scrubbing geometry, media, and scrubbing liquor are the design variables that must be optimized for optimal efficiency of the unit. The spent scrubbing liquor is typically sent for treatment before discharge.

Spray Chambers. Spray scrubbers are useful in collecting particulate matter. The device uses spray nozzles to atomize the liquid. Particles collide with the droplets and are entrained. Because of the simple design of the system, the scrubbing liquid can be recycled and still be effective, even with relatively high suspended solids concentrations. A mist eliminator is necessary for this type of device.

Packed Towers. Packed towers are essentially spray chambers that contain packing material. The packing material provides a high surface area for optimizing mass transfer of the particulates to the liquid. Packed towers are typically used to remove gaseous contaminants from airstreams. The packing may serve as a mist eliminator, so a special device for this purpose may not needed with some configurations.

Orifice Scrubbers. Orifice scrubbers use the carrier gas velocity and turbulence to promote dispersal of the scrubbing liquid and contact of the contaminated gas stream with the dispersed liquid. These devices make use of internal geometric designs that attempt to supply scrubbing liquid uniformly across the cross section of the gas flowing through the unit. In the turbulent contacting zone, the intensely turbulent motions provide violent contact between particulate matter and the larger scrubbing liquid droplets, which possess high inertia.

Although there are numerous geometries and internal designs for orifice scrubbers, the venturi scrubber, which operates at very high velocities, is a common design. This technology uses a venturi oriface (an inlet with a constricted throat to create pressure differntial) and a cyclone separator.

Combustion

Combustion, also known as *incineration,* converts matter entrained in exhaust gases to a less toxic form, principally carbon dioxide and water. Equipment of this type is mostly direct flame or catalytic combustion.

Direct Flame Applications

Flares. Flares are used for oxidation of combustible gases that are present in a stream that is within the range of flammability.Flares are typically used in petrochemical plants, refineries, chemical processing plants (for emergency situations), and research/pilot plants for these types of processes. Design features include gas stream parameters and velocity, addition of auxiliary fuel and/or air (optional), fuel–air mixing, ratio of hydrogen to carbon in the materials being combusted, flare height, heat and light emitted from the flare, and heat recovery.

Direct Flame Combustion. Direct flame combustors (incinerators) may or may not use auxiliary fuel and heat recovery to ensure efficient combustion to minimize operation costs. The key factors for the success of this type of combustor include sufficient residence time of the gas in the chamber, sufficient fuel, and sufficient mixing. Generally, a heat exchanger is used to preheat incoming gases to a temperature high enough to allow for as complete combustion as possible. Applications include removal of combustible particulate matter, particularly of particulate matter that is of submicro size.

Catalytic Combustion

Catalytic combustors use a catalyst composed of platinum, palladium, or similar metal to accelerate the rate of combustion and minimize the requirements for auxiliary fuel. A typical system includes a preheat burner, exhaust fan, catalyst, heat exchanger, and safety mechanisms. Catalitic combustors are used where concentrations of hydrocarbons are relatively low. In some cases, the airstream is incompatible with this method of treatment, because selected constituents can poison the catalyst and render it ineffective. Further, the systems are incompatible with airstreams containing particulates unless a precleaning device is incorporated into the system.

REFERENCES

American Society for Testing and Materials (1991). *Annual Book of ASTM Standards,* Vol. 11.03, "Atmospheric Analysis; Occupational Health and Safety," ASTM, Philadelphia.

Clayton, G. D., and F. E. Clayton, eds. (1991). *Patty's Industrial Hygiene and Toxicology,* Vol. 1, Part A, 4th ed., Wiley, New York.

Environmental Protection Agency (1987a). *Industrial Source Complex (ISC2) Dispersion Model User's Guide,* Vols. 1 and 2 (EPA/450/4-92/008a and EPA/450/4-92/008b). U.S. EPA, Washington, DC.

——— (1987b). *UNAMAP: User's Network for the Applied Modeling of Air Pollution,* EPA/600/D-87/330, prepared by T. E. Pierce and D. B. Turner, Atmospheric Sciences Research Laboratory, Office of Research and Development, Research Triangle Park, NC.

——— (1988). *Compendium of Methods for the Determination of Toxic Organic Compounds in Ambient Air,* EPA/600/4-89/017, Research Triangle Park, NC.

——— (1989). *Advanced Methodologies for Sampling and Analysis of Toxic Organic Chemicals in Ambient Outdoor, Indoor, and Personal Respiratory Air,* EPA/600/J-89/315, Atmospheric Research and Exposure Assessment Laboratory, Research Triangle Park, NC.

——— (1991). *National Air Quality and Emissions Trends Report,1990,* EPA 450/4-91-023, Office of Air Quality Planning and Standards, Research Triangle Park, NC.

Hanna, S. R., and P. J. Drivas (1987). *Guidelines for Use of Vapor Cloud Dispersion Models,* Center for Chemical Process Safety, American Institute of Chemical Engineers, New York.

National Institute for Occupational Safety and Health (1984). *NIOSH Manual of Analytical Methods,* 3rd ed., prepared by M. Millson and R. Delon Hull, NIOSH, Cincinnati, OH.

BIBLIOGRAPHY

Bernarde, M. (1992). *Global Warning...Global Warming,* Wiley, New York.

Bubenick, D. V., ed. (1984). *Acid Rain Information Book,* Noyes Data Corporation, Park Ridge, NJ.

Ehrenfeld, J. R., et al. (1986). *Controlling Volatile Emissions at Hazardous Waste Sites,* Noyes Data Corporation, Park Ridge, NJ.

Environmental Protection Agency (1984). *Compendium of Methods for the Determination of Toxic Organic Compounds in Ambient Air,* EPA 600/4-84-041, U.S. EPA, Research Triangle Park, NC.

——— (1986). *Supplement to the Compendium of Methods for the Determination of Toxic Organic Compounds in Ambient Air,* EPA 600/4-87-006, prepared by R. M. Riggin et al., Quality Assurance Division, Environmental Monitoring Systems Laboratory, U.S. EPA, Research Triangle Park, NC.

——— (1988). *Second Supplement to Compendium of Methods for the Determination of Toxic Organic Compounds in Ambient Air,* prepared by W. T. Winbery, N. T. Murphy, and R. M. Riggin, Atmospheric Research and Exposure Assessment Laboratory, U.S. EPA, Research Triangle Park, NC.

——— (1991). *The Clean Air Act Section 183(d) Guidance on Cost-Effectiveness,* EPA-450/2-91-008, Office of Air Quality Planning and Standards, Office of Air and Radiation, U.S. EPA, Research Triangle Park, NC.

——— (1991). *Evaluation of High Volume Particle Sampling and Sample Handling Protocols for Ambient Urban Air Mutagenicity Determinations,* EPA/600/D-91/128, Health Effects Research Laboratory, U.S. EPA, Research Triangle Park, NC.

Wuebbles, D. J. (1991). *Primer on Greenhouse Gases,* Lewis Publishers, Boca Raton, FL.

Godish, T. (1990). *Air Quality,* 2nd ed., Lewis Publishers, Boca Raton, FL.

Lodge, J. P., ed. (1989). *Methods of Air Sampling and Analysis,* 3rd ed., Lewis Publishers, Boca Raton, FL.

National Research Council (1991). *Rethinking the Ozone Problem in Urban and Regional Air Pollution,* National Academy Press, Washington, DC.

Stensvaag, J.-M. (1991). *The Clean Air Act of 1990,* Wiley, New York.

16

FACILITY ENVIRONMENTAL ASSESSMENTS

During the past several years, managers of industrial facilities have found it advantageous to assess facility operations in terms of regulatory compliance and good management practices. A comprehensive environmental assessment program can aid in detecting potential regulatory problems early, before they become noncompliance issues. In addition, a self-auditing program may be beneficial during regulatory prosecution, as stated in the introduction to a guidance document prepared by the U.S. Department of Justice, dated July 1, 1991: "It is the policy of the Department of Justice to encourage self-auditing, self-policing and voluntary disclosure of environmental violations by the regulated community by indicating that these activities are viewed as mitigating factors in the Department's exercise of criminal environmental enforcement discretion" (*Environmental Law Reporter* 1991).

Once developed, the assessment program also can be used by the waste-generating facility to evaluate vendored operations for regulatory compliance. Target vendors might include off-site waste treatment and disposal facilities, waste brokers, and recycling vendors. In this chapter we outline a methodology for developing an organized approach to hazardous materials and hazardous waste compliance assessments.

REVIEW OF PERMITS AND OPERATIONS

The first step to beginning an environmental assessment is to perform a permit review and an operations field check. If the permits are several years old and if the facility is one that has been upgraded recently with new chemical processes, there may be discrepancies in the permits that need to be addressed. Some key items that should be included during a permits review are presented in Table 16.1. Other items to be reviewed that may not be related directly to

TABLE 16.1 Items to Review During a Permit Assessment

Equipment Review Items

- Process equipment and associated exhausts
- Hazardous waste tanks and certifications by a registered professional engineer
- Chemical ventilation systems in all buildings and process centers
- Equipment preventive maintenance schedules

Process and Operations Review Items

- Hazardous materials storage, treatment, and disposal operations
- Chemical use, including individual chemicals and quantities
- Chemical dispensing or waste consolidation activities that are performed under local exhaust, including chemicals and quantities
- Identification of waste streams and wastewater treatment efficiencies, including end-of-pipe discharges
- Identification of lab, bench-scale, or pilot operations
- Occurrence (or recurrence) of compliance problems and corrective actions

Review of Changes to Process or Equipment

- Conversions of nonhazardous waste tanks to hazardous waste tanks and certification by a registered professional engineer
- Closure of storage and treatment tanks or surface impoundments
- Conversion of a warehouse or storeroom space to a hazardous materials or hazardous waste storage area
- Relocation of process centers
- Increases in production rates, resulting in greater chemical use
- Changes in workday in terms of adding overtime hours or additional shifts
- Changes in on-site vendor operations

a permit or facility changes, but that affect the overall regulatory compliance or potential environmental liability of a facility, are presented in Table 16.2.

Depending on the size of the facility, the permit and operations review could take several days. All discrepancies or inadequacies found should be addressed with management for corrective action.

DOCUMENTATION REVIEW

Along with permits and general operational compliance with regulatory standards, proper documentation is another important area to be assessed during a facility assessment or self-audit. All documents and plans should be up to date and should reflect current operating rates and practices. Some of the documents that might be required of an industrial facility are listed in Table

TABLE 16.2 Other Items to Review That May Not Be Related Directly to a Permit

Items Related to Safety and Security

- Adequacy of fire suppression systems, in terms of changes in chemicals, quantities, and location of operations
- Security of the site, individual buildings, hazardous materials and waste storage areas, and docks
- Designation of adequate evacuation routes to be used in case of an emergency

Items Related to Spill Containment

- Adequacy of docks and load/unload spill containment to contain catastrophic failures
- Adequacy of tank system spill containment, including those containing product materials and waste, to contain catastrophic failures
- Adequacy of spill containment in container storage areas

Items Related to Runoff Management

- Identification and proper control of storm drains near docks, loading and unloading stations, and tank systems and identification of other nearby migration paths to surface or groundwater
- Runoff controls or diversions to contain a release within the property boundaries
- Identification of land use adjacent to the facility and within a 2- to 5-mile radius

Other Items

- Control of fill material brought onsite for construction or other purposes and certification that the material is not contaminated
- Control of contractor chemical and waste activities and other controls to protect the environment

16.3. As can be seen from the table, documentation requirements can be extensive. In addition to assessing the completeness of the documentation, the organization of the documents for easy retrieval can also be evaluated.

FIELD REVIEW FOR REGULATORY COMPLIANCE AND GOOD MANAGEMENT PRACTICES

No facility environmental assessment is complete until a field review is conducted. This review, or physical inspection, is necessary to ensure that compliance programs and good management procedures are being followed in manufacturing areas, chemical and waste control areas, and in other parts of the facility. This part of the assessment is perhaps best handled with a standard checklist that can be duplicated and filled out for every major process center, laboratory, and chemical and waste support area. Table 16.4 details some major categories and items that might be included on a checklist.

TABLE 16.3 Deocumentation That Could Be Required to Meet Current Regulations

Regulation	Documents That Could Be Required
RCRA	RCRA Part A, PRCA Part B, waste analysis plan, waste analysis records and waste determinations, inspection plan, emergency and contingency plan, groundwater monitoring plan, closure plan, postclosure plan, training plan, financial bond or certification, tank assessments and certifications, closure documentation, postclosure documentation, groundwater monitoring data and reports, training records, a list of job titles that include hazardous waste operations and names of employees occupying those job positions, manifests and land disposal restriction notifications from the past three years, waste inventory, waste container and tank inspection logs, biennial or annual reports to EPA or equivalent, waste minimization reports, the facility operating record, fugitives monitoring records, records of noncompliance release reporting, documentation of incidents where the contingency plan was invoked, registration of petroleum underground storage tanks
OSHA	Hazard communication plan and training records, lead training records, other training records for handling chemicals such as asbestos, industrial hygiene plan for chemical laboratories, personal monitoring records, respirator training and fit test records, respirator inspection logs, medical surveillance records, laser taining records, hazardous waste operations and emergency response training records and certifications, documentation of tasks requiring personal protection, documentation of proper removal of asbestos, documentation of location of ionizing radiation producing equipment; documentation of process hazard reviews
SARA	Notice to SERC and LEPC of chemicals subject to section 312 reporting, including a list of chemicals or MSDS sheets, notice of applicability, tier I and/or tier II reports, toxic release inventory reports, records of noncompliance release reporting
TSCA	Premanufacturing notification submittals, documentation of allegations of adverse effects of chemicals, health and safety data reports pertaining to new chemicals, production records of TSCA listed chemicals, PCB storage and disposal logs, documentation of location of asbestos-containing materials
Clean Water Act	Wastewater discharge monitoring data and reports, spill prevention control and countermeasure plan, pretreatment compliance reports, local POTW limits compliance reporting or certification, notification to local POTW of hazardous waste discharges to sewer system, stormwater monitoring data and reports, stormwater pollution prevention plans, documentation of best management practices, whole effluent toxicity test data and reports, records of noncompliance reporting

TABLE 16.3 (*Continued*)

Regulation	Documents That Could Be Required
Clean Air Act	Construction and operating permits, emissions monitoring data and reports, new source performance (NSPS) test results and other NSPS data and records, hours of operation, records of control system efficiency, chemical consumption tracking logs, production tracking logs, fuel oil or natural gas consumption tracking logs, fugitive monitoring records, emissions inventory
Nuclear Regulatory Commission	Training records, quality control plan, records of defects or noncompliance with quality control plan, waste analysis plan and records, manifests and disposal records, groundwater monitoring records, compliance with other facility permit requirements

The checklist can be customized to the facility activities and can be used as a periodic self-test by the area manager. An example of a waste management checklist for waste accumulation areas is presented in Table 16.5. Additional checklists would be necessary for chemical-using departments and departments that require special safety measures, such as those using lasers.

REVIEW OF CONTRACTOR ACTIVITIES

Often overlooked, but very important in terms of overall hazardous materials and hazardous waste regulatory compliance, is control of contractor activities. Although many contractors are aware of basic regulatory requirements, they may not be aware of the specifics of the facility's permits or of the facility's general procedures for chemical and waste handling. In particular, contractors who bring chemicals on site and who generate waste should be candidates for periodic field inspections, just like other chemical using/waste generating areas of the facility.

Just as a checklist is useful for process centers, labs, and chemical and waste support areas, a checklist can be useful in identifying items to assess during a contractor review. These may differ slightly from the facility checklist, but the basic compliance requirements will be similar. Examples of categories and items that might be included in a contractor inspection checklist are listed in Table 16.6.

The frequency of these inspections will depend on the number of contractors on site and turnover of the contractor staff. Usually, the facility's contractor coordinator can perform these inspections during job setup. Additional "spot" inspections may be necessary for new contractors who are not accustomed to the procedures and practices of the facility.

TABLE 16.4 Categories and Items to Include on a Field Inspection Checklist

Category	Inspection Items
Labeling	Chemical and waste containers are labeled for content and hazards; hazardous waste containers are dated clearly as to when the waste container was filled; tanks and containers containing hazardous waste are marked "Hazardous Waste"; Tanks are labeled for quantity, content, and hazard; gas cylinders are labeled for content and hazard and have "In Use" or other appropriate tag; radioactive waste is labeled with appropriate decals; biomedical waste is labeld with appropriate decals; pipes are labeled for content; asbestos-containing materials are labeled to indicate such; tank trucks and other on-site carriers are placarded properly; chemical and waste containers are labeled with DOT labels prior to shipment
Chemical information	A written hazard communication plan is available to all employees; a written hazardous materials response plan is available to all employees; MSDSs or the equivalent are available in the work area for all chemicals used in the process center, chemical support area, or lab; documentation shows employees have been trained on all hazards in the area
Signs	Exit signs are placed properly and lit at all times; emergency evacuation routes are posted with location of fire extinguishers and emergency response equipment; emergency contact numbers are posted on all phones; personal protective equipment signs are posted on doors of process centers requiring such protection; "No Eating, Drinking, and Smoking" signs are posted in lead-using and other chemical areas; "Danger, Keep Out" signs are posted at permitted hazardous waste operation sites; "No Smoking" signs are posted in flammable areas; "Keep Out, Asbestos-Containing Materials" signs are posted during asbestos removal projects; PCB warning signs are posted where PCB-containing equipment is used or PCB waste is stored
Ionizing radiation sources	On-site inventory of ionizing radiation sources matches inventory of laser safety officier
Personal protection	Requirements for personal protection are documented at process center; inventory of personal protective equipment in area meets criteria documented in personal protection plan; personal protective equipment is being used, cleaned, and inspected, as specified
Local exhaust/ ventilation	Alarm set points are calibrated on schedule; makeup air vents are open; system balancing has been performed on schedule
Storage and containment	Chemical and waste containers are segregated properly and stored in cabinets or in containment pans; chemical tanks are segregated properly with vaults, berms, or other mechanisms to ensure that incompatibles cannot mix if a failure occurs; containments are clean, dry, and in good physical condition; required inspection logs are available to indicate inspection of waste tanks, containers, and associated containments; inspection logs are available indicating inspection for a visible sheen before releasing rainwater from fuel oil or other containments
Fire protection	All fire suppression systems are tested on schedule; all portable fire extinguishers are checked regularly to ensure availability

TABLE 16.5 Waste Management Checklist for Waste Accumulation Areas

Department Managers: This questionnaire must be filled out monthly. If you have questions, call Environmental Engineering on extension 6924. Completed checklists should be reviewed with employees and kept in the department's chemical safety notebook.

1. Is the accumulation date written on all hazardous waste containers?
   ____ yes ____ no (If no, do so immediately, using the date when the container was first used.)
2. Is the label "Hazardous Waste" attached to all containers that receive hazardous waste?
   ____ yes ____ no (If no, attach immediately. If labels are not available in the area, call chemical distribution on extension 6870.)
3. Is the label "Nonhazardous Waste" attached to all containers that receive normal trash and waste not contaminated with hazardous materials?
   ____ yes ____ no (If no, attach immediately. If labels are not available in the area, call chemical distribution on extension 6870.)
4. Are the hazards of the materials in the hazardous waste containers identified (i.e., rags contaminated with solvent, discarded lab chemicals, gloves contaminated with isopropyl alcohol, etc.)
   ____ yes ____ no (If no, do so immediately. If you are not sure of the hazard, call the waste management engineer on extension 6954.)
5. Has the waste been in the area more than 90 days?
   ____ yes ____ no (If yes, call waste pickup immediately on extension 6872.)
6. Has the waste been in the area more than 60 days?
   ____ yes ____ no (If yes, arrange for pickup before next month's waste management audit or before 90 days.)
7. Are weekly inspections performed on waste containers to assess condition of containers, spill containment, labeling, proper segregation, accumulation time, and other appropriate items?
   ____ yes ____ no (If no, call the waste management/engineer on extension 6954 immediately to report deviations.)
   Forms for the inspections can be viewed on-line under "Chemical Library" and can be printed directly or ordered from the waste management engineer.)
8. Are waste handling procedures available in the area, and have all employees been trained on these procedures?
   ____ yes ____ no (If no, correct immediately.)
9. Are procedures being followed to minimize the amount of waste generated in the area?
   ____ yes ____ no (If no, institute procedures immediately. If there are no procedures in the area, call the waste management engineer on extension 6954.)

Comments:

____________________	____	____________________
Manager's Signature	Date	Department Number

TABLE 16.6 Categories and Items to Include in a Contractor Inspection Checklist

Category	Inspection Items
Training	Documentation is available that demonstrates the contractor has given employees appropriate training, including hazard communication training and other job-specific training, such as hazardous waste operation training, personal protective equipment training, and asbestos removal training.
Labeling	Contractor chemical and waste containers are labeled with contractor name (in case of leakage or other problems), content, and hazard; contractor gas cylinders are labeled with contractor name, content, and hazard and have "In Use" or another appropriate tag.
Chemical information	Contractor chemicals are approved before use onsite and the contractor has MSDSs or the equivalent available for review by the facility's contractor coordinator or other persons working in the area; contractor employees have access to MSDS or equivalent for chemicals in the workplace.
Signs	The contractor has the work area roped off and signs posted with the contractor's name and contact number; "No Smoking, Hard Hat Area", and other safety signs are posted outside the roped off area; any other signs required for specialty jobs, such as asbestos removal, are posted.
Permits	Contractor has obtained necessary control permits such as confined space entry permits, permits for powdered-actuated tools, and welding or other hot-work permits.
Ionizing radiation sources	Contractors have received approval and have documented training before bringing on site and using ionizing radiation equipment; contractors are approved and have demonstrated training before operating ionizing radiation equipment
Personal protection	Required personal protection is identified and worn by the contractor; appropriate workplace monitoring is performed.
Local exhaust/ ventilation	Contractors are approved to use existing local exhaust systems for dispensing operations; contractors are approved to set up a portable local exhaust system.
Storage and containment	Chemicals and wastes are segregated properly in work area; adequate spill containment is used for temporary storage of drums and other containers.
Fire protection	Fire extinguishers are available and adequate for suppressing fires in the work area.

ADDITIONAL REVIEW ITEMS

For a complete facility assessment, there are several other items that could be reviewed during an environmental assessment. These include:

- Facility plans for chemical reductions and/or chemical substitutions with less toxic chemicals
- Waste minimization goals and schedules
- Facility plans for fugitives management
- Process hazard reviews and other risk assessments
- Incident investigation process and documentation
- Preventive maintenance plans and schedules
- Personal monitoring plan
- Stack sampling plan
- Spill or release documentation and reduction plan
- Facility assessment documentation, including corrective actions and schedules for subsequent assessments

Goals or schedules can become outdated quickly, so these items should be reviewed at least annually. Discrepancies to plans should be investigated and resolved as soon as possible.

ENVIRONMENTAL ASSESSMENTS FOR REAL ESTATE TRANSACTIONS

In the past decade, there has been emphasis on environmental investigations prior to the purchase of real estate. Typically, this assessment is performed based on ASTM Standard E1527 (ASTM 1994). Information obtained in performing this type of assessment (typically termed a *phase I environmental site assessment*) is detailed in Table 16.7.

TABLE 16.7 Information Obtained when Performing a Phase 1 Environmental Site Assessment

Project Information

- Project property name, address, county/state, and location description
- Property owner name, address, and phone
- Key site manager name and phone
- Identification of parties involved, such as lenders, brokers, tenants, and attorneys, including names, addresses, and phone numbers
- Property description, lot size, building(s) size, building(s) age, and legal description
- Past property uses and name and address of prior property owner
- Adjoining properties to north, northeast, east-southeast, south, southwest, west, and northwest
- Known contamination on properties in the vicinity of the site

TABLE 16.7 (*Continued*)

Records Review

- Review federal environmental record sources, including the following databases: NPL, CERCLIS, RCRA TSD, RCRA generators, and ERNS
- Review state environmental record sources, including state lists of hazardous waste sites, state landfill and/or solid waste disposal site lists, state leaking UST lists, and state registered UST lists (property and adjoining properties)
- Review local record sources, such as the department of health/environmental division, fire department, planning/zoning department, building permit/inspection department, local/regional pollution control agency, local/regional water quality control agency, local electric utility companies (with respect to PCBs); include contact person's name and phone number
- Review local lists, such as lists of landfill/solid waste disposal sites, lists of hazardous waste/contaminated sites, lists of registered underground storage tanks, record of emergency release reports, and records of contaminated public wells
- Review standard historical sources, such as aerial photographs, fire insurance maps, property tax files, recorded land title records (mandatory), national wild life maps for designation of wetlands
- Review physical setting sources, such as topographic map (mandatory), hydrology, and hydrogeology maps
- Interview persons familiar with the property, such as key site manager and major occupant of property

Site Reconnaissance

- General information about activities at site, such as raw materials, processes, products, and services; identification of any property contamination and any major environmental concerns, including areas of regulatory uncertainty; identification of environmental incidents or problems on surrounding properties; identification of community complaints
- Air issues, such as air permits, emissions inventory, fugitive emissions, and radon
- Existence of friable and nonfriable asbestos and lead-based paint
- Existence of polychlorinated biphenyls (PCBs) in transformers, capacitors, or other potential PCB-containing equipment located on the property; identification of leaking equipment; identification of equipment using hydraulic fluid, such as elevators and hydraulic lifts; evidence of PCB contamination
- Existence (past and present) of aboveground and underground storage tanks (USTs), including tank capacity, substance stored, installation date, registration, corrosion protection, leak detection, tank construction material, results of testing, date taken out of operation, name of contractor who performed removal; visible evidence or records of leaks or failures
- Information pertaining to hazardous materials handling, including chemicals handled on site, location, manufacturer, and quantities; condition of containers; description of storage areas
- Information pertaining to hazardous waste storage, treatment, and disposal, including EPA identification numbers, processes that generate hazardous waste, waste types, and quantities; facility RCRA status; storage information, including number and size of drums

TABLE 16.7 (*Continued*)

- Operations pertaining to treatment, disposal, reclamation, and recycling; permit numbers; transporters and storage and disposal facilities currently used for hazardous waste; inactive land disposal sites owned or operated by the facility; identification of waste disposal sites for which the facility has been identified as a potential contributor; identification of any spill cleanup/soil remediation projects
- Condition of hazardous waste containers; description of storage area for hazardous waste; identification of hazardous waste disposed with solid waste; identification of visible evidence of hazardous waste disposal problems
- Description of biological or medical waste generated on property
- Description of solid waste handling operations, including how solid waste is collected and disposed of
- Description of waste oil–handling operations, including how waste oil is collected and the ultimate disposition of waste oil, such as recycling, reuse, incineration, fuel blending, etc.; description of container and tank type, capacity, generation rate, and disposal transporter
- Wastewater treatment issues, including industrial process discharge sources, approximate flow rates, type and number of discharge points, and permits and issuing agency for each discharge; description of pretreatment operations associated with industrial process discharges
- Stormwater discharges, including approximate volume, area drained, number of discharge points and permits
- Description of groundwater monitoring programs onsite or near the site, including number and location of wells, latest sample date, frequency of sampling; description of remediation programs on site or near the site
- Description of sanitary discharges and on-site treatment; identification of septic tanks both operating and abandoned, including location and size
- Identification of oil/water separators connected to the sanitary system, including clean-out records
- Description of water wells on site for drinking, process, or cooling, both operating and abandoned, including location and date last used; identification of water source for site; last test of potable water source
- Description of surface water, standing water, saturated surfaces, ditches, and drainage depressions located on site; floodplain information; description of direction of runoff and run-on
- Issues concerning adjoining properties, including high-risk operations such as service stations, dry cleaners, chemical manufacturing facilities, etc.; potential for groundwater contamination from adjoining properties; potential for contaminated stormwater run-on from adjoining properties
- Inclusion of information about photographs, including date, time, location, and description of photo, and direction faced when taking photo
- Professional's comments, conclusions, and recommendations

REFERENCES

American Society for Testing and Materials (1994). *Environmental Site Assessments – Phase I: Assessment, Process, Practice,* E1527, ASTM, Philadelphia.

Environmental Law Reporter, (1991). "Factors in Decisions on Criminal Prosecutions for Environmental Violations in the Context of significant Voluntary Compliance or Disclosure Efforts by the Violator," *Environmental Law Reporter* (21 ELR 35400). November.

BIBLIOGRAPHY

Government Institutes (1985). *Good Laboratory Practice Compliance Inspection Manual,* GI, Rockville, MD.

——— (1989). *Environmental Audits,* 6th ed., GI, Rockville, MD.

——— (1991). *OSHA Field Operations Manual,* 4th ed., GI, Rockville, MD.

Greeno, J. L., G. S. Hedstrom, and M. A. DiBerto (1988). *The Environmental, Health, and Safety Auditor's Handbook,* Arthur D. Little, Inc., Cambridge, MA.

Ortolano, L. (1984). *Environmental Planning and Decision Making,* Wiley, New York.

Tarantino, J. A. (1992). *Environmental Liability Transaction Guide: Forms and Checklists,* Wiley, New York.

U.S. Army Corps of Engineers (1990). *Environmental Review Guide for Operations,* Construction Engineering Research Laboratory, U.S. Army COE, Champaign, IL.

Water Environment Federation (formerly Water Pollution Control Federation) (1989). *Environmental Audits: Internal Due Diligence,* WEF, Washington, DC.

17

POLLUTION PREVENTION

Pollution prevention, or P2 for short, has become an integral part of environmental regulations. Throughout the 1980s, interest in pollution prevention grew steadily as a result of increasing environmental awareness and concern, tightening waste regulations, and liabilities incurred for contaminated sites. In 1990, pollution prevention became national policy when Congress passed the Pollution Prevention Act.

The ultimate goal of pollution prevention is to reduce or eliminate pollution at its source. Where source reduction is not feasible, recycling should be considered. Where recycling is not feasible, proper treatment of a waste is required. Any remaining residuals must be disposed of properly in an environmentally safe manner.

As a subset to pollution prevention, many companies are implementing programs that focus on design for the environment. These programs emphasize the development of environmentally friendly products.

THE NEED FOR POLLUTION PREVENTION

The Toxic Release Inventory (TRI) reports were first required by EPA in 1988 for chemical releases that occurred in 1987. The reports detailed releases (including permitted releases) to air, land, and water for over 300 compounds from thousands of companies. Once the data were compiled and published, the public was shocked at the amounts of chemicals that were being discharged. The need for pollution prevention became a national focus. In 1991, EPA listed 17 toxic chemicals that were targeted for immediate reduction, and added over 300 additional chemicals to the list during the mid-1990s. The initial list of 17 chemicals is presented in Table 17.1.

TABLE 17.1 EPA's Initial List of Chemicals Targeted for Reduction

Benzene	Methyl ethyl ketone
Cadmium	Methyl isobutyl ketone
Carbon tetrachloride	Nickel
Chloroform	Tetrachloroethylene
Chromium	Toluene
Cyanide	1,1,1-Trichloroethane
Dichloromethane	Trichloroethylene
Lead	Xylene
Mercury	

DESIGNING AND IMPLEMENTING A P2 PROGRAM

Steps for designing and implementing a P2 program are summarized in Table 17.2. Below is a discussion of some of the key elements of P2.

Commitment

Pollution prevention has obvious rewards: potential cost savings, reduction in waste quantities or toxicities, and improved company image. However, other aspects of pollution prevention hinder many projects: reluctance to change from tried-and-true methods or processes (particularly when such changes can affect product quality); time, expense, and difficulty in studying alternatives; initial capital expense; and structural or space constraints at existing locations. Consequently, implementation of significant pollution prevention projects usually requires a commitment from those persons with the authority to make things happen—that is, management. But employee commitment is important, too. Therefore, it is not surprising that the most successful pollution prevention projects are those that have the commitment of both management and employees.

Management can start by developing a pollution prevention policy or mission statement that gives a clear idea of the company's goals, objectives, and approach. As such, the policy or statement can be used as a general gauge to assess progress. ISO 14001, an international standard for environmental management, requires top management to set policy, the chairman or equivalent in most companies. Employees set goals and objectives along with management. The company's pollution prevention policy should be communicated to all employees so that they understand the company's commitment and how they, as employees, play a part in it.

Setting clear, measurable pollution prevention goals is important to show commitment. These goals should be set beyond mere compliance with environmental regulations; however, they must be realistic and achievable or else

TABLE 17.2 Steps for Design and Implementation of a P2 Program

Get Top Management Support for P2

- Demonstrate top management commitment for P2 through policy.
- Demonstrate endorsement of the policy by all levels of management.
- Communicate policy for P2 to all employees.

Develop a P2 Program

- Designate a P2 coordinator.
- Develop a P2 team.
- Develop a written P2 plan.
- Set strategies and goals.

Involve Employees

- Solicit and reward employee suggestions for waste reduction.
- Ensure that every employee who has the potential to affect waste reduction activities understands that his or her contribution to pollution prevention is important.

Characterize Process

- Develop process flow diagrams.
- Develop material balance.
- Identify process units that generate waste.

Assess Wastes and Identify Opportunities

- Identify waste streams.
- Prioritize waste streams to be reduced or eliminated.
- Generate reduction and elimination options.
- Analyze source reduction and source control options.

Analyze Costs

- Determine full cost of waste generation, including costs incurred in producing and handling the waste, as well as treatment and disposal costs.
- Develop economic matrix to determine economic feasibility of projects.
- Consider benefits, such as reduced long-term liability, reduced worker exposure to toxic chemicals, and improved community relations.
- Establish cost-allocation system to "charge" the process or department that generates the waste.

Identify and Implement P2 Options

- Assess technical feasibility.
- Evaluate economics of project.
- Implement projects.

TABLE 17.2 (*Continued*)

Evaluate Program
• Assess commitment.
• Assess progress.
Sustain Program
• Reemphasize economic benefits.
• Rotate members of pollution prevention team.
• Provide refresher training.
• Reward success.
• Publicize success.

Source: Information from Freeman (1995).

the company stands to lose credibility when it cannot meet its stated goals. Management should be prepared to provide the necessary resources—personnel, information, and money—to achieve its pollution prevention goals. Setting goals without making it possible to achieve them is only giving lip service.

Management can show its commitment to its employees by valuing employee input and rewarding employees who help achieve pollution prevention goals. Environmental staff, who are usually on the front line of pollution prevention because of compliance pressures, should be able to communicate directly to management. Environmental managers should have a presence in upper management. Examples of actions that can encourage pollution prevention in the workplace are presented in Table 17.3.

Evaluating Performance

Tracking progress in pollution prevention allows a company to see how its stated goals are being met, to document its commitment to its employees and the public, to provide feedback to ongoing and future projects, and to justify cost or environmental benefits. Because many larger companies are required to report chemical releases annually under the TRI, they can use these data to track and document reductions. Other regulatory reports can provide waste data for tracking, such as RCRA hazardous waste elimination and biennial reports, state or local waste reports, waste manifests, wastewater discharge monitoring reports, and air emission inventory reports. Other sources of data are environmental audit reports, waste analyses and waste profiles, material inventory and use records, continuous emission monitoring, production records, and industrial sector waste surveys. Results of pollution prevention efforts should be reported to management to provide feedback on what does and does not work and where the greatest potential benefits are.

TABLE 17.3 Actions to Encourage Pollution Prevention in the Workplace

General

- Enlist support for pollution prevention from top management. Have them demonstrate that support by providing a written policy statement.
- Appoint an unsinkable champion and give that champion the power to implement.
- Keep hammering away at the nonbelievers—be patient and persistent.
- Believe in P2 yourself.
- Identify and publicize low-tech or retrotechnology options in place of high-impact processes.
- Seek a fundamental understanding of the sources of waste.
- Focus on optimizing the use of resources consumed in the process.
- Use pollution prevention as a competitive strategy in public relations and to attract a better caliber workforce.
- Devote adequate resources (people, money, and energy) to pollution prevention.
- Set up a structure for recycling that complements source reduction.

Culture

- Establish a clear pollution prevention goal that everyone in the organization feels empowered to put into practice.
- Reestablish the culture of not wasting.
- Instill a philosophy of continuous improvement.
- Link zero discharge, total quality management, and pollution prevention into a working program.
- Publicize pollution prevention accomplishments.
- Convince all personnel that they have a role to play in pollution prevention; no one is exempt from pollution prevention, no matter what the job.
- Share money saved through pollution prevention with the originator(s) of the idea.
- Incorporate pollution prevention into performance evaluations for middle management.
- Establish teams to promote pollution prevention, receive ideas and solicit suggestions, evaluate projects, and champion implementation.
- Have top management personally hand out all pollution prevention awards; participate in state and regional awards programs.

Workplace Education

- Produce innovative and exciting training for teaching pollution prevention concepts.
- Have primary contractors train and assist subcontractors to practice pollution prevention.
- Educate each person as to what pollution prevention means.

Decision Making

- Use expert systems and process simulation modeling to develop effective pollution prevention strategies.
- Use and develop life-cycle studies for processes and products.
- Develop and implement methods for measuring progress to support continued pollution prevention investments.

TABLE 17.3 ***(Continued)***

- Develop performance-based specifications rather than prescriptive or design specifications.
- Set a goal: Think of a bubble around your facility—nothing comes out but finished product.
- Plan your waste reduction work and work your waste reduction plan.
- Use expert systems to assist the design of new processes that avoid pollution in the first place.
- Design products with zero ultimate waste potential, with cradle-to-grave functionality.
- Incorporate pollution prevention into the development of new products and processes.
- Assign life-cycle responsibility to production management, linked to cost and liability.
- Charge the true cost of the waste created to the operating unit and make operating units responsible for liabilities, management, and costs of waste streams.
- Develop analytical tools for accountants and financial managers to recognize full environmental costs of unwanted environmental programs.

Outreach

- Establish an industry round table to communicate and share methods for compliance and low-impact processes.
- Develop corporate–public partnerships to protect local resources and to explore solutions to difficult problems.
- Develop pollution prevention in all locations of the company on an equal basis. Treat foreign facilities with the high level of expectation applied to domestic facilities.
- Establish a process (newsletter, etc.) for communicating progress.

Communication

- Develop an internal information system to exchange good ideas and technical knowledge within your business unit, division, or company.
- Shift from paper systems to paperless communication (e.g., e-mail, electronic bulletin boards).
- Create effective, friendly, human information networks for effective pollution prevention.
- Encourage policies that allow plant engineers and managers to spend ample time on the factory floor.

Recyling and Resource Conservation Examples

- Xeriscape—use low-water-consumption drought-resistant native plants and low-impact irrigation in corporate landscaping.
- Make precycled writing tablets out of paper used on one side.
- Establish a chemical inventory program to reduce redundant purchases, track chemical use more efficiently, and avoid disposal of unused material.
- Change computer printer to print only "flag" sheet plus manuscript, not additional sheets.
- Use two-sided copying.

Source: AIPP (http://www.envirosense.com/aipp/48_pg.html).

Identifying Chemicals of Concern

When seeking to reduce the quantity of hazardous waste or at least reduce the degree of hazard associated with such wastes, wastes should be characterized by hazard and hazardous constituents. There are many regulations that focus on different types of hazardous chemicals, materials, or wastes, and these can be reviewed for those chemicals that might be of concern and would be candidates for pollution prevention efforts. For example, EPA has challenged industry to reduce wastes containing any of 17 specific toxic chemicals under the TRI 33/50 program. A "list of lists" can be developed from the various regulations as a database for pollution prevention projects. All personnel involved with pollution prevention should be made aware of these lists so that they can focus on important chemicals or characteristics. Such information is particularly important to those persons in research, development, and manufacturing because they are the ones that are most likely to have detailed information on products and waste streams, as well as the expertise to suggest modifications and alternatives. Examples of regulated chemicals are found in the RCRA hazardous waste regulations, the TRI reporting regulations, CERCLA lists of hazardous and extremely hazardous substances, CAA lists of hazardous and highly hazardous chemicals, and the CAA list of ozone-depleting substances.

Process Modifications

Process modifications naturally are easiest to make during the process development and design phase. Retrofitting and modifications of an existing process are hampered by ongoing operations, available space, and existing layout. Understandably, they can be disruptive and expensive—two big drawbacks that most people like to avoid. Even so, many companies have made such modifications and have been rewarded not only with pollution reductions, but with tangible and immediate cost savings. Process modifications, in both the design/development stage and in existing operations, are normally quite specific to the particular process. Therefore, in this section we discuss process modifications in a broad and general sense.

Material substitution or improvements should be considered for feedstock and secondary materials used in the process. Such process changes can include switching to a less toxic or less corrosive material. Using a higher-purity material or developing a more selective catalyst may reduce by-product formation, improve product yield, and reduce the transfer of unreacted materials into the waste stream.

Other substitutions include using less toxic and volatile solvents and cleaners or those that are not ozone depleting. For example, alternative cleaners and solvents include aqueous and emulsion cleaners, detergent cleaners, and terpene-based chemicals. EPA has developed or is working on various guides to alternative solvents. One is called SAGE and is for solvents in the TRI

33/50 program. Other guides being developed are CAGE for coating processes and AGE for adhesives. EPA maintains a solvent substitution database on the Internet (see "Enviroene" later in this chapter) including the following data systems:

- *Solvent Alternatives Guide (SAGE):* a logic tree system that can be used to evaluate a current operating scenario and then identify possible surface cleaning alternative chemistries and processes that best suit the defined operating and material requirements.
- *Hazardous Solvent Substitution Data System (HSSDS):* an on-line information system on alternatives to hazardous solvents and related subjects. HSSDS contains product information, material safety data sheets, and related information.
- *Department of Defense (DOD) Pollution Prevention (P2) Technical Library:* an electronic resource maintained by the Naval Facilities Engineering Service Center (NFESC). To foster technology transfer with other organizations, the NFESC has made this library available by direct access or by search facilities.
- *DOD Ozone Depleting Chemical/Substance Information–U.S. Air Force Ozone Depleting Chemicals (ODC) Information Exchange:* provides information exchange to the Air Force weapons system community to aid in complying with federally mandated ODC reduction goals; courtesy of the Brooks Air Force Base Human Systems Center. The DOD ODS MIL-SPEC database contains a listing of military and federal specifications that may require the use of class I ODSs. This database includes information on each document, including identity of the ODS, how it is used, whether non-ODS alternatives are specified, potential substitutes for the ODS called out in the documents, and modification and cancellation information. The U.S. Navy CFC & Halon Clearinghouse provides users of ODSs with a central point of contact for information, data, and expertise on Navy ODS policy.
- *Solvent Handbook Database System (SHDS):* a database providing access to environmental and safety information on solvents used in maintenance facilities and paint strippers. The database also contains empirical data from laboratory testing.
- *National Center for Manufacturing Sciences (NCMS) Solvent Alternatives Database and NCMS Materials Compatibility Database:* contributed by the NCMS, which is a large manufacturing consortium of 175 members, including some of the largest manufacturers in the United States.

When designing chemical reactors with pollution prevention in mind, it is important to ensure proper mixing to maximize product yield and minimize the formation of by-products. Unwanted by-products can result from poor mixing, fluctuating residence times, impurities in process materials and feedstocks, hot spots on catalysts, and catalytic effects on construction materials. Reactors should be designed to drain completely and clean easily.

Piping designed for pollution prevention should have minimal run and be able to drain completely. It is particularly important to look at the collection of individual streams (both process materials and wastes) in segregated piping because the trend in environmental regulations is moving from management of commingled materials (end-of-pipe) further into the process area before streams are combined. The number of valves and fittings should be minimized. Any drain lines, sample lines, and vents should be routed to recycle, treatment, or disposal as appropriate. Piping should be avoided that drains unto the process pad (later collected in process washdown) or into open trenches or ditches.

Process control can be designed and optimized to maximize product yield and minimize waste streams. Process controls can include instrumentation, remote monitoring and data collection, statistical analysis, and inventory management. Methods of analysis, such as statistical process control (SPC), which are commonly used in total quality management programs to assess product quality can be used in innovative ways to control process chemicals. SPC in conjunction with an automated control system allows the process to be optimized using real-time data, which can restrict chemical usage to much tighter tolerances. Material inventory management can be optimized to reduce the wastage of unused or expired chemicals.

Good Housekeeping and Preventive Maintenance

Good housekeeping and preventive maintenance not only can help reduce waste generation but can also help provide a safe and clean working environment. Simple housekeeping tasks should include inspection of material storage areas, spill prevention and control, segregation of incompatible or inappropriate materials, and container labeling. Preventive maintenance tasks should include regularly scheduled checks or maintenance, record keeping for inspections and repairs, testing of overflow alarms and release detection equipment, replacing damaged or leaking containers, and maintaining up-to-date equipment manuals.

Implementing good housekeeping practices and effective preventive maintenance requires not only employee training, but cooperation as well. Employees need to understand and value the importance of a clean and well-maintained workplace and how this avoids waste generation. Management provides the training and workplace environment; however, that itself is not enough—management must lead by example and enforcement of company policies and practices. Those employees directly involved in housekeeping and maintenance should to be asked for their input on how to make a task easier while getting the job done.

Reuse, Reclamation, and Recycling

Most people are familiar with recycling in their everyday life. Household and office waste that is commonly collected for recycling includes aluminum cans, white ledger paper, newspaper, cardboard, glass, metal cans, and plastic. Recycling is also common at commercial and industrial facilities. Many processes incorporate recycling because it conserves raw materials and reduces waste generation. Some of the most common recycling processes are solvent recovery and wastewater recycling.

Common processes used in solvent recovery include distillation, steam stripping, carbon adsorption, and membrane separation (described in Chapter 12). Solvent recovery may be closed-loop and integral to the manufacturing process. Alternatively, spent or contaminated solvents may be removed from the process for recovery on site or off site. In some cases, a solvent that is lightly contaminated from use in one process can be reused in another process where a lesser quality solvent is acceptable.

In a similar fashion, process water that is slightly contaminated, such as high-purity rinse waters, may be reused where process requirements are less stringent. Effluent from a facility's wastewater treatment plant may be recycled back to a process area or to a cooling tower for makeup water, minimizing water use as well as wastewater discharges into surface waters. Treated sanitary (nonindustrial) wastewater may be reused as drinking water by using it to recharge groundwater supplies (health concerns and public perceptions about wastewater have inhibited direct reuse).

Oil recovery is common at many industrial facilities, particularly petroleum refineries and chemical manufacturers, because of the types of materials that are handled and because high concentrations of oil are unacceptable to the biological wastewater treatment systems. The oil in wastewaters is usually removed in some type of oil–water separator and/or dissolved air floatation unit. Recovered oil may be recycled back to the process or burned for its fuel value on site. It may also be sent off site for material or energy recovery. Waste oil from the maintenance and cleaning of equipment and vehicles can also be recovered.

Wastewaters containing metals may be treated to recover the metals, depending on both the value and concentration of the particular metal. For example, the silver in photographic wastewaters can be recovered. Metals in wastewaters may be recovered by chemical precipitation, ion exchange, membrane separation, or electrowinning.

Solvents, metals, oil, and water are some of the most common recovered or recycled materials. There are many others, of course, some that are specific to a particular process and others that have a more universal application. Some of these applications use wastes that otherwise have few options except disposal. For example, soil contaminated with petroleum products or metals may be converted into asphalt-type paving materials, inorganic residues from supercritical oxidation may be incorporated into bricks, and foundry sand may be used as daily cover for landfills or in construction of dikes, roads, and parking lots. Ideas for recycling materials are constantly evolving.

Prioritizing Projects

When a facility is assessing possible pollution prevention projects for more than one process or several alternatives for a single process, it helps to have a way to prioritize projects. There are many different ways to approach the problem; however, most follow a typical sequence: characterizing material streams, identifying pollution prevention alternatives, ranking the alternatives, and finally, choosing an alternative.

The most effective pollution prevention projects are those that receive input from people who are knowledgeable about the process or will be affected by it. Usually, representatives from the various areas at the site will form a pollution prevention team that will develop and evaluate the pollution prevention projects. Members of the team may include, as appropriate, environmental, health, and safety; production; engineering; research and development; maintenance; purchasing and accounting; legal; and information systems. Outside consultants may be added to the team to provide special expertise or to represent an unbiased (non-company) view.

The first step in characterizing material streams associated with a process is to identify them. The types of information that might be included in the characterization are listed in Table 17.4. The information that is collected for each stream should be summarized in a standard form to make sure that characterizations are complete and to make comparisons among streams easier.

Next, pollution prevention alternatives should be listed for each material stream. Source reduction techniques should be given special consideration in keeping with the goal of pollution "prevention." Recycling, treatment, and disposal alternatives should also be listed. Some judgment will be needed to screen out those alternatives that are obviously unsuitable for one reason or another (project scope, technical feasibility, budget constraints, high cost with little benefit). Care has to be taken not to strike down alternatives too quickly

TABLE 17.4 Information for Material Stream Assessments

- Source (from which part of process)
- State or federal waste identification code
- Physical characteristics (solid, liquid, gas, semisolid, etc.)
- Generation rate
- Hazardous properties (toxic, flammable, reactive, corrosive, etc.)
- Hazardous constituents and quantities released (i.e., TRI reports)
- Other health, safety, or environmental issues
- Applicable regulations (existing, pending, or potential)
- Management costs (treatment, disposal, transport, analysis, etc.)

so that no truly promising alternative will be overlooked. Alternatives that make the final list will then undergo a more intensive review.

Criteria that may be used to rank alternatives include:

- Reduction in quantity generated
- Reduction in hazards
- Persistence of hazardous constituents and assimilative capacity of environmental media
- Effect on product quality
- Regulatory compliance
- Technical feasibility
- Performance track record
- Ease of implementation
- Time to implement
- Effect on existing operations
- Capital cost
- Operating cost
- Management cost
- Process material cost
- Liability and insurance cost
- Training requirements

Each criterion may be assigned a weighting factor to represent its importance relative to the others. Pollution prevention alternatives can then be assigned scores for each criterion. Using the weighted sum method, the score for each criterion is multiplied by its weighting factor. The weighted scores are summed for each alternative, representing its overall score.

Another method for comparing alternatives is to use leverage. *Leverage* is defined as a function of priority and feasibility:

$$\text{leverage} = \text{priority} \times \text{feasibility}$$

Feasibility can be viewed as the possible percentage reduction based on engineering studies of technical feasibility. Priority is a function of potential environmental impact and feasibility is the probability of minimizing the stream. *Priority* can be defined further as:

$$\text{priority} = \text{environmental units} \times \text{fate factor}$$

where *environmental units* are assigned to a stream as a function of its mass and its toxicity:

$$\text{environmental units} = \text{mass} \times \text{toxicity}$$

Toxicity is determined by health data such as an LD_{50}, permissible exposure limit; leachability, such as the TCLP value; or other data uniformly applied to all the streams being ranked. Other factors can be included in the toxicity value, such as hazardous constituents on CERCLA, SARA, or other lists, and contributions to the greenhouse effect or ozone depletion.

The *fate factor* should account for the potential impact of the stream on health and the environment based on how it is managed. Streams that are recycled or that are incinerated will be given a lower fate factor than those that are released to the air or surface water. The hierarchy of fate factors is source reduction, recycling, incineration, treatment or hazard reduction, discharge to air, water, and land.

Leverage can also be defined in terms of cost where *management cost* is substituted for *priority* in the equations above.

DESIGN FOR ENVIRONMENT

A subset of P2, design for environment (DfE) typically has product development as its focus. A illustration of where DfE fits in relation to other environmental initiatives under the P2 umbrella is presented in Figure 17.1. The potential for DfE initiatives typically is identified through a life-cycle assessment or analysis (LCA). This analysis looks at a product from its inception to final disposition. In LCA, a product may begin with the acquisition of raw materials, followed by manufacturing or development of the product, then recycling or maintenance of the product, and finally, waste management of the product or its residuals. LCA is useful because in addition to looking at impacts on the ecosystem and human health, it also looks at impacts on resources such as raw materials or energy.

LCA generally is conducted in three phases: inventory, impact assessment, and improvement assessment. In the inventory phase, the pollution prevention team first defines its goals and how broad the life cycle of the product will be. The life-cycle stages that may be included are marketing, research and development, raw material acquisition, material manufacture, production, packaging, distribution, consumer use, and disposal. After the scope of the life cycle is defined, the use and generation rates of the different material and energy streams are estimated.

The second phase of LCA, impact assessment, itself consists of three phases: classification, characterization, and valuation. In the classification phase, inventoried items are assigned to affect categories of environmental stressors. In the characterization phase, the values assigned to inventoried items are converted into impact descriptors such as direct measure and equivalency factors. In the third phase, valuation, impacts are compared by methods that generally incorporate some type of weighting factors applied to dissimilar impacts to produce a single value.

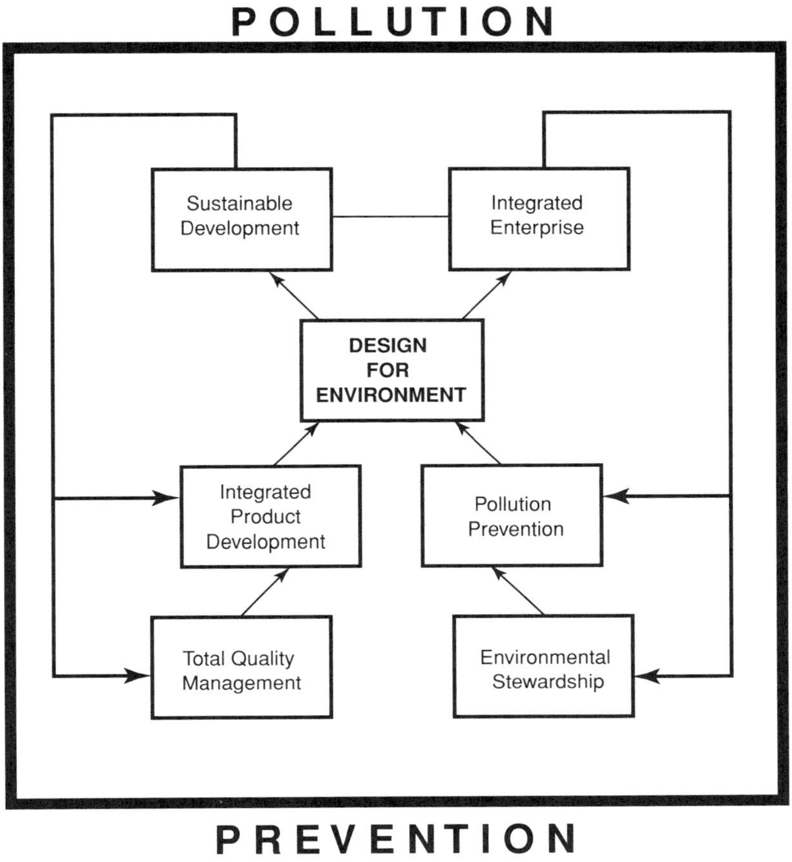

Figure 17.1 Design for environment in relation to other environmental initiatives. (Adapted from Fiksel, 1996.)

Following impact assessment is the improvement assessment phase of LCA. In this phase, options are identified that will improve negative impacts associated with a product. These options are evaluated and compared. Some typical DfE guidelines are presented in Table 17.5.

PREVENTION OF POLLUTION FROM HOUSEHOLD HAZARDOUS WASTE

Many private citizens are interested in practicing pollution prevention in their own lives, particularly with materials that are hazardous. Cities and communities can help by distributing information to let people know how they can reduce their use of hazardous materials that they may end up discarding. For those materials that are discarded, it helps to know how to dispose of them

TABLE 17.5 DfE Guidelines

Design for Material and Waste Recovery

- Avoid composite materials.
- Specify recyclable materials.
- Use recyclable packaging.
- Design for waste recovery and reuse.

Design for Component Recovery

- Design reusable containers.
- Design for refurbishment.
- Design for remanufacture.

Design for Disassembly

- Optimize disassembly sequence.
- Design for easy removal.
- Avoid embedded parts.
- Simplify interfaces.
- Reduce product complexity.
- Reduce the number of parts.
- Design for multifunctional parts.
- Utilize common parts.

Design for Source Reduction

- Reduce product dimensions.
- Specify lighter-weight materials.
- Design thinner enclosures.
- Reduce mass of components.
- Reduce packaging weight.
- Use electronic documentation.

Design for Separability

- Facilitate identification of materials.
- Use fewer types of materials.
- Use similar or compatible materials.

Design for Energy Conservation

- Reduce energy use in production.
- Reduce device power consumption.
- Reduce energy use in distribution.
- Use renewable forms of energy.

Design for Material Conservation

- Design multifunctional products.
- Specify recycled or renewable materials.
- Use remanufactured components.

TABLE 17.5 (*Continued*)

- Design for product longevity.
- Design for closed-loop recycling.
- Design for packaging recovery.

Design for Chronic Risk Reduction

- Reduce production releases to the environment.
- Avoid hazardous or toxic substances.
- Avoid ozone-depleting chemicals.
- Use water-based technologies.
- Assure product biodegradability.
- Assure waste disposability.

Design for Accident Prevention

- Avoid caustic and/or flammable materials.
- Provide pressure relief mechanisms.
- Minimize leakage potential.

properly. Communities can also arrange for special hazardous waste collection days (annually or more frequently) to give their citizens an incentive to get rid of accumulated materials as well as a means of disposing of them safely. Ideas for reducing household hazardous waste are summarized in Table 17.6.

TABLE 17.6 Recommendations for Reducing Household Hazardous Waste

General Recommendations

- If available, use an alternative product that is less toxic or less hazardous.
- Before buying a product, be certain that it can do the job so that it won't be thrown out partially or completely unused.
- Buy only the amount of product that is actually needed or will be used within a reasonable time period so that extra or expired quantities don't have to be thrown out.
- Use the product correctly so that it has the best chance of doing its job properly.

Specific Recommendations

- *Household cleaners.* Use less hazardous cleaners such as baking soda, vinegar, borax, detergents, and lemon juice.
- *Paints.* Use latex or water-based paints, recycled paints made from mixing discarded paints (prepared during waste collection drives), natural earth pigment finishes, limestone-based whitewash, and casein-based paints.
- *Pesticides.* Reduce need for fungicides by keeping areas clean, dry, and not overwatered. Use naturally derived pesticides such as pyrethrum, rotenone, sabadilla, nicotine, and insecticidal soap.

P2 PROGRAMS AND INFORMATION SOURCES

There are many programs and information sources available on the subject of pollution prevention, many of which are government based or supported. Brief descriptions of some of these resources (listed alphabetically) are described in this section.

American Institute for Pollution Prevention (AIPP)

The American Institute for Pollution Prevention (AIPP) is a nonprofit organization of individuals representing industry trade associations and professional societies. AIPP is devoted to fostering and enhancing sustainable pollution prevention solutions to environmental problems. Sponsor organizations are the EPA and DOE. AIPP's goals are:

- To serve as a bridge for communication by providing information on pollution prevention resources available from its members and others
- To promote policies and provide for the transfer of information to encourage the culture shifts necessary for implementation of pollution prevention technologies throughout society
- To proactively set future directions for pollution prevention through cooperative and collaborative efforts among industry, government, and the public.

Enviroene

Enviroense, funded by EPA and the Strategic Environmental Research and Development Program, contains a wide collection of pollution prevention information accessible through the World Wide Web (http://envirosense.com) and an EPA bulletin board. It contains or is linked to pollution prevention resources and programs, including several of the resources described elsewhere in this section. Enviroene contains a host of information on pollution prevention, including technical information on grants, international resources, training opportunities, information notebooks and case studies for specific industries, and solvent substitution (known as the *Solvent Umbrella*). Abstracts and information texts can be searched by keywords.

Pollution prevention notebooks are available for 18 industrial sectors. Each notebook contains a comprehensive environmental profile, industrial process information, pollution prevention techniques, pollutant release data, regulatory requirements, compliance and enforcement history, innovative programs, and names of contacts for further information.

"Guided tours" of the Web database are also being developed for industrial sectors to help users find information. The tour or content guides include frequently asked questions (FAQs), visual content maps, subject indices, and core document lists.

National Pollution Prevention Center for Higher Education

In 1991, the EPA created the National Pollution Prevention Center for Higher Education (NPPC) to collect, develop, and disseminate educational materials on pollution prevention. The NPPC is located at the University of Michigan. The NPPC is a collaborative effort among business and industry, government, nonprofit organizations, and academia. The NPPC develops pollution prevention educational materials for college faculty. These materials are designed so that the principles of pollution prevention may be incorporated into existing or new courses. They contain resources for professors as well as assignments for students. Educational materials are available in a variety of disciplines and offer background material, case studies, problem sets, and multimedia resources to use in the classroom. The NPPC's research program develops new frameworks, principles, and tools for achieving sustainable development through industrial ecology and life-cycle assessment and design.

The NPPC offers an internship program, professional education and training, and conferences. Professional education courses are offered through the University of Michigan College of Engineering Continuing Education program and special courses for industry and government. The internship program promotes hands-on pollution prevention experience to undergraduate and graduate students in engineering, business, natural resources, and other fields of study. Students work with faculty mentors to develop educational materials based on their internships. Conferences and workshops organized by the NPPC are designed to identify and discuss critical issues in pollution prevention education and research.

National Pollution Prevention Roundtable

The National Pollution Prevention Roundtable (NPPR), located in Washington, D.C., is a membership organization that provides a national forum for pollution prevention. Voting membership in the roundtable includes state, local, and tribal government pollution prevention programs. Affiliate members include representatives from federal agencies, nonprofit groups, and private industry. The roundtable serves its members by:

- Providing access to the latest information on legislative and regulatory pollution prevention developments
- Preparing and distributing information on pollution prevention technologies and technical assistance practices
- Serving as a national pollution prevention information clearinghouse containing publications of state, local, and related programs

- Sponsoring semiannual national meetings that attract the largest gathering of pollution prevention experts from across the country

The NPPR publishes a national directory, called "P2 Yellow," of state, regional, and local pollution prevention programs that includes information, areas of expertise, contact names, and type description of each program.

P2Info

The DOE provides a clearinghouse of information on pollution prevention technologies and vendors call P2Info. The clearinghouse is operated by Pacific Northwest Laboratory (see related resource below).

Pacific Northwest Pollution Prevention Research Center

The Pacific Northwest Pollution Prevention Research Center (PPRC) is a nonprofit organization serving EPA Region 10 (Alaska, Idaho, Oregon, and Washington) and British Columbia. PPRC publishes a bimonthly newsletter called *Pollution Prevention Northwest* that covers topics such as industry-specific research and state pollution prevention programs. Industry-specific information is available for the fields of electronics, forestry, sulfite pulp processing, fish processing, wood products, and oil and gas. PPRC also maintains an online Pollution Prevention Research Projects Database.

Pollution Prevention Information Clearinghouse

EPA's Pollution Prevention Information Clearinghouse (PPIC) offers EPA documents and fact sheets about pollution prevention, including case studies and the Pollution Prevention Directory. PPIC also provides a referral service for technical questions.

Waste Reduction Innovative Technology Evaluation

EPA's Waste Reduction Innovative Technology Evaluation (WRITE) program focuses on pilot projects in pollution prevention performed in cooperation with state and local governments. Program priorities are new source reduction and recycling technologies. Goals of the program include identifying and evaluating new or improved economically favorable technologies for pollution prevention in industries presently experiencing waste problems. The program is designed to assist federal, state, and local governments, as well as small and midsized industries, by providing performance and cost information on pollution prevention technologies.

Waste Reduction Resources Center

The Waste Reduction Resources Center (WRRC) is a clearinghouse for pollution prevention information, focusing on industries in EPA Regions 3 and 4 (Alabama, Delaware, the District of Columbia, Florida, Georgia, Kentucky, Maryland, Mississippi, North Carolina, Pennsylvania, South Carolina, Tennessee, Virginia, and West Virginia). WRRC contains articles, case studies, and technical reports covering issues such as economic analyses, process descriptions, pollution prevention techniques, and implementation strategies. Upon request by an industry within its area, WRRC will prepare reports and information packets for specific facility or waste problems. These industries can also obtain on-site technical training assistance and pollution prevention audits. For facilities outside its area, WRRC will provide case summaries, contacts and referrals, and vendor information.

REFERENCES

Fiksell, J. (1996). *Design for the Environment,* McGraw-Hill, New York.

Freeman, H. M. (1995). "Developing and Maintaining a Pollution Prevention Program," in *Industrial Pollution Prevention Handbook,* McGraw-Hill, New York.

BIBLIOGRAPHY

Chemical Manufacturers Association (1991). *CMA Pollution Prevention Resource Manual,* CMA, Washington, DC.

Hanus, D. (1995). "Assessments Aid Waste Minimization Efforts," *Pollution Engineering,* vol. 27, no. 8.

Luper, D. (1995). "Process, Design Considerations Minimize Waste," *Pollution Engineering,* vol. 27, no. 6.

Mooney, G. A. (1992). "Pollution Prevention: Shrinking the Waste Stream," *Pollution Engineering,* vol. 24, no. 3.

Quinn, B. (1995). "Beyond the Big Stick: EPA as Business Aide," *Pollution Engineering,* vol. 27, no. 8.

——— (1995). "Finding the Right Recipes for Pollution Prevention," *Pollution Engineering,* vol. 27, no. 6.

——— (1995). "Info Sources Help Develop the Substance in the Middle," *Pollution Engineering,* vol. 27, no. 12.

——— (1995). "Panning for Gold," *Pollution Engineering,* vol. 27, no. 5.

Texas Natural Resource Conservation Commission (1996). "Household Hazardous Wastes: Alternatives and General Storage Directions," brochure, TNRCC, Austin, TX.

Water Environment Federation (undated). "Household Hazardous Waste: What You Should and Shouldn't Do," brochure, WEF, Alexandria, VA.

Woodside, G. (1993). *Hazardous Materials and Hazardous Waste Management: A Technical Guide,* Wiley, New York.

Woodside, G. and D. S. Kocurek (1997). *Environmental, Safety, and Health Engineering,* Wiley, New York.

——— (1994). *Resources and References: Hazardous Waste and Hazardous Materials Management,* Noyes Data Corporation, Park Ridge, NJ.

PART IV

INCIDENT AND EMERGENCY MANAGEMENT

18

PROCESS SAFETY MANAGEMENT

Section 304 of the CAAA requires the Secretary of Labor, in conjunction with the Administrator of EPA, to promulgate, pursuant to the Occupational Safety and Health (OSH) Act of 1970, a chemical process safety standard to prevent accidental releases of chemicals that could pose a threat to employees. This standard, delineated under 29 CFR 1910.119 and finalized on February 24, 1992, set forth requirements for an organized approach to process safety management (PSM). In addition, on May 24, 1996, the EPA finalized the last portions of its Accidental Release Prevention Provisions in 40 CFR Part 68, which are essentially commensurate to OSHA's PSM standard, with the exception that EPA's regulation requires off-site accidental release analysis in addition to the hazard assessment. Also, threshold quantities that invoke coverage under the standards differ in some cases. Examples of listed chemicals and threshold quantities are presented in Table 18.1.

Elements included in both standards, as specified in the CAAA, are process hazard assessment, employee education and training, documentation of operating and safety information, equipment maintenance and testing, emergency preparedness planning, and other management aspects. The standards apply primarily to manufacturing industries and other high-use chemical sectors. Examples of types of industries and other sectors that might be covered—depending on amounts of chemicals stored at the facility and other factors—are listed in Table 18.2. Covered employers are those companies that manufacture, store, or use regulated toxic and reactive chemicals in listed quantities. Flammable liquids and gases stored or used are covered if the amount is 10,000 pounds or greater. The standards do not apply for some specific activities, as defined in the regulations.

TABLE 18.1 Examples of Chemicals and Threshold Quantities Listed in OSHA's Process Safety Management Standard and EPA's Risk Management Program

Chemical Name	CAS Number	OSHA Threshold Quantity (lb)	EPA Threshold Quantity (lb)
Ammonia (anhydrous)	7664417	10,000	10,000
Arsine	7784421	100	1,000
Boron trichloride	10294345	2,500	5,000
Boron trifluoride	7637072	250	5,000
Dibrane	19287457	100	2,500
Hydrogen chloride (anhydrous)	7647010	5,000	5,000
Trichlorosilane	10025782	10,000	5,000

Source: 29 CFR 1910.119 and 40 CFR Part 68, Subpart F.

BASIC ELEMENTS OF A PROCESS SAFETY MANAGEMENT PROGRAM

Process Safety Information

All employers who have a process safety management program must compile written process safety information about the hazards of the highly hazardous chemicals used or produced by the process, information about the technology of the process, and information about the equipment used in the process. This information is necessary for process hazard analysis. Types of information to be included are as follow:

- Hazard information for each highly hazardous chemical—such as toxicity, permissible exposure limits, physical data, corrosivity data, thermal

TABLE 18.2 Examples of Manufacturing and Other Sectors That Are Likely to Be Covered Under the OSHA and EPA Standards

Manufacturing Industries	Other Sectors
Chemical manufacturing	Natural gas liquids
Chemical products	Farm product warehousing
Transportation equipment	Electric
Primary metals	Gas
Fabricated metal products	Sanitary services
Others	Wholesale trade
	Pyrotechnics/explosives manufacturers

and chemical stability data, and hazardous effects of inadvertent mixing of different materials

- Information pertaining to the technology of the process—such as a block flow diagram or simplified process flow diagram, process chemistry, maximum intended inventory, safe upper and lower limits for items such as temperatures, pressures, flows, or compositions, and an evaluation of the consequences of deviations, including those affecting the safety and health of employees
- Information pertaining to the equipment in the process—such as materials of construction, piping and instrument diagrams (P&IDs), electrical classification, relief system design and design basis, ventilation system design, design codes and standards employed, material and energy balances for processes built after May 26, 1992, and safety systems such as interlocks, detection of suppression system

Employee Involvement

Section 304 of the CAAA requires that employers involve employees in developing and implementing the process safety management program elements and hazard assessments. A written plan for methods of accomplishing this should be included in the employer's process safety management program. Existing health and safety programs can be used, as applicable, to ensure that employees are properly involved. In addition, the employer might want to initiate programs such as employee involvement teams, employee safety partnerships, and other programs in order to fulfill this obligation.

Process Hazard Analysis

The process hazard analysis (PHA) is a thorough, organized, and systematic approach used to identify, evaluate, and control the hazards of processes involving highly hazardous chemicals. The process must address or include the following:

- Hazards of the process
- Identification of any previous incident that had a potential for catastrophic consequences in the workplace
- Engineering and administrative controls applicable to the hazards and theirinterrelationships, such as appropriate application of detection methodologies to provide early warning of releases. Acceptable detection methods might include process monitoring and control instrumentation with alarms, and detection hardware such as hydrocarbon sensors
- Consequences of failure of engineering and administrative controls
- Facility siting

- Human factors
- Qualitative evaluation of a range of the possible safety and health effects on employees in the workplace if there is a failure of controls

Methodologies for performing a process hazard analysis are well documented and have been used throughout industry for many years. An excellent review of methods for performing a process hazard assessment is presented in *Guidelines for Hazard Evaluation Procedures* (AIChE 1992). Both OSHA and EPA require the use of one or more of the following methodologies (or equivalent methodology appropriate for the particular process) for reviewing processes involving highly hazardous chemicals over threshold amounts:

- *What-if.* This methodology is used for relatively uncomplicated processes. "What if..." questions are used to evaluate the effects of component failures at each process or material handling step.
- *Checklist.* A checklist approach is used for more complex processes. During this analysis, an organized checklist is developed and certain aspects of the process are assigned to committee members having the greatest experience or skill in evaluating those aspects. Operator practices and job knowledge are audited in the field, and suitability of equipment and materials of construction is studied. In addition, process chemistry and control systems are reviewed and operating and maintenance records are audited. A checklist evaluation typically precedes use of the more sophisticated processes.
- *What-if/checklist.* This combination methodology is a comprehensive approach that combines the creative thinking of a selected team of specialists with the methodical focus of a prepared checklist. The review team methodically examines the operation from receipt of raw materials to delivery of the finished product to the customer's site. Information acquired from this method can be used in training of operating personnel.
- *Hazard and operability study* (*HAZOP*). HAZOP is a formal method of systematically investigating each element of a system for all potential deviations from design conditions. Parameters such as flow, temperature, pressure, and time are reviewed in conjunction with material strength, locations of connections, and other design aspects. Piping and instrument designs (or plant model) are analyzed critically during this process for potential problems that could arise in each vessel or pipeline in the process. Potential causes for failure are evaluated, consequences of failure are reviewed, and safeguards against failure are assessed for adequacy.
- *Failure mode and effect analysis.* This technique is a methodical study of component failure. The analysis includes all system diagrams and reviews all components in the system that could fail, such as pumps, instrument transmitters, seals, temperature controllers, and other components. Consequences of failure, hazard class, probability of failure, and

detection methods are evaluated, and multiple concurrent failures are assessed.

- *Fault tree analysis.* This methodology is a qualitative or quantitative model of all undesirable outcomes from a specific event. It includes catastrophic occurrences such as explosion, toxic gas release, and rupture. A graphic representation of possible sequences results in a diagram that looks like a tree with branches. Assessment of probability using failure rate data is used to calculate probability of occurrence.

Table 18.3 presents examples of types of processes that might be suitable for each of the above-mentioned methodologies.

Other recognized hazard analysis methods (EPA 1987) are used frequently to augment the six methodologies described above. These include:

- *Preliminary hazard analysis.* This methodology is used for hazard identification during the preliminary phase of plant development. It is especially useful for new processes where there is limited past experience.
- *Human error analysis.* This analysis is a systematic evaluation of the factors that influence the performance of human operators, maintenance

TABLE 18.3 Examples of Applications Suitable for Selected Process Hazard Analysis Methodologies

Methodology	Applications
What-if	Applicable to all parts of a facility, such as bulk chemical systems, storage areas, and processing or manufacturing areas; useful during process development, pre-startup, and operation
Checklist	Applicable to equipment, materials, and procedures in a chemical process facility; useful during design, construction, startup, operation, and shutdown
What-if/checklist	Applicable to all parts of the facility, including equipment and operations of a process; useful during design, construction, and operation
Hazard and operability study	Applicable to systems and equipment with a high degree of operating variability; useful during late design and operation phases
Failure mode and effect analysis	Applicable to individual system and effect components; useful during design, construction, and operation
Fault tree analysis	Applicable to equipment and system design features and operational procedures that are interrelated; useful during design and operation phases

Source: Information from EPA (1987).

staff, technicians, and other personnel. Human errors that are likely to occur and that could cause an incident are identified.

- *Event tree analysis.* This analysis considers operator or safety system response to the initiation of an event. Incident consequences from varied responses (or failures in response) are assessed.
- *Cause–consequence analysis.* This methodology uses a blend of fault tree and event tree analysis to diagram the interrelationships between accident outcomes and their basic causes.
- *Dow and Mond indices.* These indices provide an empirical method for ranking the risks in a chemical process plant. Penalties are assigned to process materials and conditions that can cause accidents, while credits are assigned to plant safety procedures that can mitigate the effects of an accident. These indices are useful in the early stages of plant design.
- *Probabilistic risk assessment.* This assessment measures the overall risk through numerical evaluation of both accidental consequences and probabilities. This method generally is used to compare risks when alternative designs exist or to assess risk reduction strategies.
- *Safety review.* This review is used as a comprehensive facility inspection to identify facility conditions or procedures that might allow an accident or incident.

Other preliminary or screening analyses that can be used during process hazard analysis include review of chemical reactions caused by mixing of materials, evaluation of momentum forces such as pressure fluctuations or water hammers, and evaluation of physical characteristics such as liquid levels, temperature, and mass flow rates.

The team conducting the PHA needs to understand the methodology used, and the team leader should be fully knowledgeable about the methodology and the PHA process. The team should have expertise in the various areas of process technology, process design, operating procedures and practices, instrumentation, routine and nonroutine tasks, safety and health, and other relevant areas, as the need dictates. At least every five years after the completion of the initial process hazard analysis, the process hazard analysis must be updated and revalidated. Documentation pertaining to the PHA and updates or revalidation must be kept on file for the life of the facility.

Operating Procedures

Part of a process safety management program includes written operating procedures, consistent with the process safety information, that provide clear instructions for safely conducting activities involved in each covered process. Elements to be addressed in the procedures include the following:

- *Delineation of steps for each operating phase:* including initial startup; normal operations; temporary operations; emergency shutdown, including the conditions under which emergency shutdown is required and the assignment of shutdown responsibility to qualified operators to ensure that emergency shutdown is executed in a safe and timely manner; emergency operations; normal shutdown; and startup during turnaround or after an emergency shutdown
- *Operating limits:* including upper and lower boundaries of operating limits; consequences of deviation to these limits; steps required to correct or avoid deviation
- *Safety and health considerations:* including properties of, and hazards presented by, the chemicals used in the process; precautions necessary to prevent exposure, including engineering controls, administrative controls, and personal protective equipment; control measures to be taken if physical contact or airborne exposure occurs; quality control for raw materials and control of hazardous chemical inventory levels; any special or unique hazards; safety systems (e.g., interlocks, detection or suppression systems) and their functions

Training

Each employee presently involved in operating a process covered under process safety management standards, including maintenance and contractor employees, must be trained in an overview of the process and in its operating procedures. The training must include emphasis on the specific safety and health hazards, emergency operations including shutdown, and safe work practices applicable to the employee job tasks. If an employee was already operating the process on the effective date of the standards, the employer had the option, in lieu of initial training, of certifying in writing that the employee has the required knowledge, skills, and abilities to safely carry out the duties and responsibilities specified in the operating procedure.

Refresher training must be provided at least every three years, or more often if necessary, to ensure that the employee understands and adheres to the current operating procedures of the process. The employer, in consultation with the employees involved in operating the process, must determine the appropriate frequency of refresher training. Documentation of the training must be kept, including employee's identity, the date of training, and means used by the employer to verify that the employee understood the training.

Contractors

The process safety management program applies to contractors performing maintenance or repair, turnaround, major renovation, or specialty work on or adjacent to a covered process. It does not apply to contractors providing incidental services that do not influence process safety, such as janitorial, food and drink, laundry, delivery, or other supply services.

Employers who use contractors to perform work in and around processes that involve highly hazardous chemicals have to establish a screening process so that they hire and use only contractors who accomplish the desired job tasks without compromising the safety and health of any employees at the facility. Employer and contractor responsibilities with respect to process safety management are presented in Table 18.4.

Pre-Startup Safety Review

A pre-startup safety review must be conducted before any covered new process facilities or modified facilities—when the modification has been signif-

TABLE 18.4 Employer and Contractor Responsibility with Respect to Process Safety Management

Employer Responsibilities

- *Ensure safe work environment:* including obtaining and evaluating information regarding the contract employer's safety performance and programs; developing and implementing safe work practices consistent with standards to control the entrance, presence, and exit of contract employers and contract employees in covered process area; and evaluating periodically the performance of contract employers in fulfilling their obligations under the standards.
- *Provide information to contractor employer and employees:* including information of the known potential fire, explosion, or toxic release hazards related to the contractor's work and the process; and the applicable provisions of the emergency action plan.
- *Maintain a contract employee injury and illness log related to the contractor's work in the process areas.*

Contractor Employer Responsibilities

- *General safety requirements:* including ensuring that the contract employees are trained in the work practices necessary to perform their job safely; ensuring that the contract employees are instructed in the known potential fire, explosion, or toxic release hazards related to their job and the process, and in the applicable provisions of the emergency action plan; ensuring that each contract employee follows the safety rules of the facility, including the required safe work practices.
- *Provide information to employer:* including information of any unique hazards presented by the contract employer's work; information about safety training; other safety-related information.
- *Documentation:* including documentation that each contract employee has received and understood the training required by the standard by preparing a record that contains the identity of the contract employee, the date of training, and the means used to verify that the employee understood the training.

icant enough to require a change in the process safety information—can be brought on line. The pre-startup safety review includes confirmation that construction and equipment is in accordance with design specification, as well as assuring that safety, operating, maintenance, and emergency procedures are in place and are adequate. For new facilities, the pre-startup review ensures that a PHA has been performed and that recommendations have been resolved or implemented before startup, and for modified facilities, any changes, other than "in-kind replacements," to the facilities must go through management of change procedures. This includes updates to P&IDs, operating procedures and instructions, training, and other aspects of the PSM standard (see "Management of Change" later in this chapter.)

Mechanical Integrity of Equipment

Employers must establish and implement written procedures to maintain the ongoing integrity of process equipment, including pressure vessels and storage tanks, piping systems, relief and vent systems and devices, emergency shutdown systems, pumps, and controls such as monitoring devices and sensors, alarms, and interlocks. Each employee involved in maintaining the mechanical integrity of the equipment must be trained in the process hazards and the procedures applicable to the employee's job tasks. Inspection and testing must be performed on the process equipment using the following criteria:

- Procedures must follow recognized and generally accepted good engineering practices.
- Frequency of inspections and tests of process equipment must be consistent with applicable manufacturer's recommendations, state and federal regulations, and good engineering practices, and must be performed more frequently if determined to be necessary by prior operating experience.
- Testing documentation must include the date of the inspection or test, the name of the person who performed the inspection or test, the serial number or other identifier of the equipment on which the inspection or test was performed, a description of the inspection or test performed, and the results of the inspection or test.

Equipment deficiencies outside the acceptable limits defined by the process safety information must be corrected before further use, unless other steps are taken to ensure safe operation. In these cases the deficiencies must be corrected in a safe and timely manner. In constructing new plants and equipment, the employer must ensure that equipment as it is fabricated is suitable for the process application for which it will be used. Appropriate field checks and inspections should be performed to ensure that equipment is installed properly and is consistent with design specifications, engineering drawings, and the manufacturer's instructions.

Hot-Work Permit

The employer is responsible for issuing a hot-work permit for hot-work operations conducted on or near a covered process. The permit must document that the fire prevention and protection requirements specified in the standards are met. Further information, such as date of authorization for hot work and identity of the object on which the hot work is to be performed, must be included. The permit should be kept on file until the work is complete, or, as a good management practice, filed at the facility for a period of time designated by the safety professional.

Management of Change

Employers covered under the process safety management standards must establish and implement written procedures to manage changes to process chemicals, technology, equipment, and procedures, as well as changes to facilities that affect a covered process. Considerations to be addressed in the procedures include the technical basis for the proposed change, the impact of the change on employee safety and health, modifications to operating procedures, necessary time period for the change, and authorization requirement for the change.

Employees who operate a process and maintenance and contract employees whose job tasks will be affected by a change in the process must be informed of, and trained in, the change prior to startup of the process or startup of the affected part of the process. Updates to information, procedures, and other documents should also be made before startup.

Incident Investigation

Any incident that resulted in or reasonably could have resulted in a catastrophic release of a highly hazardous chemical in the workplace must be investigated by the employer. The incident investigation must be initiated as promptly as possible, but not later than 48 hours. An incident investigation team must be established and must include at least one person knowledgeable of the process, and if a contractor is involved, a contract employee must also participate. A report must be prepared at the conclusion of the investigation which includes the date and time of the incident, the date the investigation began, a description of the incident, factors contributing to the incident, and recommendations resulting from the investigation. The report must be reviewed by all affected personnel whose job tasks are relevant to the incident findings, including contract employees, where applicable. The incident investigation must be retained for five years. In some cases, if the release is also regulated by other agencies or authorities, investigation and reporting must meet these requirements, which may be different and possibly more stringent.

Emergency Planning and Response

Emergency planning and response are required for the entire facility. The plan must include procedures for handling catastrophic and small releases. In some cases, the hazardous waste and emergency response provisions under 29 CFR 1910.120(a). (p). and (q) also apply. Emergency planning and response at the facility and community level is discussed in Chapter 19.

Compliance Audits

To ensure that the process safety management program is effective, employers must certify that they have evaluated compliance with the provisions of the process safety management program every three years. The compliance audit must be conducted by at least one person knowledgeable in the process. A findings report must be written, and the employer must determine and document an appropriate response to each of the findings of the compliance audit. Corrected deficiencies must be documented, and the most recent two compliance audit reports should be kept on file.

Trade Secrets

Employers must make available all information necessary to comply with PSM to those persons responsible for compiling the process safety information, those developing the PHA, those responsible for developing the operating procedures, and those performing incident investigation, emergency planning and response, and compliance audits, without regard to the possible trade secret status of such information. However, nothing in PSM precludes the employer from requiring those persons to enter into confidentiality agreements not to disclose the information.

ACCIDENTAL RELEASE PREVENTION PROVISIONS

As mentioned earlier in the chapter, EPA's accidental release prevention provisions essentially mirror OSHA's process safety management program, with the exception that EPA's standard requires a risk management plan (RMP) that includes offsite consequence analysis—or accidental release modeling analysis—in addition to the hazard process assessment described previously. In the following sections we detail this additional requirement, which is delineated in 40 CFR Part 68, Subpart G.

General Overview

Depending on the type and serverity of the chemical processes at a facility, owners of covered processes must follow one of three programs outlined in the regulations, with the more "at risk" processes requiring the most comprehensive off-site consequence analysis. Eligibility requirements for the three programs are presented in Table 18.5.

Owners of covered processes—regardless of which program the process fall under—must submit a single RMP that includes an executive summary, a completed registration form, required off-site consequence analysis or analyses, the five-year accident history of the process, the emergency response program, and certification. Some additional information pertaining to prevention is required for programs 2 and 3.

TABLE 18.5 Eligibility Requirements for Programs 1, 2, and 3 under 40 CFR Part 68

Program 1 Eligibility Requirements

- For five years prior to the submission of the risk management plan, the process has not had an accidental release of a regulated substance where exposure to the substance, its reaction products, overpressure generated by an explosion involving the substance, or radiant heat generated by a fire involving the substance led to off-site death, injury, or response or restoration activities for an exposure of an environmental receptor.
- The distance to a toxic or flammable endpoint[a] for a worst-case release assessment is less than the distance to any public receptor, including residential populations, institutions, parks and recreational aras, and major commercial, office and industrial buildings.
- Emergency response procedures have been coordinated between the stationary source and local emergency planning and response organizations.

Program 2 Eligibility Requirements

- Does not meet the eligibility requirements of program 1 or 3.

Program 3 Eligibility Requirements

- The covered process does not meet the requirements of Program 1.
- The process is in any of the following SIC codes:

—SIC Code 2611	Pulp mills
—SIC Code 2812	Alkalies and chlorine
—SIC Code 2819	Industrial inorganic chemicals
—SIC Code 2821	Plastic materials, synthetic resins, and nonvolcanizable elastomers
—SIC Code 2865	Cyclic organic crudes and intermediates, and organic dyes and pigments
—SIC Code 2869	Industrial organic chemicals
—SIC Code 2873	Nitrogenous fertilizers
—SIC Code 2879	Pesticides and agricultural chemicals
—SIC Code 2911	Petroleum refining

- The process is subject to the OSHA process safety management standard

[a]Toxic or flammable endpoints are defined in 40 CFR 68.22.

Accidental Release Modeling Analysis

Accidental release modeling analysis must be conducted for each program, as follows:

- *Program 1 requirements.* Analyze one worst-case release scenario for each process covered.
- *Programs 2 and 3 requirements.* Analyze one worst-case release scenario to represent all regulated toxic substances held above the threshold quantity and one worst-case release scenario to represent all regulated flammable substances held above the threshold quantity; additional worst-case release scenarios for a hazard class are required if a worst-cast release from another covered process at the stationary source potentially affects public receptors differently from those potentially affected by the worst-case release scenario already developed.

A worst-case release is defined as a 10-minute release from the largest vessel (or process) with passive or administrative mitigation measures (e.g., concrete dike) under the worst-case meteorological condition. A worst-case scenario may include a passive safety control system, but not an active system (e.g., water curtain). because of the failure possiblity of an active safety measure. Off site consequences to be considered include consequences of toxics and flammables—with explosion, radiant heat/exposure time, and lower flammablility limit being subset consequences of flammables.

Data elements required for accidental release source modeling analysis are presented in Table 18.6. As can be seen from the table, the data consist of substance properties, system condition, environmental and meteorolgical factors, and source parameters.

Source Considerations

In addtion to considering the source parameters detailed in Table 18.6, the source should be characterized as to whether an accidental release may involve single-phase or two-phase releases of a pure compound or a mixture. A *single-phase release* consists of only gas *or* liquid, whereas a *two-phase release* consists of a gas *and* a liquid. An example of the latter is a release of pressurized liquefied gas as a result of a rupture of the container. In this case, the release can form both gas and a liquid.

Another consideration to evaluate when perfoming accidental release modeling is the possiblity of choked flow. When compounds are stored under high pressure, a release from a valve failure will exhibit choked flow until the storage pressure drops to approximately twice the ambient pressure. For modeling purposes, the maximum gas jet velocity is the sonic velocity of the release material. Other considerations for input into the model include the possiblity of flashing of a liquefied gas, nonideal solutions for multicomponent releases, and mitigation measures.

TABLE 18.6 Data Elements Required for Accidental Release Modeling

Substance Properties	
Molecular weight	Boiling point
Vapor pressure	Latent heat
Heat capacities	
System Conditions	
Quantity of substance held in a system	Release-point opening area
System pressure	Spill area
System temperature	Safety measurements
Environmental and Meteorological Factors	
Anemoneter height	Ground temperature
Stability class	Wind speed
Ambient temperature	Humidity
Source Parameters	
Source height	Emission duration
Source diameter (area)	Vapor cloud temperature
Emission rate	Vapor density
Emission velocity	Vapor heat capacity

REFERENCES

Federal Register, 55 FR 29150, July 17, 1990.

Environmental Protection Agency (1987). *Prevention Reference Manual: User's Guide Overview for Controlling Accidental Releases of Air Toxics,* EPA/600/8-87/028, prepared by Radian Corporation for Office of Research and Development, Air and Energy Engineering Research Laboratory, Research Triangle Park, NC.

——— (1991). *Environmental Protection Agency Accidental Release Questionnaire,* OMB 2050-0065, U.S. EPA, Washington, DC.

BIBLIOGRAPHY

American Institute of Chemical Engineers (1988a). *Guidelines for Safe Storage and Handling of High Toxic Hazard Materials,* prepared by Arthur D. Little, Inc., and Richard LeVine for Center for Chemical Process Safety, AIChE, New York.

——— (1988b). *Guidelines for Vapor Release Mitigation,* prepared by R. W. Prugh and R. W. Johnson for Center for Chemical Process Safety, AIChE, New York.

——— (1992). *Guidelines for Hazard Evaluation Procedures,* 2nd ed., Center for Chemical Process Safety, AIChE, New York.

American Institute of Chemical Engineers–Center for Chemical Process Safety (1993). *Guidelines for Auditing Process Safety Management Systems,* AIChE–CCPS, New York.

——— (1993). *Guidelines for Engineering Design for Process Safety,* AIChE–CCPS, New York.

——— (1994). *Guidelines for Chemical Process Documentation,* AIChE–CCPS, New York.

——— (1994). *Guidelines for Evaluating the Characteristics of Vapor Cloud Explosions, Flash Fires, and Bleves,* AIChE–CCPS, New York.

——— (1994). *Guidelines for Implementing Process Safety Management Systems,* AIChE–CCPS, New York.

——— (1994). *Guidelines for Process Safety Fundamentals for General Plant Operations,* AIChE–CCPS, New York.

Chemical Manufacturers Association (1985). *Community Awareness and Emergency Response Program Handbook,* CMA, Washington, DC.

——— (1985). *Process Safety Management (Control of Acute Hazards).* CMA, Washington, DC.

——— (1986). *Site Emergency Response Planning Handbook,* CMA, Washington, DC.

——— (1987). *An Analysis of Risk Assessment Methodologies for Process Emissions from Chemical Plants,* CMA, Washington, DC.

——— (1989). *Evaluation Process Safety in the Chemical Industry: A Manager's Guide to Quantitative Risk Assessment,* CMA, Washington, DC.

Davis, D. S., et al. (1989). *Accidental Releases of Air Toxics: Prevention, Control and Mitigation,* Noyes Data Corporation, Park Ridge, NJ.

Drivas, P. J., (1995). "A Review of Source Emission Models for Accidental Releases," *Proceedings of the 88th Annual Meeting of the Air and Waste Management Association,* 95-WA54A.03, AWMA, Pittsburgh, PA.

Environmental Protection Agency (1992). *Workbook of Screening Techniques for Assessing Impacts of Toxic Air Pollutants (Revised).* EPA-454/R-92-024, U.S. EPA, Research Triangle Park, NC.

——— (1993). *Guidance on the Application of Refined Dispersion Models to Hazardous Toxic Air Pollutant Releases,* EPA-454/R-93-002, U.S. EPA, Research Triangle Park, NC.

——— (1993). *Guiding Principles for Chemical Accident Prevention, Preparedness, and Response,* Office of Solid Waste and Emergency Response, U.S. EPA, Washington, DC.

——— (1996). *RMP Off-Site Consequence Analysis Guidance,* U.S. EPA, Research Triangle Park, NC.

Hanna, S. R., and P. J. Drivas (1987). *Guidelines for Use of Vapor Cloud Dispersion Metals,* AIChE–CCPS, New York.

Kelly, R. B. (1989). *Industrial Emergency Preparedness,* Van Nostrand Reinhold, New York.

National Response Team (1987). *Hazardous Materials Emergency Planning Guide,* National Response Team of the National Oil and Hazardous Substances Contingency Plan, Washington, DC.

Occupational Safety and Health Administration (1992). *Standard for Process Safety Management of Highly Hazardous Chemicals,* 29 CFR 1910.119, OSHA, Washington, DC.

——— (1993). *Process Safety Management Guidelines for Compliance,* OSHA 313, OSHA, U.S. Department of Labor, Washington, DC.

Office of Management and Budget (1988). "Using Community Right to Know: A Guide to a New Federal Law," *OMB Watch,* Washington, DC.

19

EMERGENCY PLANNING AND RESPONSE

Accidental releases of hazardous constituents and the prevention of these releases have been a focus in U.S. and international societies for several decades. Incidents like those that have occurred in Chernobyl, Seveso, and Bophal have made the public aware that there are risks associated with activities that use hazardous materials. As a result, legislation aimed at decreasing or eliminating hazard potential and developing or improving emergency response planning has been developed in the United States, as well as internationally through the European Economic Community and other organizations.

EPA regulations that address the issue of hazard assessment and emergency response planning were developed as part of the Emergency Planning and Community Right-to-Know Act (EPCRA) incorporated into Title III of the Superfund Amendments and Reauthorization Act of 1986 (known as SARA Title III). These regulations required each community to set up a formal Local Emergency Planning Committee (LEPC), which has the mission of assessing hazards in the area and developing an emergency response plan. The LEPC includes representation from locally elected officials, the fire department, police, the civil defense unit, hospitals, health and first-aid groups, local environmental entities, broadcast and print media, industry, and citizens from the community. This extensive representation allows input into the planning process from all affected parties in the local area. In addition, a State Emergency Response Commission (SERC) is required in every state to aid in emergency response coordination. In addition, OSHA passed a rule that addresses process safety management of highly hazardous chemicals, which is described in Chapter 18.

Any unplanned release of hazardous materials or hazardous waste must be handled by trained responders, as set forth by OSHA in 29 CFR §1910.120. The foundation of emergency response, as outlined by the NFPA, is the incident command system, which is used widely for incident management. The incident command system provides structure, coordination, and effectiveness in emergency situations, thus enhancing the safety of all responding personnel. In this chapter we cover the topics of emergency planning and response and includes information about incident mitigation.

LOCAL EMERGENCY RESPONSE PLANNING

The EPCRA or SARA Title III regulations required all LEPCs to develop a plan for emergency response to incidents that could occur in a designated local district such as a county or township. The deadline for the plans to be submitted to the EPA and the SERC was October 1988. Updates to the original plan are required if there are changes in local hazards or emergency response procedures.

Developing a Plan

Several guidance documents are available on the subject of local emergency planning, including:

- *Hazardous Materials Emergency Planning Guide,* published by the National Response Team (NRT) of the National Oil and Hazardous Substances Contingency Plan
- *Criteria for Review of Hazardous Materials Emergency Plans,* also published by the NRT
- *Guide for Development of State and Local Emergency Operations Plans,* published by the Federal Emergency Management Agency

Developing a thorough, workable emergency response plan can be a large task, depending on the number of facilities and amounts of chemicals stored in the local area. Elements to be included in a local emergency response plan are summarized in Table 19.1.

Assessing Local Hazards

An integral part of developing a local emergency response plan is a hazard analysis. A hazard analysis is a systematic method of identifying hazards that could affect the community and types of emergencies that could occur as a result of these hazards. Basic elements included in the analysis as identified in EPA's emergency planning guide, *Technical Guidance for Hazard Analysis* (EPA 1987), are: hazards identification, vulnerability analysis, and risk analysis. Utilizing all three elements in a hazard analysis allows for evaluation of:

TABLE 19.1 Elements to Include in a Local Emergency Response Plan

Planning Factors

- Identification and description of all facilities in the district that possess extremely hazardous substances
- Identification of other facilities that may contribute to risk in the district or may be subject to risks as a result of being within close proximity of facilities with extremely hazardous substances
- Documentation of methods for determining that a release of extremely hazardous substances has occurred and the area of population likely to be affected by a release
- Other optional information, such as findings from the hazard analysis, geographical features, demographical features, and other planning information

Concept of Operations

- Designation of community emergency coordinator and facility emergency coordinators, who will make determinations necessary to implement the plan

Emergency Notification Procedures

- A description of procedures for providing reliable, effective, and timely notification by the facility coordinators and community emergency coordinator to persons designated in the plan and to the affected public that a release has occurred
- Other optional information such as emergency hotline numbers and lists of names and numbers of organizations and agencies that are to be notified in the event of a release; optional description of methods to be used by facility emergency coordinators to notify community and state emergency coordinators of a release

Direction and Control

- Descriptions of methods and procedures to be followed by facility owners and operators and local emergency and medical personnel to respond to a release of extremely hazardous substances
- Optional information identifying organizations and persons who provide direction and control during the incident, including chain of command

Resource Management

- A description of emergency equipment and facilities in the community, and identification of persons responsible for such equipment and facilities
- Optional list of all personnel resources available for emergency response

Health and Medical

- Descriptions of methods and procedures to be followed by facility owners and operators and local emergency and medical personnel to respond to a release of extremely hazardous substances
- Optional information on major types of emergency medical services in the district and neighboring districts, including emergency medical services, first aid, triage, ambulance service, and emergency medical care

TABLE 19.1 ***(Continued)***

Personal Protection of Citizens

- Descriptions of methods in place in the community and in each of the affected facilities for determining areas likely to be affected by a release
- Optional description of methods for indoor protection of the public

Personal Protective Measures and Evacuation Procedures

- A description of evacuation plans, including those for precautionary evacuations and alternative traffic routes
- Optional information on precautionary evacuations of special populations and information on mass care facilities that provide food, shelter, and medical care to relocated populations

Procedures for Testing and Updating the Plan

- Descriptions of methods and schedules for exercising the emergency response plan

Training

- Descriptions of the training programs, including schedules for training of local emergency response and medical personnel

Optional Information (Recommended for Inclusion)

- General information, including a description of essential information to be recorded in an actual incident and signatures of LEPC chairperson and other officials and industry representatives endorsing the plan
- Instructions for plan use and record of amendments, including listings of organizations and persons receiving the plan or plan amendments, and other data about the dissemination of the plan
- Description of communication methods among responders
- Identification of warning systems and emergency public notification
- Description of methods used for public information and community relations, prior to any emergency
- Descriptions of procedures for responders to enter and leave the incident area, including safety precautions, medical monitoring, sampling procedures, and designation of personal protective equipment
- Identification of major tasks to be performed by firefighters, including a listing of fire response and HAZMAT personnel
- A description of the command structure for multiagency/multijurisdictional incident management systems
- Descriptions of major law enforcement tasks related to responding to releases, including security-related tasks
- Descriptions of methods to assess areas likely to be affected by an ongoing release
- A description of agencies responsible for providing emergency human services
- A description of the chain of command for public works actions and a listing of major tasks
- A description of major containment and mitigation activities for major types of HAZMAT incidents

TABLE 19.1 (*Continued*)

- Descriptions of major methods for cleanup
- A list of reports required following and incident and methods of evaluating response activities
- Other information outlined by the National Response Team as appropriate for inclusion in an emergency plan

Source: Information from NRT (1988).

- The potential for a situation to cause injury to life or damage to the environment
- The susceptibility of life or property to such injury or damage
- The probability that such injury or damage will occur

There are many complexities associated with an inclusive analysis, particularly for large industrialized areas; thus this type of analysis may be too costly or impractical to perform. For these cases an analysis of major hazards may be sufficient.

Hazard Identification. This part of the hazard analysis includes information about facilities in the area and transportation situations that have the potential for causing injury to the public or damage to the environment. To develop this information, hazardous materials that are (or could be) maintained in the area should be considered. Facilities that store or use hazardous materials might include:

- Industrial facilities
- Storage facilities/warehouses
- Public works facilities such as water and wastewater treatment plants
- Hospitals, education, and governmental facilities
- Trucking, rail, air, and other transportation terminals
- Waste disposal and treatment facilities
- Nuclear facilities

In addition, transportation corridors and types of hazardous hemicals likely to be transported through the area via highways, railways, and waterways should be included in the hazard identification.

Vulnerability Analysis. This part of the analysis identifies local area vulnerabilities that are susceptible to damage if a release occurs. Normally, worst-case conditions are assumed when making this assessment. Aspects to consider include:

- Location of release or spill

- Potential size of release or spill
- Wind direction and speed
- Population zones and locations of schools, nursing homes, and other institutions
- Location of essential life support areas within the affected zone, such as power plants, major transportation corridors, and water supplies
- Environmentally sensitive areas within the zone, such as wildlife refuges and endangered species habitat

For releases to air, the use of dispersion models can aid in predicting the zone of impact and contaminant concentrations. For releases to the ground, geographical maps can aid in predicting the zone of impact if the release reaches rivers or streams. Potential groundwater contamination should also be reviewed.

Risk Analysis. During this part of the analysis, the probability of a major release occurring, along with the type of harm to people or damage to property expected from a release, is evaluated. To understand risks appropriately, the facility of concern should be contacted to assess the following:

- Safety features of equipment, including leak detection systems, automatic shutoff features, fail-safe systems, and alarm capabilities for early warning
- Training programs for safe handling techniques
- Spill prevention and countermeasure plans
- Facility emergency plans
- Completion of process hazard analyses for highly toxic chemical processes and/or chemical bulk storage areas

For planning on a community-wide scale, the hazard analysis can provide an overall understanding of hazards in the area. Populations affected, transportation corridors affected, incident probability, and incident severity must all be evaluated to help a community prioritize its emergency preparedness and response activities.

Hazard analysis specific to processes is also used to evaluate potential risk in the community. Methodologies for this type of analysis vary and are presented in Chapter 18.

EMERGENCY RESPONSE

Effective emergency response and mitigation can prevent an unplanned release from becoming a disaster. In the following section we provide information about the incident command system, setting up response zones, and techniques that can be used to mitigate the incident.

Incident Command System

The incident command system requires emergency responders to perform in assigned roles during the incident. The incident commander is at the head of the command system, and all decisions pertaining to the incident are made by this person, with advice and counsel from qualified experts. This ensures a strong central command, which is necessary to keep the incident response organized and safe. The number and types of personnel needed to respond to an incident will depend on the size and complexity of the incident. Necessary personnel and their functions for proper management of an incident are documented by NFPA and others. Examples of the types of personnel that typically are needed during an incident and types of assignments or duties expected of these personnel are listed in Table 19.2.

In addition to these resources, other "behind the scenes" personnel will be needed to carry out other duties, including:

- Bringing water and food to the area for people involved in the incident

TABLE 19.2 Personnel Who Might Be Needed During a Hazardous Materials Incident

Personnel	Responsibilities
Incident commander	Establish and manage the incident response plan; allocate resources, assign activities, manage information, and ensure response is completed per plan Sector officers Manage geographical areas, including hazardous materials response teams in that area, as needed; provide specific functions, such as serving as safety officer
Command staff	Assist incident commander and/or sector officers; gather data such as chemical information, weather data, and status of other operations
Police or security	Manage the location and activities of the general public
Hazardous materials response teams	Directly manage and mitigate the hazardous materials release under the direction of a specified team leader
Communications personnel	Perform central communication function for incident, including requests of additional emergency resources; notify proper agencies of the release; answer media questions
Technical information specialist	Provide expertise directly to the incident commander in terms of chemical information and exposure hazards

Source: Information from NFPA (1992) and Noll et al. (1988).

- Acquiring additional emergency equipment, such as shovels, blankets, raingear, and other items that are needed
- Acquiring decontamination equipment and other equipment
- Calling families of the responders to inform them of the status of their spouse's return home
- Performing other errands as the situation warrants

Levels of training for hazardous materials (HAZMAT) incident responders is defined in 29 CFR §1910.120. Requirements for each of the five OSHA-defined levels of response are outlined in Table 19.3. Requirements for personnel such as medical professionals and communications personnel are not cited in the OSHA regulations, but training should be adequate to perform duties successfully.

Command Post and Zoning During the Incident

The incident command post is the first area to be established during a hazardous materials incident, and the position of this post is determined by wind factors and materials released. Setting up the command post $\frac{1}{2}$ to 2 miles from the incident is not uncommon. Once the command post is set, other zones can be set up by appropriate responders. Figure 19.1 illustrates the appropriate zones that should be established prior to an response to the emergency. As can be seen in the figure, the zones include the support zone, the limited access zone, and the restricted or hazard zone. The designated activities in each of these zone are as follows:

- *Support zone*. The command post is within this zone, and communication equipment for dispatching and calling outside agencies is maintained within the command post. Staging of equipment and resources occurs in this zone, and key technical personnel such as the safety officer, medical support, and the communications specialist are stationed in this zone. Backup resources to the responders also reside in this zone until needed, and decontaminated equipment and personnel exit through this zone. Nonessential personnel such as news media should remain outside this zone, and the communications specialist should exit the support zone to communicate with these personnel.
- *Limited access zone* (*contamination reduction corridor*). Within this zone are personnel who are integral to managing the incident. Persons who are awaiting entry into or returning from the restricted zone are staged in this zone. Decontamination activities, including personnel and equipment decontamination, occur in this zone. There is one point of entry into the restricted zone and one point of entry into and out of the limited access zone that are established early and enforced. Protective equipment is required in this zone, and the HAZMAT team leader and/or other control officer maintains communication equipment.

TABLE 19.3 OSHA Requirements for HAZMAT Training

First Responder Awareness Level (e.g., security or police)

- Must demonstrate a basic understanding of what hazardous materials are and the risks associated with them in an incident. This includes understanding the potential outcomes associated with an emergency created when hazardous materials are present.
- Must be able to recognize the presence of hazardous materials in an emergency and, if possible, identify the hazardous materials.
- Must demonstrate an understanding of the role of the first-responder awareness person in the site emergency plan. Must also be able to realize the need for additional resources and make appropriate notifications to the communication center.

First Responder Operations Level (e.g., responders stationed in the decontamination zone or people who contain the leak at a distance using diking or other techniques)

- Must demonstrate competency in requirements of first responder awareness level.
- Must demonstrate knowledge of the basic hazard and risk assessment techniques, and must have a basic understanding of hazardous materials terms.
- Must be able to select and use proper personal protective equipment (PPE).
- Must know how to perform basic control, containment, and/or confinement operations within the capabilities of the resources and PPE of the unit. Must know how to implement basic decontamination procedures as well as understand relevant operating procedures and termination procedures.

Hazardous Materials Technician (e.g., responders mitigating the incident)

- Must demonstrate competency in requirements of first responder operations level.
- Must know how to implement the site emergency response plan.
- Must know how to identify, classify, and verify known and unknown materials by using field survey instruments.
- Must be able to function within an assigned role in the incident command system.
- Must know how to select and use specialized chemical PPE provided to the hazardous materials technician.
- Must understand hazard and risk assessment techniques.
- Must be able to perform advanced control, containment, and confinement operations within the capabilities of the resources and PPE available to the unit, and must understand decontamination procedures.
- Must understand termination procedures.
- Must understand basic chemical and toxicological terminology and behavior.

Hazardous Materials Specialist (e.g., safety officer, chemical specialist, or other technical information specialist)

- Must demonstrate competency in requirements of hazardous materials technician.
- Must know how to implement the local emergency response plan, as well as understand the state emergency response plan.
- Must understand in-depth hazard and risk techniques.
- Must be able to determine and implement decontamination procedures.

TABLE 19.3 (*Continued*)

- Must have the ability to develop a site safety and control plan.
- Must understand chemical, radiological, and toxicological terminology and behavior.

On-Scene Incident Commander

- Must demonstrate competency in requirements of first responder awareness level and first responder operations level.
- Must know and be able to implement the site incident command system.
- Must know how to implement the site emergency response plan and the local emergency response plan.
- Must know of the state emergency response plan and the federal regional response team.
- Must know and understand the hazards and risks associated with employees working in chemical protective clothing.
- Must know and understand the importance of decontamination procedures.

Source: Adapted from 29 CFR 1910.120(q)(6).

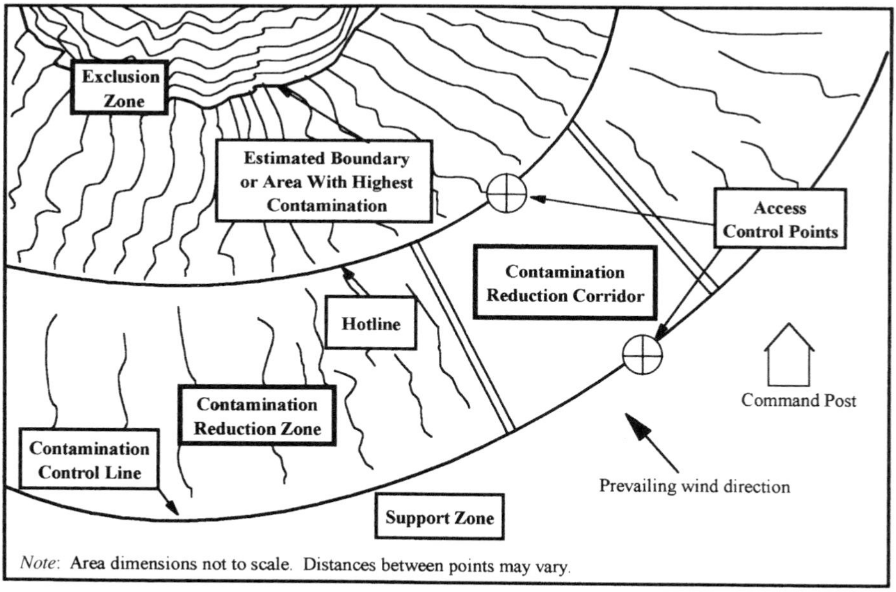

Figure 19.1 Zoning used during emergency response and mitigation.

- *Restricted (exclusion) zone.* Personnel entering this zone—often termed the hazard zone or hot zone—are those who are responding to and mitigating the incident. Two-way communication equipment is carried by the responding personnel. Full protective equipment (usually SCBA and fully encapsulating suit) is worn in this zone.

Use of precisely marked zones ensures that activities can be carried out safely and in a methodical way. In addition, the area directly outside the support zone should be protected by security or the police, and only a limited number of persons who are prespecified—such as the media or local or state officials—should be allowed near the support zone for briefings by designated support personnel.

Techniques for Incident Mitigation

General. Techniques for incident mitigation are determined by the materials involved in the incident. Priorities during the incident always include:

- *Stopping the release.* This can be done by turning off flow valves, plugging or patching the hole, or capping the release mechanism. In the case of fire, it must be suppressed first by means of water or foam suppressants before other activities can take place.
- *Stopping the spread of material.* This can be done by closing storm drain valves, by diking or berming the area, or by diverting the flow to a confined area.
- *Cleanup of the released material in a timely fashion so as to limit seepage into waterways or groundwater.* This can be done using adsorbents, neutralizers, earth moving equipment, and other cleanup techniques.

Factors that are considered in selecting the proper incident management and mitigation techniques include:

- *Chemical hazard (if known).* This includes physical properties such as specific gravity, flash point, vapor pressure, corrosivity reactivity, and toxicity.
- *Physical situation.* This includes aspects such as container type, other chemicals in area, release type and rate, and LEL readings or the existence of fire.
- *Worker/community hazard.* This includes things such as distance of the incident to community and potential impact via air or stream, the need for evacuation of workers and/or community members, and the need for response from outside agencies such as state or national emergency response teams.

In all cases, incident management techniques should be selected so as to disallow or reduce, to the extent possible, contamination of area soils and waterways from runoff associated with the incident.

In-Plant Incidents. Since approximately three-fourths of reportable incidents occur at fixed facilities, it is appropriate during a hazard assessment to focus on methods that mitigate potential for these types of incidents. Process hazard analysis is one of the best methods of pinpointing factors that can cause (or avert) a disastrous incident or accident (see Chapter 18). Other release mitigation techniques can be used and include items such as prerelease controls,

TABLE 19.4 Examples of Release Mitigation Techniques

Category	Specific Mitigation Measures
Prerelease controls	Preventive maintenance of equipment; regular equipment inspections; regular equipment testing; comprehensive safety audit; preinstallation assessment of equipment designs; process controls for operations monitoring; alarm capability for early warning of problems or out-of-spec conditions; regular upgrading of equipment; investigation of all spills and releases and assessment of similar equipment for the same problems; development of standard operating procedures; pre-startup equipment checks; release prevention equipment
Prerelease protection equipment	Containment for tanks, process equipment, overhead piping; neutralization capabilities; flares or incinerators; adsorbers; scrubbers; spray curtains; emergency equipment, including spill cleanup supplies, firefighting equipment, air monitors, and other necessary equipment
Management activities	Employee safety training, hazard communication training, emergency response training, certification of operators on equipment or system; membership in community emergency planning groups; development of a release control program; development of an accident/incident investigation program; participation in research/conferences; development of a safety loss prevention program; formalized corrective action process for deviation from rules; formalized notification procedures for accidental releases; initiation of a program to improve system designs
Safety systems and procedures	Backup systems; redundant systems; value lockout program; automatic shutoffs; bypass and surge systems; manual overrides; interlocks; alarms; formalized safety procedures; formalized testing program for all safety equipment; formalized color coding/labeling program

Source: Information from EPA (1991).

prerelease protection equipment, management activities, and safety systems and procedures. Table 19.4 provides information about these release mitigation techniques.

Release Mitigation Techniques During Transport of Hazardous Materials. The potential for a release of hazardous material during transport is a concern of many cities and communities, since railways, waterways, and federal or state highways are not controlled locally. EPA data show that approximately 25% of the incidents surveyed are in-transit incidents, with incidents involving trucks being the most common, followed by railcars, water vessels, and pipelines (EPA 1985).

Mitigating techniques for the transport of hazardous chemicals and wastes include proper maintenance of trucks and railcars, proper maintenance of highways and railroad tracks, and adequate training for the transport operators. Proper shipping papers, which define the chemicals in transit, are a necessity in the event of a release. These include bills of lading for highway transportation, waybills for rail shipments, and dangerous cargo manifests for shipment by water. In addition, manifests are required when shipping hazardous waste, PCBs, and low-level radioactive waste. The Chemical Manufacturer's Association operates a chemical information service, CHEMTREC, which was established during the 1970s. This service provides a toll-free hotline to persons needing spill response and other emergency information.

Mitigating techniques have been established for the transport of radioactive wastes in terms of waste packaging requirements. Type A and type B packages for transporting intermediate and large quantities of radioactive materials, respectively, are specified by international radioactive transportation regulations (DOE 1990) as well as by DOE in 10 CFR Part 71. These packages must pass tests in terms of free drop, compression, penetration, water immersion, and extended heating to ensure that they will not be damaged in the event of a transportation accident.

REFERENCES

International Fire Service Training Association (1988). *Hazardous Materials for First Responders,* Fire Protection Publications, Oklahoma State University, Stillwater, OK.

Noll, G. G., M. S. Hildebrand, and J. G. Yvorra (1988). *Hazardous Materials: Managing the Incident,* Fire Protection Publications, Oklahoma State University, Stillwater, OK.

National Fire Protection Association (1987). *Technical Guidance for Hazard Analysis,* Washington, DC.

——— (1989). *Standard for Professional Competence of Responders to Hazardous Materials,* NFPA 472, NFPA, Quincy, MA.

——— (1990). *Fire Department Incident Management System,* NFPA 1561, NFPA, Quincy, MA.

——— (1991). *Environmental Protection Agency Accidental Release Questionnaire,* OMB # 2050-0065, Washington, DC.

——— (1992a). *Hazardous Materials Response Handbook,* Gary Tokle, ed., NFPA, Quincy, MA.

——— (1992b). *Recommended Practice for Responding to Hazardous Materials Incidents,* NFPA 471, NFPA, Quincy, MA.

National Response Team (1988), "Criteria for Review of Hazardous Materials Emergency Plans," Washington, DC.

BIBLIOGRAPHY

Chemical Manufacturers Association (1987). *Teamwork: Safe Handling of a Hazardous Materials Incident,* CMA, Washington, DC. videotape.

——— (1990). "CHEMNET Brochure," CMA, Washington, DC.

——— (1992). "National Chemical Response and Information Center (NCRIC) Brochure," CMA, Washington, DC.

——— (1992). "CHEMTREC Brochure," CMA, Washington, DC.

——— (1992). "Chemical Referral Center Brochure," CMA, Washington, DC.

Department of Energy (1991). *OSHA Training Requirements for Hazardous Waste Operations,* DE92 004780, DOE/EH-0227P, Office of Environment, Safety and Health, U.S. DOE, Washington, DC.

Environmental Protection Agency (1985). *Field Standard Operating Procedures for Establishing Work Zones F.S.O.P. 6,* OSWER Directive 9285.2-04, Office of Emergency and Remedial Response, U.S. EPA, Washington, DC.

——— (1988). *EPA Standard Operating Safety Guides,* OSWER Directive 9285.1-01C, Office of Emergency and Remedial Response, Environmental Response Team, U.S. EPA, Washington, DC.

——— (1989). *EPA Health and Safety Audit Guidelines,* EPA/540/G-89/010, OSWER Directive 9285.8-02, Office of Solid Waste and Emergency Response, Emergency Response Division, U.S. EPA, Washington, DC.

Fire Protection Publications (1983). *Incident Command System,* Oklahoma State University, Stillwater, OK.

Theodore, L., J. P. Reynolds, and F. B. Taylor (1989). *Accident and Emergency Management,* Wiley, New York.

Varela, J., ed. (1996). *Hazardous Materials Handbook for Emergency Response,* Van Nostrand Reinhold, New York.

APPENDICES

APPENDIX A

LIST OF ACRONYMS

AA	Atomic absorption
ACGIH	American Conference of Governmental Industrial Hygienists
ACL	Alternate concentration limit
AEC	Atomic Energy Commission
AIChE	American Institute of Chemical Engineers
AIChE-CCPS	American Institute of Chemical Engineers–Center for Chemical Process Safety
AIHA	American Industrial Hygiene Association
ANPR	Advanced notice of proposed rulemaking
ANSI	American National Standards Institute
API	American Petroleum Institute
ASHRAE	American Society of Heating, Refrigeration, and Air-Conditioning Engineers
ASME	American Society of Mechanical Engineers
ASTM	American Society for Testing and Materials
BACT	Best Available Control Technology
BTEX	Benzene, toluene, ethylbenzene, and xylene
C	Ceiling
CAA	Clean Air Act
CAAA	Clean Air Act Amendments
CERCLA	Comprehensive Environmental Response, Compensation, and Liability Act
CFC	Chlorofluorocarbon
CFR	Code of Federal Regulations
CGI	Combustible gas indicator
CPC	Chemical protective clothing
CWA	Clean Water Act
DFE	Design for environment

DOE	Department of Energy
DOT	Department of Transportation
EMS	Environmental management system
EPA	Environmental Protection Agency
FID	Flame ionization detector
FIFRA	Federal Insecticide, Fungicide, and Rodenticide Act
FML	Flexible membrane liner
FR	Federal register
FTIR	Fourier transform infrared spectrometer
GC	Gas chromatograph
GC/MS	Gas chromatograph/mass spectrometer
HAZMAT	Hazardous materials
HAPS	Hazardous air pollutants
HLW	High-level waste (radioactive)
HMTA	Hazardous Materials Transportation Act
HONS	Hazardous organic air pollutants
HSWA	Hazardous and Solid Waste Amendments
ICS	Incident command system
IDLH	Immediately dangerous to life or health
IR	Infrared analyzer
ISC	Industrial source complex
ISO	International Organization for Standardization
LAER	Lowest achievable emission rate
LEL	Lower explosive limit
LEPC	Local Emergency Planning Committee
LLW	Low-level waste (radioactive)
MACT	Maximum achievable Control technology
MQL	Method quantitation limit
MS	Mass spectrometer
MSDS	Material safety data sheet
MSHA	Mine Safety and Health Administration
NAAQS	National ambient air quality standards
NACE	National Association of Corrosion Engineers
NEPA	National Environmental Policy Act
NESHAP	National Emission Standards for Hazardous Air Pollutants
NFPA	National Fire Protection Association
NIOSH	National Institute for Occupational Safety and Health
NPDES	National pollution discharge elimination system
NSPS	New source performance standards
NRC	National Response Center
NRC	Nuclear Regulatory Commission
NTIS	National Technical Information System
NVLAP	National Voluntary Laboratory Accreditation Program
ODC	Ozone-depleting chemical
OPA	Oil Pollution Act

OSHA	Occupational Safety and Health Administration
OSH Act	Occupational Safety and Health Act
OVA	Organic vapor analyzer
PCBs	Polychlorinated biphenyls
PEL	Permissible exposure limit
PI	Plasticity index
PID	Photoionization detector
PMN	Premanufacture notice
POTW	Publicly owned treatment works
PPE	Personal protective equipment
PQL	Practical quantitation limit
RACT	Reasonably available control technology
RCRA	Resource Conservation and Recovery Act
REL	Recommended exposure limit
SARA	Superfund Amendments and Reauthorization Act
SCBA	Self-contained breathing apparatus
SDWA	Safe Drinking Water Act
SE	Synchronous excitation (fluorescence spectroscopy)
SERC	State Emergency Response Commission
SOCMI	Synthetic organic chemical manufacturing
SPC	Statistical process control
SPCC	Spill prevention, control and countermeasure
STEL	Short-term exposure limit
SWE	Single-wavelength excitation (fluorescence spectroscopy)
TCLP	Toxicity characteristic leaching procedure
TLV-TWA	Threshold limit value-time weighted average
TQM	Total quality management
TRU	Transuranic waste (radioactive)
TSCA	Toxic Substance Control Act
TWA	Time-weighted average
UEL	Upper explosive limit
UIC	Underground injection control
UL	Underwriters' Laboratories
UNAMAP	User's Network for the Applied Modeling of Air Pollution
USEPA	U.S. Environmental Protection Agency
USGS	U.S. Geological Survey
UV	Ultraviolet
VPP	Voluntary Protection Program
WRAP	Waste reduction assessments program
WREAFS	Waste reduction evaluations at federal sites
WRITE	Waste reduction innovative technology evaluation

APPENDIX B

SELECTED STANDARDS OF THE AMERICAN NATIONAL STANDARDS INSTITUTE

Standards Related to Radiation Monitoring

ANSI N13.1-1969(R1993)—Guide to Sampling Airborne Radioactive Materials in Nuclear Facilities

ANSI N13.2-1969(R1982)—Administrative Practices in Radiation Monitoring (A Guide for Management)

ANSI N13.3-1969(1981)—Dosimetry for Criticality Accidents

ANSI N13.5-1972(R1989)—Performance and Specification for Direct Reading and Indirect Reading Pocket Dosimeters for X- and Gamma-Radiation

ANSI N13.6-1966(R1989)—Practice for Occupational Radiation Exposure Records System

ANSI N13.7-1983(R1989)—Criteria for Photographic Film Dosimeter Performance

ANSI N13.11-1993—Criteria for Testing Personnel Dosimeter Performance

ANSI N13.14-1994—Internal Dosimetry Standards for Tritium

ANSI N13.15-1985—Dosimetry Systems, Performance of Personnel Thermoluminescence

ANSI N13.27-1981(R1992)—Dosimeters and Alarm Ratemeters, Performance Requirements for Pocket-Sized Alarm

ANSI N42.6-1980(R1991)—Interrelationship of Quartz-Fiber Electrometer Type Exposure Meters and Companion Meter Chargers

ANSI N42.14-1991—Calibration and Usage of Germanium Detectors for Measurement of Gamma-Ray Emission of Radionuclides

ANSI N42.15-1990—Performance Verification of Liquid-Scintilliation Counting Systems

ANSI N42.17B-1989(R1994)—Performance Specifications for Health Physics Instrumentation—Occupational Airborne Radioactivity Monitoring Instrumentation

ANSI N42.18-1980(R1991)—Specification and Performance of On-Site Instrumentation for Continuously Monitoring Radioactivity in Effluents

ANSI N317-1980(R1991)—Performance Criteria for Instrumentation Used for Inplant Plutonium Monitoring

ANSI N322-1977(R1991)—Inspection and Test Specification for Direct and Indirect Reading Quartz Fiber Pocket Dosimeters

ANSI N545-1975(R1993)—Performance Testing and Procedural Specifications for Thermoluminescence Dosimetry: Environmental Applications

ANSI/IEEE 309-1970(R1991)—Test Procedures for Geiger–Müller Counters

ANSI/IEEE 325-1986—Test Procedures for Germanium Gamma-Ray Detectors

ANSI/IEEE 398-1972 (R1991)—Standard Test Procedures for Photo-Multipliers for Scintillation Counting and Glossary for Scintillation Counting Field

Standards Related to Transportation of Radioactive Materials

ANSI N14.1-1995—Nuclear Materials—Uranium Hexafluoride—Packaging for Transport

ANSI N14.5-1987—Radioactive Materials—Leakage Tests on Packages for Shipment

ANSI N14.6-1993—Radioactive Materials—Special Lifting Devices for Shipping Containers Weighing 10,000 Pounds (4,500 kg) or More

ANSI N14.19-1986—Ancillary Features of Irradiated Fuel Shipping Casks

ANSI N14.24-1985(R1985)—Domestic Barge Transport of Highway Route Controlled Quantities of Radioactive Materials

ANSI N14.27-1986(R1993)—Carrier and Shipper Responsibilities and Emergency Response Procedures for Highway Transportation Accidents Involving Truckload Quantities of Radioactive Materials

Other Standards Related to Radioactive Materials

ANSI/ASME N509-1989—Nuclear Power Plant Air-Cleaning Units and Components

ANSI/ASME N510-1989R(1995)—Testing of Nuclear Air-Cleaning Systems

ANSI/ASME N626-1990—Qualifications and Duties for Authorized Nuclear Inspection Agencies and Personnel

ANSI/ASME N626.3-1993—Qualifications and Duties of Specialized Professional Engineers

ANSI/ASME NQA-1-1994—Quality Assurance Requirements for Nuclear Facility Applications

ANSI/ASME NQA-3-1989—Quality Assurance Program Requirements for the Collection of Scientific and Technical Information for Site Characterization of High-Level Nuclear Waste Repositories

ANSI/ASME OM-1990—Operation and Maintenance of Nuclear Power Plants

Standards Related to Personnel Protective Equipment

ANSI/ISEA 101-1993—Size and Labeling Requirements for Limited-Use and Disposable Coveralls

ANSI/NFPA 1971-1991—Protective Clothing for Structural Fire Fighting

ANSI Z87.1-1989—Practice for Occupational and Educational Eye and Face Protection

ANSI Z88.2-1992—Respiratory Protection

ANSI Z89.1-1986—Protective Headwear for Industrial Workers

ANSI Z358.1-1990—Emergency Eyewash and Shower Equipment

Standards Related to Ventilation

ANSI Z9.2-1979(1991)—Fundamentals Governing the Design and Operation of Local Exhaust Systems

ANSI Z9.4-1985—Ventilation and Safe Practices of Abrasive Blasting Operations

ANSI/ASHRE 41.2-1987(R1992)—Standard Methods for Laboratory Air Flow Measurement

ANSI/ASHRE 41.3-1989—Method for Pressure Measurement

ANSI/ASHRE 41.6-1994—Method of Measurement of Moist Air Properties

ANSI/ASHRE 55-1992—Thermal Environmental Conditions for Human Occupancy

ANSI/ASHRE 62-1989—Ventilation for Acceptable Indoor Air Quality

ANSI/ASHRE 62a-1991—Ventilation for Acceptable Indoor Air Quality

ANSI/ASHRE 110-1995—Method of Testing Performance of Laboratory Fume Hoods

ANSI/ASHRE 111-1988—Practices for Measurement, Testing, Adjusting, and Balancing of Building Heating, Ventilation, Air-Conditioning, and Refrigeration Systems

ANSI/ASHRE 113-1990—Method of Testing for Room Air Diffusion

ANSI/ASHRE 114-1986—Energy Management Control Systems Instrumentation

ANSI/ASHRE/IEEE 90A-1-1988—New Building Design, Energy Conservation

ANSI/ASHRE/IES 100.4 (1984)—Energy Conservation in Existing Facilities—Industrial

Standards Related to Tank and Piping Systems

ANSI/ASME B31G-1991—Manual for Determining the Remaining Strength of Corroded Pipelines

ANSI/ASME B31.3-1993—Chemical Plant and Petroleum Refinery Piping

ANSI/ASME B31.4-1992—Liquid Transportation Systems for Hydrocarbons, Liquid Petroleum Gas, Anhydrous Ammonia, and Alcohol

ANSI/ASME B31.8-1995—Gas Transmission and Distribution Piping Systems

ANSI/ASME B31.11-1989—Slurry Transportation Piping Systems

Standards Related to Waste Treatment and Pollution Prevention Equipment

ANSI/ASME PTC 33-1978(R1991)—Performance Test Code for Large Incinerators

ANSI/ASME PTC 33a-1980(R1987)—Performance Test Code for Large Incinerators

ANSI/ASME PTC 40-1991—Flue Gas Desulfurization Units

ANSI/ASME QRO-1-1994—Qualification and Certification of Resource Recovery Facility Operators

ANSI/ASME SPPE-1-1994—Quality Assurance and Certification of Safety and Pollution Prevention Equipment Used in Offshore Oil and Gas Operations

ANSI/ASME SPPE-2-1994—Accreditation of Testing Laboratories for Safety and Pollution Prevention Equipment Used in Offshore Oil and Gas Operations

APPENDIX C

SELECTED STANDARDS OF THE AMERICAN SOCIETY FOR TESTING AND MATERIALS

Standards Related to Air Monitoring

ASTM D1356-95A(1995)—Standard Definitions of Terms Relating to Atmospheric Sampling and Analysis

ASTM D1357-95—Standard Practice for Planning the Sampling of the Ambient Atmosphere

ASTM D1704-95—Test Method for Determining the Amount of Particulate Matter in the Atmosphere by Measurement of the Light Absorptance of a Filter Sample

ASTM D2009-65(1990)—Practice for Collection by Filtration and Determination of Mass, Number, and Optical Sizing of Atmospheric Particulates

ASTM D-2913-87(1991)—Standard Test Method for Mercaptan Content of the Atmosphere

ASTM D2914-1995—Test Method for Sulfur Dioxide Content of the Atmosphere (West–Gaeke Method)

ASTM D3162-94—Test Method for Carbon Monoxide in the Atmosphere (Continuous Measurement by Nondispersive Infrared Spectrometry)

ASTM D3249-95—Recommended Practice for General Ambient Air Analyser Procedures

ASTM D3608-91—Test Method for Nitrogen Oxides (Combined) Content in the Atmosphere by the Greiss–Saltzman Reaction

ASTM D3686-95—Standard Practice for Sampling Atmospheres to Collect Organic Compound Vapours (Activated Charcoal Tube Adsorption Method)

ASTM D3687-95—Standard Practice for Analysis of Organic Compound Vapours Collected by Activated Charcoal Tube Adsorption Method (Using Gas/Liquid Chromatography)

ASTM D3824-95—Standard Test Method for Continuous Measurement of Oxides of Nitrogen in the Ambient or Workplace Atmosphere by the Chemiluminescent Method

ASTM D4096-91—Test Method for Determination of Total Suspended Particulate Matter in the Atmosphere (High-Volume Sampler Method)

ASTM D5149-90—Test Method for Ozone in the Atmosphere: Continuous Measure by Ethylene Chemiluminescence

ASTM D4240-83(1989)—Standard Test Method for Airborne Asbestos Concentration in Workplace Atmosphere

ASTM D4490-90—Standard Practice for Measuring the Concentration of Toxic Gases or Vapours Using Detector Tubes

ASTM D4532-92—Test Method for Respirable Dust in Workplace Atmospheres

ASTM D4597-92—Practice for Sampling Workplace Atmospheres to Collect Organic Gases or Vapours with Activated Charcoal Diffusional Samplers

ASTM D4599-90—Standard Practice for Measuring the Concentration of Toxic Gases or Vapours Using Length-of-Stain Dosimeter

ASTM D4844-88(1993)—Guide for Air Monitoring at Waste Management Facilities for Worker Protection

ASTM D4861-94A—Standard Practice for Sampling and Analysis of Pesticides and Polychlorinated Biphenyls in Indoor Atmospheres

ASTM D4947-94—Practice for Chlordane and Heptachlor Residues in Indoor Air

ASTM D5015-95—Test Method for pH of Atmospheric Wet Deposition Samples by Electrometric Determination

ASTM E1370-90—Guide to Air Sampling Strategies for Worker and Workplace Protection

ASTM F328-80(1989)—Practice for Determining Counting and Sizing Accuracy of an Airborne Particle Counter Using Near-Monodisperse Spherical Particulate Materials

ASTM G91-92—Practice for Monitoring Atmospheric SO_2 Using Sulfation Plate Technique

Standards Related to Testing of Protective Fabrics

ASTM D751-95—Test Methods for Coated Fabrics

ASTM D1003-92—Test Method for Haze and Luminous Transmittance of Transparent Fabrics

ASTM D1043-92—Test Method for Stiffness Properties of Plastics as a Function of Temperature by Means of a Torsion Test

ASTM D2582-93—Test Method for Puncture Propagation Tear Resistance of Plastic Film and Thin Sheeting

ASTM D4157-92—Test Method for Abrasion Resistance of Textile Fabrics (Oscillatory Cylinder Method)

ASTM F739-91—Test Method for Resistance of Protective Clothing Materials to Penetration by Liquids or Gases Under Conditions of Continuous Contact

ASTM F903-90—Test Method for Resistance of Protective Clothing Materials to Penetration by Liquids

ASTM F1001-89(1993)—Guide for Selection of Chemicals to Evaluate Protective Clothing Materials

ASTM F1052-87(1991)—Practice for Pressure Testing of Gas-Tight Totally-Encapsulating Chemical Protective Suits

Standards Related to Radiation Protection

ASTM E1167-87—Guide for a Radiation Protection Program for Decommissioning Operations

ASTM E1168-87—Guide for Radiological Protection Training for Nuclear Facility Workers

Standards Related to Cathodic Protection

ASTM G8-90—Test Method for Cathodic Disbonding of Pipeline Coatings

ASTM G19-88—Standard Test Method for Disbonding Characteristics of Pipeline Coatings by Direct Soil Burial

ASTM G42-90—Method for Cathodic Disbonding of Pipeline Coatings Subjected to Elevated or Cyclic Temperatures

ASTM G80-88(1992)—Test Method for Specific Cathodic Disbonding of Pipeline Coatings

Standards Related to Tank Testing

ASTM A275/A 275M-94—Method for Magnetic Particle Examination of Steel Forgings

ASTM A754-79(1990)—Test Method for Coating Thickness by X-ray fluorescence

ASTM C868-85(1990)—Standard Test Method for Chemical Resistance of Protective Linings

ASTM C982-88(1992)—Guide for Selecting Components for Generic Energy Dispersive X-ray Fluorescence (XRF) Systems for Nuclear Related Material Analysis

ASTM C1118-89(1994)—Guide for Selecting Components for Wavelength-Dispersive X-ray Fluorescence (XRF) Systems

ASTM D471-95—Test Method for Rubber Property—Effect of Liquids

ASTM D3491-92—Standard Methods of Testing Vulcanizable Rubber Tank and Pipe Lining

ASTM E125-63(1993)—Reference Photographs for Magnetic Particle Indications on Ferrous Castings

ASTM E165-95—Practice for Liquid Penetrant Examination

ASTM E432-91—Guide for the Selection of a Leak Testing Method

ASTM E433-71(1993)—Reference Photographs for Liquid Penetrant Inspection

ASTM E709-95—Guide Magnetic Particle Examination

ASTM E750-88—Practice for Characterizing of Acoustic Emission Instrumentation

ASTM E1002-94—Method for Testing for Leaks Using Ultrasonics

ASTM E1003-84(1994)—Method for Hydrostatic Leak Testing

ASTM E1065-92—Guide for Evaluating Characteristics of Ultrasonic Search Units

ASTM E1067-89—Practice for Acoustic Emission Testing of Fiberglass Reinforced Plastic Resin (FRP) Tanks/Vessels

ASTM E1139-92—Practice for Continuous Monitoring of Acoustic Emission from Metal Pressure Boundaries

ASTM E1172-87(1992)—Practice for Describing and Specifying a Wavelength-Dispersive X-ray Spectrometer

ASTM E1208-94—Test Method for Fluorescent Liquid Penetrant Examination Using the Lipophilic Post-emulsification Process

ASTM E1209-94—Test Method for Fluorescent Penetrant Examination Using the Water-Washable Process

ASTM E1210-94—Test Method for Fluorescent Penetrant Examination Using the Hydrophilic Post-emulsification Process

ASTM E1211-87(1992)—Practice for Leak Detection and Location Using Surface-Mounted Acoustic Emission Sensors

ASTM E1219-94—Test Method for Fluorescent Penetrant Examination Using the Solvent-Removable Process

ASTM E1220-92—Test Method for Visible Penetrant Examination Using the Solvent-Removable Process

ASTM G1-90(1994)—Recommended Practice for Preparing, Cleaning, and Evaluating Corrosion Test Specimens

ASTM G4-95—Guide for Conducting Corrosion Coupon Tests in Plant Equipment

ASTM G5-94—Reference Test Method for Making Potentiostatic and Potentiodynamic Anodic Polarization Measurements

ASTM G15-93—Terminology Relating to Corrosion and Corrosion Testing

ASTM G41-90(1994)—Practice for Determining Cracking Susceptibility of Metals Exposed Under Stress to a Hot Salt Environment

ASTM G46-94—Recommended Practice for Examination and Evaluation of Pitting Corrosion

ASTM G48-92—Testing for Pitting and Crevice Corrosion Resistance of Stainless Steels and Related Alloys by Use of Ferric Chloride

ASTM G49-85(1990)—Recommended Practice for Preparation and Use of Direct Tension Stress Corrosion Test Specimen

ASTM G50-76(1992)—Recommended Practice for Conducting Atmospheric Corrosion Tests on Metals

ASTM G51-95—Test Method for pH of Soil for Use in Corrosion Testing

ASTM G62-87(1992)—Test Method for Holiday Detection in Pipeline Coatings

ASTM G71-81(1992)—Practice for Conducting and Evaluating Galvanic Corrosion Tests in Electrolytes

ASTM G78-95—Guide for Crevice Corrosion Testing of Iron Base and Nickel Base Stainless Alloys in Seawater and Other Chloride-Containing Aqueous Environments

ASTM G82-83(1993)—Guide for Development and Use of a Galvanic Series for Predicting Corrosion Performance

ASTM G90-94—Practice for Performing Accelerated Outdoor Weathering of Nonmetallic Materials Using Concentrated Natural Sunlight

ASTM G96-90—Guide for On-Line Monitoring of Corrosion in Plant Equipment (Electrical and Electrochemical Methods)

ASTM G97-89—Test Method for Laboratory Evaluation of Magnesium Sacrificial Anode Test Specimens for Underground Applications

ASTM G101-94—Guide for Estimating the Atmospheric Corrosion Resistance of Low-Alloy Steels

ASTM G104-89(1993)—Test Method for Assessing Galvanic Corrosion Caused by the Atmosphere

Other Standards Related to Tank Systems

ASTM D2487-93—Standard Test Method for Classification of Soils for Engineering Purposes

ASTM D3299-88—Specification for Filament-Wound Glass Fiber Reinforced Thermoset Resin Chemical-Resistant Tanks

ASTM D4021-92—Standard Specification for Glass-Fiber-Reinforced Polyester Underground Petroleum Storage Tanks

ASTM D4097-88—Specification for Contact Molded Glass-Fiber-Reinforced Thermoset Resin Chemical-Resistant Tanks

ASTM D4865-91—Guide for Generation and Dissipation of Static Electricity in Petroleum Fuel Systems

ASTM F670-87(1994)—Specification for Tanks—5 and 10 Gallon Lube Oil Dispensing

Standards Related to Testing of Landfill Liners

ASTM D2434-68(1994)—Test Method for Permeability of Granular Soils (Constant Head)

ASTM D4491-92—Test Methods for Water Permeability of Geotextiles by Permittivity

ASTM D4716-87—Test Method for Constant Head Hydraulic Transmissivity (In-Plane Flow) of Geotextiles and Geotextile Related Products

ASTM D4751-93—Test Method for Determining Apparent Opening Size of a Geotextile

Standards Related to Ground Water Monitoring

ASTM C150-95—Specification for Portland Cement

ASTM D1785-94—Specification for Poly Vinyl Chloride (PVC) Plastic Pipe, Schedules 40, 80, and 120

ASTM D2113-83(1993)—Practice for Diamond Core Drilling for Site Investigation

ASTM F480-94—Specification for Thermoplastic Water Well Casing Pipe and Couplings Made in Standard Dimension Ratio (SDR)

Standards Related to Soils Investigation

ASTM D420-93—Recommended Practice for Investigating and Sampling Soil and Rock for Engineering Purposes

ASTM D421-85(1993)—Practice for Dry Preparation of Soil Samples for Particle-Size Analysis and Determination of Soil Constants

ASTM D422-63(1990)—Method for Particle-Size Analysis of Soils

ASTM D1452-80(1990)—Practice for Soil Investigation and Sampling by Auger Borings

ASTM D1586-84(1992)—Method for Penetration Test and Split-Barrel Sampling of Soils

ASTM D1587-94—Method for Thin-Walled Tube Sampling of Soils

ASTM D2216-90—Test Method for Laboratory Determination of Water (Moisture) Content of Soil and Rock

ASTM D2217-85(1993)—Practice for Wet Preparation of Soil Samples for Particle-Size Analysis and Determination of Soil Constants

ASTM D2487-93—Classification of Soils for Engineering Purposes

ASTM D2488-93—Practice for Description and Identification of Soils (Visual-Manual Procedure)

ASTM D4318-93—Test Method for Liquid Limit, Plastic Limit, and Plasticity Index of Soils

ASTM D4452-85(1990)—Methods for X-ray Radiography of Soil Samples

ASTM D5079-90—Practices for Preserving and Transporting Rock Core Samples

APPENDIX D

SELECTED STANDARDS OF THE NATIONAL FIRE PROTECTION ASSOCIATION

Standards Related to Fire Protection Systems

NFPA 11 (1994)—Low Expansion Foam and Combined Agent Systems
NFPA 11A (1994)—Medium and High Expansion Foam Systems
NFPA 11C (1995)—Mobile Foam Apparatus
NFPA 12 (1993)—Carbon Dioxide Extinguishing Systems
NFPA 12A (1992)—Halon 1301 Fire Extinguishing Systems
NFPA 13 (1994)—Installation of Sprinkler Systems
NFPA 15 (1990)—Water Spray Fixed Systems
NFPA 16 (1995)—Installation of Deluge Foam-Water Sprinkler and Foam-Water Spray Systems
NFPA 16A (1994)—Installation of Closed-Head Foam-Water Sprinkler Systems
NFPA 17 (1994)—Dry Chemical Extinguishing Systems
NFPA 25 (1995)—Inspection, Testing, and Maintenance of Water Based Fire Protection Systems

Standards Related to Identification of Fire Hazards of Chemicals and Materials

NFPA 49 (1994)—Hazardous Chemicals Data
NFPA 321 (1991)—Basic Classification of Flammable and Combustible Liquids

NFPA 325 (1994)—Fire Hazard Properties of Flammable Liquids, Gases, and Volatile Solids
NFPA 491M (1991)—Hazardous Chemical Reactions
NFPA 495 (1992)—Explosives Materials Code
NFPA 704 (1990)—Identification of Fire Hazards of Materials

Standards Related to Storage of Hazardous Materials

NFPA 30 (1993)—Flammable and Combustible Liquids Code
NFPA 30B (1994)—Manufacture and Storage of Aerosol Products
NFPA 40E (1993)—Storage of Pyroxylin Plastics
NFPA 43B (1993)—Storage of Organic Peroxide Formulations
NFPA 43D (1994)—Storage of Pesticides
NFPA 55 (1993)—Standard for Storage, Use and Handling of Compressed and Liquefied Gases in Portable Cylinders
NFPA 58 (1995)—Standard for Storage and Handling of Liquefied Petroleum Gases
NFPA 59 (1995)—Storage and Handling of Liquefied Petroleum Gases at Utility Gas Plants
NFPA 59A (1996)—Production, Storage, and Handling of Liquefied Natural Gas (LNG)
NFPA 231C (1995)—Rack Storage of Materials
NFPA 327 (1993)—Cleaning or Safeguarding Small Tanks and Containers Without Entry
NFPA 329 (1992)—Underground Leakage of Flammable and Combustible Liquids
NFPA 490 (1993)—Storage of Ammonium Nitrate

Standards Related to Personal Protection

NFPA 1404 (1996)—Fire Department Self-Contained Breathing Apparatus Program
NFPA 1500 (1992)—Fire Department Occupational Safety and Health Programs
NFPA 1971 (1991)—Protective Clothing for Structural Fire Fighting
NFPA 1972 (1992)—Helmets for Structural Fire Fighting
NFPA 1973 (1988)—Gloves for Structural Fire Fighting
NFPA 1974 (1992)—Protective Footwear for Structural Fire Fighting
NFPA 1976 (1992)—Protective Clothing for Proximity Fire Fighting
NFPA 1981 (1992)—Open Circuit Self-Contained Breathing Apparatus for Fire Fighters

NFPA 1991 (1994)—Vapor-Protective Suits for Hazardous Chemical Emergencies

NFPA 1992 (1994)—Liquid Splash-Protective Suits for Hazardous Chemical Emergencies

NFPA 1993 (1994)—Support Function Protective Garments for Hazardous Chemical Operations

NFPA 1999 (1992)—Protective Clothing for Emergency Medical Operations

Standards Related to Manufacturing and Handling of Hazardous Materials

NFPA 35 (1995)—Manufacture of Organic Coatings

NFPA 91 (1995)—Exhaust Systems for Air Conveying of Materials

NFPA 480 (1993)—Storage, Handling, and Processing of Magnesium

NFPA 481 (1995)—Production, Processing, Handling, and Storage of Titanium

NFPA 482 (1987)—Production, Processing, Handling, and Storage of Zirconium

NFPA 650 (1990)—Pneumatic Conveying Systems for Handling Combustible Materials

Standards Related to Fire Protection of Specific Industries or Processes

NFPA 32 (1990)—Dry Cleaning Plants

NFPA 33 (1995)—Spray Application Using Flammable and Combustible Materials

NFPA 34 (1995)—Dipping and Coating Processes Using Flammable or Combustible Liquids

NFPA 36 (1993)—Solvent Extraction Plants

NFPA 45 (1991)—Fire Protection for Laboratories Using Chemicals

NFPA 120 (1994)—Coal Preparation Plants

NFPA 654 (1994)—Prevention of Fire and Dust Explosions in the Chemical, Dye, Pharmaceutical, and Plastics Industries

NFPA 801 (1995)—Facilities Handling Radioactive Materials

NFPA 802 (1993)—Nuclear Research and Production Reactors

NFPA 803 (1993)—Light Water Nuclear Power Plants

NFPA 820 (1995)—Fire Protection for Wastewater Treatment and Collection Facilities

Standards Related to Ventilation and Exhaust Systems

NFPA 90A (1993)—Installation of Air Conditioning and Ventilating Systems
NFPA 90B (1993)—Installation of Warm Air Heating and Air Conditioning Systems
NFPA 204M (1991)—Smoke and Heat Venting

Standards Related to Life Safety

NFPA 69 (1992)—Explosion Prevention Systems
NFPA 101 (1991)—Life Safety Code
NFPA 101A (1995)—Alternate Approaches to Life Safety
NFPA 110 (1996)—Emergency and Standby Power Systems
NFPA 111 (1996)—Stored Energy Emergency and Standby Power Systems
NFPA 170 (1994)—Standard Fire Safety Symbols
NFPA 220 (1995)—Types of Building Construction
NFPA 241 (1993)—Safeguarding Construction, Alteration and Demolition Operations
NFPA 251 (1995)—Standard Methods of Tests of Fire Endurance Building Construction and Materials
NFPA 252 (1995)—Fire Tests of Door Assemblies
NFPA 253 (1995)—Test for Critical Radiant Flux of Floor Covering Systems Using a Radiant Heat Energy Source
NFPA 255 (1996)—Method of Test of Surface Burning Characteristics of Building Materials
NFPA 256 (1993)—Methods of Fire Test of Roof Coverings
NFPA 257 (1996)—Fire Tests for Window and Glass Block
NFPA 780 (1995)—Installation of Lightning Protection Systems

Standards Related to Transportation of Hazardous Materials

NFPA 306 (1993)—Control of Gas Hazards on Vessels
NFPA 385 (1990)—Tank Vehicles for Flammable and Combustible Liquids
NFPA 386 (1990)—Portable Shipping Tanks for Flammable and Combustible Liquids
NFPA 1124 (1995)—Manufacture, Transportation, and Storage of Fireworks

Standards Related to Incident Management

NFPA 471 (1992)—Responding to Hazardous Materials Incidents

NFPA 472 (1992)—Professional Competence of Responders to Hazardous Materials Incidents

NFPA 473 (1992)—Competencies for EMS Personnel Responding to Hazardous Materials Incidents

NFPA 495 (1992)—Explosive Materials Code

NFPA 902M (1990)—Fire Reporting Field Incident Manual

NFPA 903 (1992)—Fire Reporting Property Survey Manual

NFPA 904 (1992)—Incident Follow-up Report Manual

NFPA 906M (1993)—Fire Incident Field Notes

NFPA 921 (1995)—Guide for Fire and Explosion Investigations

NFPA 1033 (1993)—Professional Qualifications for Fire Investigator

NFPA 1405 (1996)—Guide for Land Based Fire Fighters Who Respond to Marine Vessel Fires

NFPA 1561 (1990)—Standard on Fire Department Incident Management System

APPENDIX E

SELECTED STANDARDS, RECOMMENDED PRACTICES, AND PUBLICATIONS OF THE AMERICAN PETROLEUM INSTITUTE

Documents Related to Fire Protection

API Publication 2009 (1995)—Safe Welding and Cutting Practices in Refineries, Gasoline Plants, and Petrochemical Plants, 6th ed.

API Publication 2021 (1991)—Fighting Fires in and Around Petroleum Storage Tanks, 3rd ed.

API Publication 2201 (1995)—Procedures for Welding or Hot Tapping on Equipment in Service, 4th ed.

API Publication 2027 (1988)—Ignition Hazards Involved in Abrasive Blasting of Atmospheric Hydrocarbon Tanks in Service, 2nd ed.

API Publication 2028 (1991)—Flame Arresters in Piping Systems, 2nd ed.

Documents Related to Tank Systems

API Recommended Practice 2X (1988)—Recommended Practice for Ultrasonic Examination of Offshore Structural Fabrication and Guidelines for Qualification of Ultrasonic Technicians, 2nd ed. (ANSI/API RP 2X-1992)

API Publication 301 (1989,1991)—Aboveground Storage Tank Survey

API Publication 306 (1991)—An Engineering Assessment of Volumetric Methods of Leak Detection in Aboveground Storage Tanks

API Publication 307 (1991)—An Engineering Assessment of Acoustic Methods of Leak Detection in Aboveground Tanks

API Standard 510 (1992)—Pressure Vessel Inspection Code: Maintenance Inspection, Rating, Repair, and Alteration, 7th ed. (ANSI/API 510 - 1992)

API Recommended Practice 520 (1993)—Sizing, Selection, and Installation of Pressure Relieving Systems in Refineries, Part 1: Sizing and Selection, 6th ed. (ANSI/API Std 520 and 520-1-1992)

API Recommended Practice 520 (1994)—Sizing, Selection, and Installation of Pressure Relieving Systems in Refineries, Part 2: Installation, 4th ed.

API Recommended Practice 521 (1990)—Guide for Pressure-Relieving and Depressuring Systems, 3rd ed. (ANSI/API RP 521-1992)

API Standard 620 (1990)—Design and Construction of Large, Welded, Low-Pressure Storage Tanks, 8th ed. (ANSI/API Std 620-1992)

API Standard 650 (1993)—Welded Steel Tanks for Oil Storage, 9th ed. (ANSI/API Std 650-1992)

API Recommended Practice 651 (1991)—Cathodic Protection of Aboveground Petroleum Storage Tanks

API Recommended Practice 652 (1991)—Lining of Aboveground Petroleum Storage Tank Bottoms (ANSI/API Std 652-1992)

API Standard 653 (1991)—Tank Inspection, Repair, Alteration, and Reconstruction (Supplement 1, January 1992) (ANSI/API Std 653-1992)

API Recommended Practice 1110 (1991)—Recommended Practice for the Pressure Testing of Liquid Petroleum Pipelines, 3rd ed.

API Recommended Practice 1621 (1993)—Recommended Practice for Bulk Liquid Stock Control at Retail Outlets, 5th ed.

API Recommended Practice 1631 (1992)—Interior Lining of Underground Storage Tanks, 3rd ed.

API Publication 1632 (1987)—Cathodic Protection of Underground Petroleum Storage Tanks and Piping Systems, 2nd ed.

API Recommended Practice 1615 (1987)—Installation of Underground Petroleum Storage Systems, 4th ed.

API Publication 2015 (1994)—Safe Entry and Cleaning of Petroleum Storage Tanks, Planning and Managing Tank Entry from Decommissioning Through Recommissioning, 5th ed. (ANSI/API Std 2015-1994)

API Recommended Practice 2350 (1987)—Overfill Protection for Petroleum Storage Tanks

Documents Related to Hazardous Materials Transportation

API Recommended Practice 1112 (1992)—Recommended Practice for Developing a Highway Emergency Response Plan for Incidents Involving Hazardous Materials

API Recommended Practice 1125 (1991)—Overfill Control Systems for Tank Barges

Documents Related to Air Quality

API Publication 4311 (1979)—NOx Emissions from Petroleum Industry Operations

API Publication 4322 (1980)—Fugitive Hydrocarbon Emissions from Petroleum Production Operations, Vols. 1 and 2

API Publication 4392 (1985)—Test Protocol for Automotive Evaporative Emissions

API Publication 4403 (1985)—Uncertainties Associated with Modelling Regional Haze in the Southwest

API Publication 4413 (1985)—Evaluation of Proposed Downwash Modifications to the Industrial Source Complex Model

API Publication 4421 (1986)—Plume Rise Research for Refinery Facilities

Documents Related to Ground Water and Soils

API Publication 1628 (1989)—A Guide to the Assessment and Remediation of Underground Petroleum Releases, 2nd ed.

API Publication 4410 (1985)—Subsurface Venting of Hydrocarbons from an Underground Aquifer

API Publication 4442 (1986)—Cost Model for Selected Technologies for Removal of Gasoline Components from Ground water

API Publication 4448 (1987)—Field Study of Enhanced Subsurface Biodegradation of Hydrocarbons Using Hydrogen Peroxide as an Oxygen Source

API Publication 4454 (1987)—Capability of EPA Methods 624 and 625 to Measure Appendix IX Compounds

API Publication 4476 (1989)—Hydrogeologic Data Base for Ground water Modelling

API Publication 4497 (1991)—Cost-Effective, Alternative Treatment Technologies for Reducing the Concentrations of Ethers and Alcohols in Ground Water

API Publication 4499 (1989)—Evaluation of Analytical Methods for Measuring Appendix IX Constituents in Ground Water

API Publication 4510 (1991)—Technological Limits of Ground Water Remediation: A Statistical Evaluation Method

API Publication 4525 (1990)—A Compilation of Field-Collected Cost and Treatment Effectiveness Data for the Removal of Dissolved Gasoline Components from Ground water

Standards Related to Process Safety

API Recommended Practice 750 (1990)—Management of Process Hazards (Reaffirmed 1995)

API Publication 2219 (1986)—Safe Operating Guidelines for Vacuum Trucks in Petroleum Service

API Recommended Practice 2220 (1991)—Improving Owner and Contractor Safety Performance

APPENDIX F

SELECTED STANDARDS OF UNDERWRITERS' LABORATORIES

Standards Related to Fire Protection

UL 8 (1995)—Foam Fire Extinguishers
UL 154 (1995)—Carbon-Dioxide Fire Extinguishers
UL 162 (1994)—Foam Equipment and Liquid Concentrates
UL 193 (1993)—Alarm Valves for Fire-Protection Service
UL 199 (1990)—Automatic Sprinklers for Fire Protection
UL 268 (1989)—Smoke Detectors for Fire Protective Signalling Systems
UL 268A (1993)—Smoke Detectors for Duct Application
UL 299 (1995)—Dry Chemical Fire Extinguishers
UL 346 (1994)—Waterflow Indicators for Fire Protective Signalling Systems
UL 401 (1993)—Portable Spray Hose Nozzles for Fire Protection Service
UL 448 (1994)—Pumps for Fire Protection Service
UL 497B (1993)—Protectors for Data Communication and Fire Alarm Circuits
UL 521 (1993)—Heat Detectors for Fire Protective Signalling Systems
UL 539 (1995)—Single and Multiple Station Heat Detectors
UL 626 (1995)—$2\frac{1}{2}$ Gallon Stored-Pressure, Water-Type Fire Extinguishers
UL 711 (1995)—Rating and Fire Testing of Fire Extinguishers
UL 753 (1995)—Alarm Accessories for Automatic Water-Supply Control Valves for Fire Protection Service
UL 864 (1991)—Control Units for Fire-Protective Signalling Systems

UL 1058 (1995)—Halogenated Agent Extinguishing System Units
UL 1093 (1995)—Halogenated Agent Fire Extinguishers
UL 1254 (1992)—Pre-engineered Dry Chemical Extinguishing System Units
UL 1767 (1995)—Early-Suppression Fast-Response Sprinklers
UL 2006 (1995)—Halon 1211 Recovery/Recharge Equipment

Standards Related to Ventilation Systems

UL 181 (1994)—Factory-Made Air Ducts and Air Connectors
UL 181A (1994)—Closure Systems for Use with Rigid Air Ducts and Air Connectors
UL 474 (1993)—Dehumidifiers
UL 555 (1995)—Fire Dampers
UL 555S (1993)—Leakage Rated Dampers for Use in Smoke Control Systems
UL 586 (1990)—High-Efficiency, Particulate Air Filter Units
UL 867 (1995)—Electrostatic Air Cleaners
UL 900 (1994)—Test Performance of Air Filter Units
UL 998 (1993)—Humidifiers
UL 1046 (1979)—Grease Filters for Exhaust Ducts
UL 1784 (1995)—Air Leakage Tests of Door Assemblies

Standards Related to Material Identification

UL 340 (1993)—Test for Comparative Flammability of Liquids
UL 746A (1995)—Polymeric Materials—Short Term Property Evaluations
UL 746B (1979)—Polymeric Materials—Long Term Property Evaluations

Standards Related to Storage of Hazardous Materials

UL 1275 (1994)—Flammable Liquid Storage Cabinets
UL 1314 (1995)—Special-Purpose Containers
UL 1853 (1995)—Nonreusable Plastic Containers for Flammable and Combustible Liquids

Standards Related to Tank Systems

UL 58 (1986)—Steel Underground Tanks for Flammable and Combustible Liquids

UL 142 (1993)—Steel Aboveground Tanks for Flammable and Combustible Liquids

UL 180 (1991)—Liquid-Level Indicating Gauges and Tank-Filling Signals for Petroleum Products

UL 353 (1994)—Limit Controls

UL 443 (1995)—Steel Auxiliary Tanks for Oil-Burner Fuel

UL 525 (1994)—Flame Arresters

UL 1238 (1975)—Control Equipment for Use with Flammable Liquid Dispensing Devices

UL 1316 (1994)—Glass-Fiber-Reinforced Plastic Underground Storage Tanks for Petroleum Products, Alcohols, and Alcohol and Gasoline Mixtures

UL 1746 (1993)—External Corrosion Protection Systems for Steel Underground Storage Tanks

Standards Related to Life Safety

UL 96 (1994)—Lightning Protection Components

UL 96A (1994)—Installation Requirement for Lightning Protection Systems

UL 263 (1992)—Fire Tests of Building Construction and Materials

UL 410 (1992)—Slip Resistance of Floor Surface Materials

UL 924 (1995)—Emergency Lighting and Power Equipment

UL 969 (1995)—Marking and Labelling Systems

UL 1715 (1994)—Fire Test of Interior Finish Material

INDEX